CW01263869

STURMGESCHUTZ & ITS VARIANTS

The Spielberger German Armor & Military Vehicles Series Vol.II

Walter J. Spielberger

Sturmgeschütz

& Its Variants

Schiffer Military/Aviation History
Atglen, PA

Scale drawings by Hilary L. Doyle
Cover artwork: Siegfried Horn, from a Walter J. Spielberger photo.

Research and technical editing by Thomas L. Jentz
Translated from the German by James C. Cable

Photographic Sources:
Bundesarchiv/Militärarchiv (BA) (37), Daimler-Benz (1), Hilary L. Doyle (17), ECPA (5), W. Fleischer (1), Icken (2) Jaugitz (1), Tom Jentz (1), Krauss-Maffei (5), Maybach (1), Karl-Heinz Münch (15), National Archives (52), W. Oswald (1), K. Sarrazin (5), W.J. Spielberger (77), Thomas (1), Tornau (3), Torina (1), Wegmann (4), Wasmuth (39)

Library of Congress Catalog Number: 92-60361.

Printed in the United States of America.
ISBN: 0-88740-398-0

This title was originally published under the title,
Sturmgeschütz Entwicklung und Fertigung der sPak,
by Motorbuch Verlag, Stuttgart.

Published by Schiffer Publishing Ltd.
4880 Lower Valley Road
Atglen, PA 19310 USA
Phone: (610) 593-1777
FAX: (610) 593-2002
E-mail: Schifferbk@aol.com.
Visit our web site at: www.schifferbooks.com
Please write for a free catalog.
This book may be purchased from the publisher.
Please include $3.95 postage.
Try your bookstore first.

In Europe, Schiffer books are distributed by:
Bushwood Books
6 Marksbury Road
Kew Gardens
Surrey TW9 4JF
England
Phone: 44 (0)181 392-8585
FAX: 44 (0)181 392-9876
E-mail: Bushwd@aol.com.

Try your bookstore first.

Contents

Foreword 9
Introduction 10
Development by the Reichswehr 13
Commercial Solution 13
Battlefield Support Vehicles 14
Army Solution 14 R.K. Schlepper 16

Sturmgeschütz: Course of Development 18
Sturmgeschütz, Ausf.A 24
Description 31
Sturmgeschütz, Ausf.B 40
Sturmgeschütz, Ausf.C 45
Sturmgeschütz, Ausf.D 48
Sturmgeschütz, Ausf.E 53
Development of the 7.5cm Sturmkanone 40 62
Description of the 7.5cm Sturmkanone 40 67
Sturmgeschütz, Ausf.F 68
Sturmgeschütz, Ausf.F/8 77
Sturmgeschütz, Ausf.G 84
Modifications introduced during production 85
Improvements in protection 90 Pistol ports 90 / Machine gun shield 90 / Gun sight guard 90 / Driver's visor 90 / Schürzen side skirts 92 / Hull armor 93 / Zimmerit protective coating 92 / Shot deflector for the commander's cupola 103 / Gun mantle 105 / Superstructure armor 106
Improvements in armament 106 Muzzle brakes 106 / Smoke grenade launcher 108 / Rotatable machine gun mount 111 / Nahverteidigungswaffe (Close quarters defense weapon) 114 / Coaxial machine gun 114
Improvements in mobility 117 Tracks 119 / Running gear components 119 / Final drives 122 / Maintenance equipment 122
Sturmhaubitze, Ausf.G 126

Sturmgeschütz IV 135
Modifications on the Sturmgeschütz IV 139

The Production Firms 151
Assembly firms: Daimler Benz AG, Werk 40, Berlin-Marienfelde 161, Altmärkische Kettenwerk GmbH (Alkett), Berlin-Tegel 163, Mühlenbau und Industrie AG (MIAG), Amme-Werk, Braunschweig 167, Fried.Krupp-Grusonwerk, AG, Magdeburg-Buckau 168
Sub-contractors of major components: Steel industry 171 / Weapons manufacturers 172/ Optics industry 172/ Transmission manufacturers 172/ Engine manufacturers 173/ Other subcontractors 174

Sturmgeschütz Production 175
15cm Sturm-Infanteriegeschütz 33 182
Sturmgeschütz (Flammenwerfer) 187
Flakpanzer for Sturmgeschütz units 188
Sturmgeschütz's for radio controlled units 190
Memorandum on the further development of the FKL-Waffe based on lessons learned from experience gained from 5-8 July 1943 in Operation "Zitadelle" 195

Battlefield Support Vehicles 198
Munitionswagen und Beobachtungswagen 198
Development of the armored variants of the leichter Zugkraftwagen 200
leichter, gepanzerter Munitionstransportwagen (Sd.Kfz.252) 201

leichter, gepanzerter Beobachtungswagen (Sd.Kfz.253) 205
Replacement of the Sturmartillerie special vehicles with the leichten Schützenpanzerwagen (Sd.Kfz.250) 208
Schwere Zugkraftwagen 18t (Sd.Kfz.9) 211
Vehicle description 213
Production of the 18t Zugkraftwagen 216
Heavy recovery vehicles authorized for Sturmgeschütz organizations 218
Tiefladeanhänger füer Panzerkampfwagen (23t) 219
Technical data of the trailer 222
Bergepanzer (Armored Recovery Vehicle) 223
Bergepanzer IV 226
Export of the Sturmgeschütz 229
Delivery of Sturmgeschütz to foreign armies 230

The Sturmartillerie in Action 233
The Western Campaign beginning May 10th, 1940 233
The Balkan Campaign beginning April 6, 1941 234
Operation "Barbarossa" from June 22, 1941 235
The Summer Offensive in the East beginning June, 1942 236
In the East from July of 1943 Operation "Zitadelle" 238
The Great Russian Offensive beginning June 22, 1944 and the Allied landing in the West 231
Sturmgeschütz with the Panzertruppe 243
Sturmgeschütz with the Waffen-SS 246
Sturmgeschütz with the Luftwaffe 247
Sturmgeschütz with the Infanterie 247
Heeres-Sturmartillerie units 250

Technical Data of the Sturmgeschütz 252

Bibliography 254

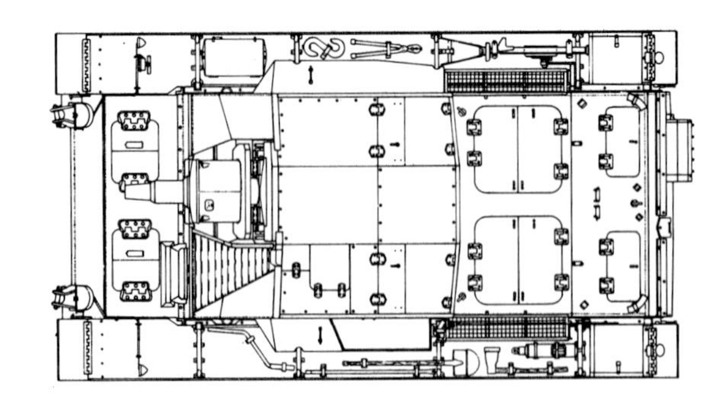

Sturmgeschütz, Ausf.A — Page 24

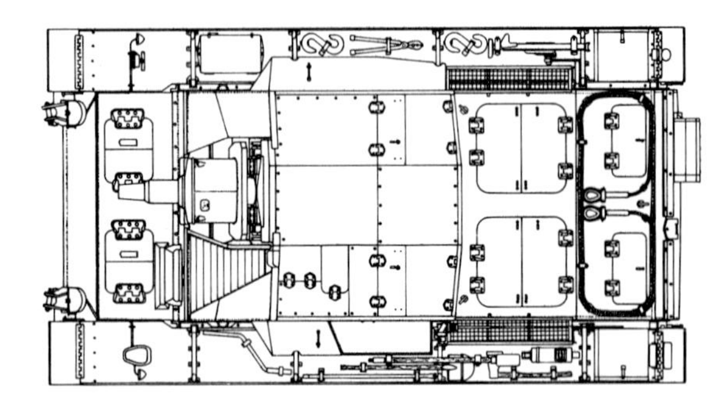

Sturmgeschütz, Ausf.B — Page 37

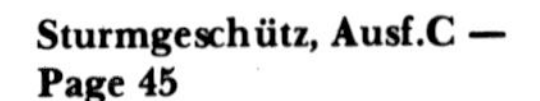

Sturmgeschütz, Ausf.C — Page 45

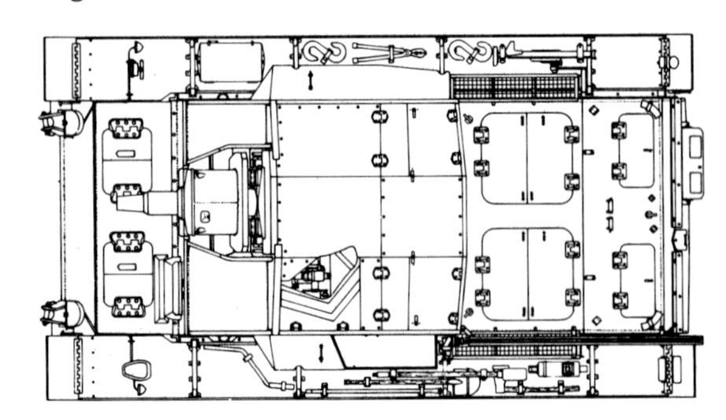

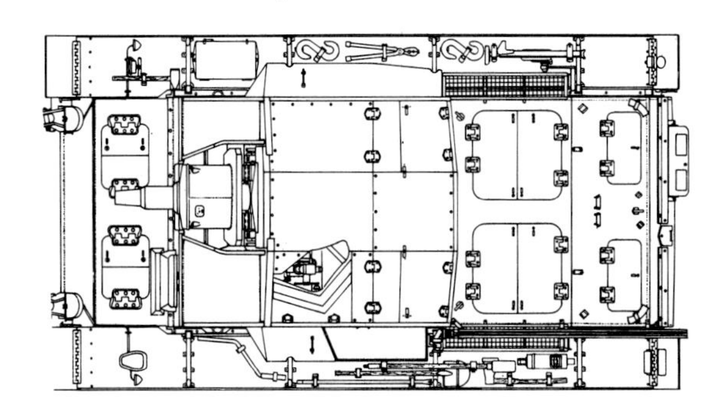

Sturmgeschütz, Ausf.D —
Page 48

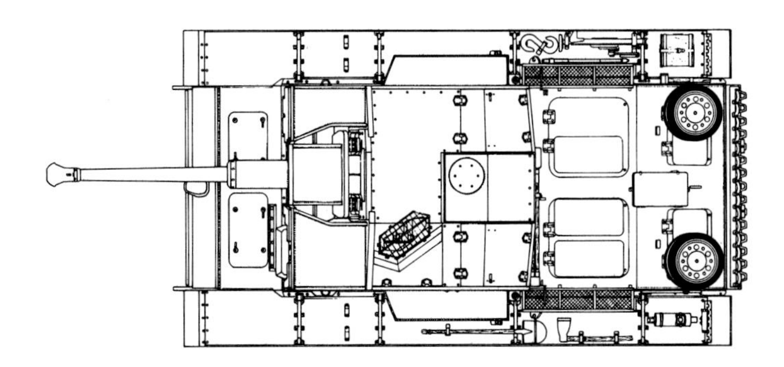

Sturmgeschütz, Ausf.F/8 —
Page 77

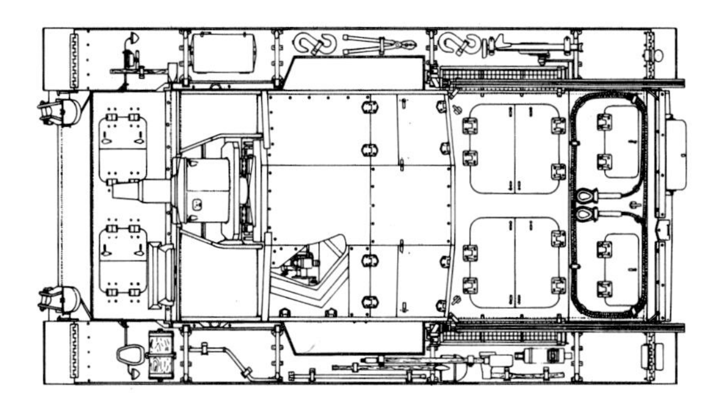

Sturmgeschütz, Ausf.E —
Page 53

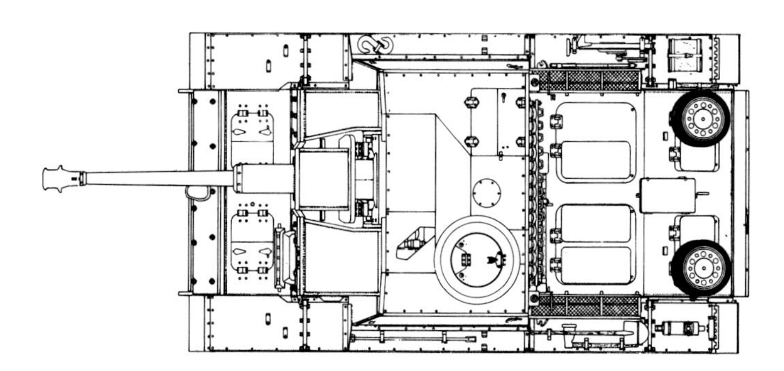

Sturmgeschütz, Ausf.G —
Page 84

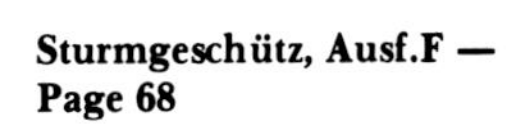

Sturmgeschütz, Ausf.F —
Page 68

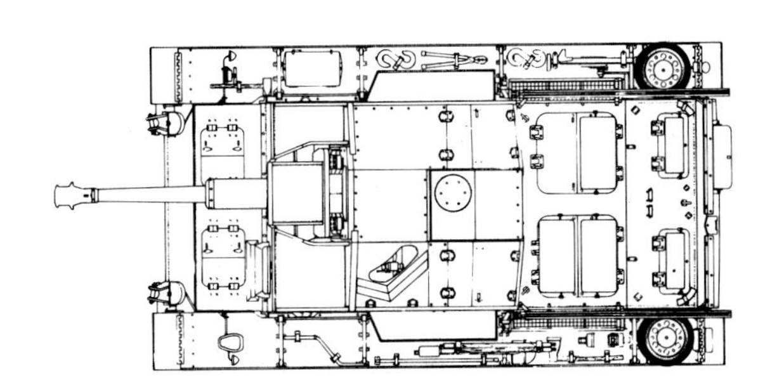

Sturmgeschütz —
Page 98

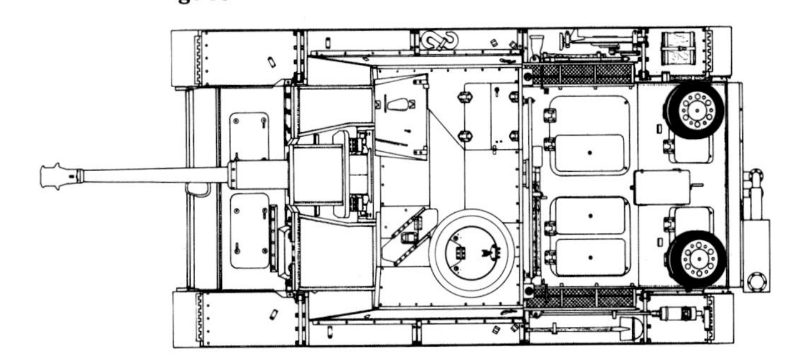

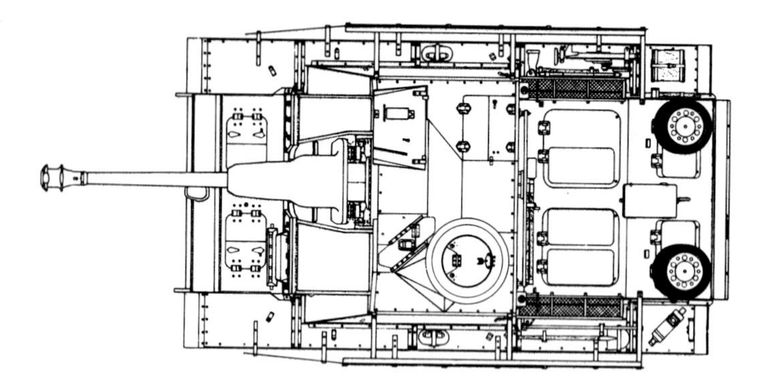

Sturmgeschütz, Ausf.G (starting 1944) —
Page 106

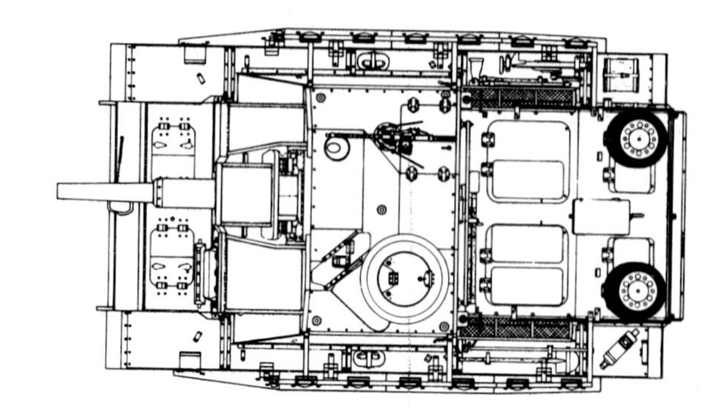

Sturmhaubitze, Ausf.G —
Page 126

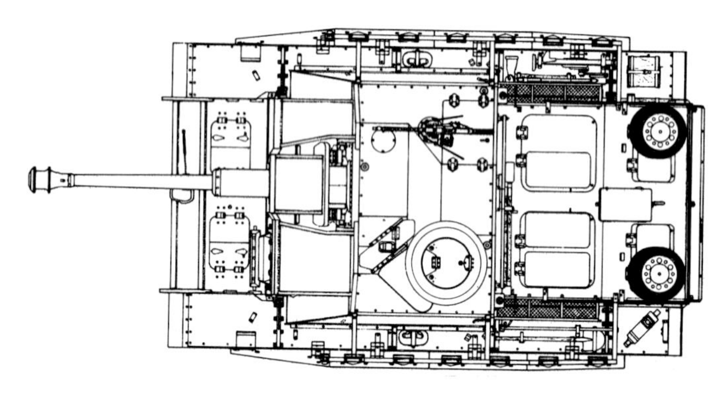

Sturmgeschütz, Ausf.G —
Page 111

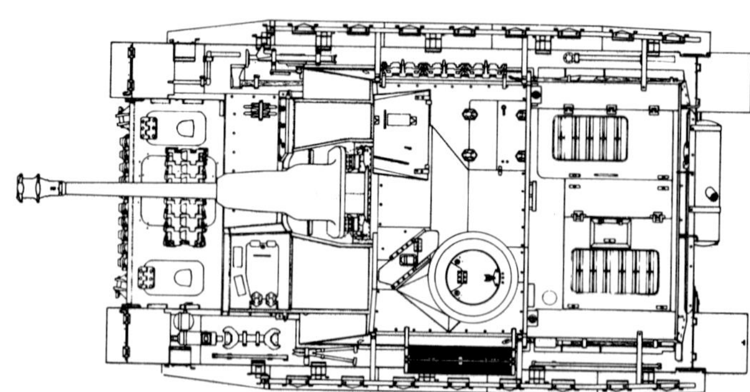

Sturmgeschütz IV —
Page 135

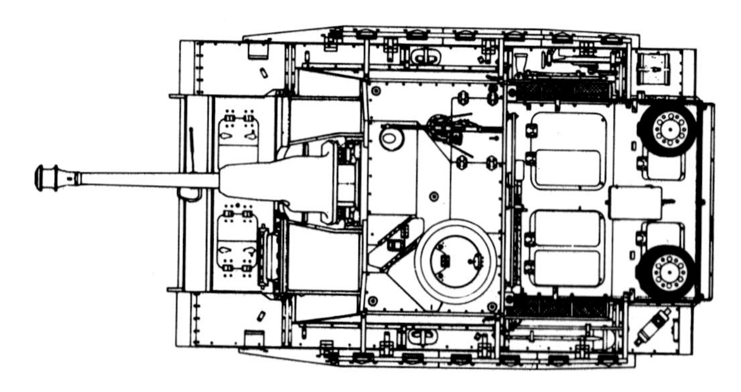

Sturmgeschütz, Ausf.G (final production) —
Page 115

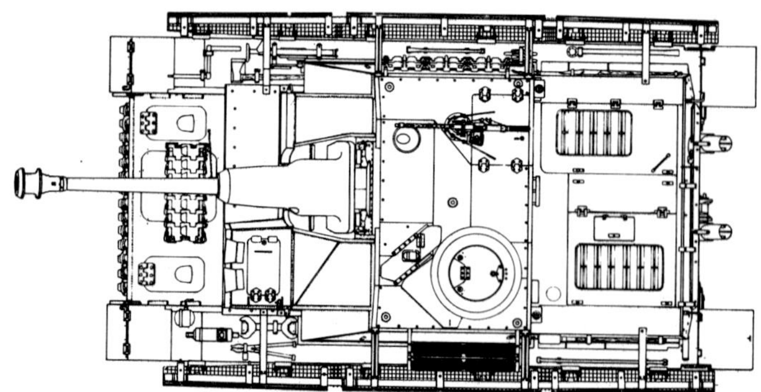

Sturmgeschütz IV (final production) —
Page 144

Foreword

The wealth of terms used by the German Wehrmacht, such as gepanzerte Selbstfahrlafette, Panzerjäger-Selbstfahrlafette, Panzerjäger, Jagdpanzer, Sturmpanzer, Sturmgeschütz, Sturmhaubitze and others still provide the interested reader with difficulty in determining the true mission of these vehicles, as well as to which branch of service they belong.

Also complicating this matter is the fact that even high echelons of the military did not clearly know the exact reasons for deployment of their weapons systems and did not fully realize their potential, therefore poor decisions and avoidable losses of valuable personnel and material were the order of the day. Additionally, the Panzertruppe had to switch from their original combat role to the engagement of the vastly superior allied Assault Forces, thereby blurring the traditional separation between the battle tanks and tank destroyer vehicles to an even greater extent.

Finally, every vehicle equipped with an armor-piercing weapon served in an anti-tank role, without taking into account its suitability for such action.

The primary purpose of this book is the depiction of a special branch: the Sturmartillerie (Assault Artillery) and its principle weapon specially developed under the cover name of "sPak": the Sturmgeschütz. Currently, no army in the world employs a similar weapons system.

In keeping with the tradition of the Militärfahrzeuge series of books, the main thrust of this book is in the area of the technical development and production of this weapon. It is to serve as a foundation and supplement for the numerous publications about the deployment and successes of the Sturmgeschütz during the Second World War.

Sturmgeschütz units were organic to the Artillerie of the German Army. and were not organized in companies, but rather batteries. The Commander was called a Geschützführer, or Gun Leader, while a loader, gunner and driver filled out the crew. Their uniforms, although cut in the style of the Panzerbesatzungen, (tank crews) were grey, not black.

Their branch color was the red of the Artillery.

For the hard-fighting Infantry, the Sturmgeschütz were often the last rescue in an emergency while confronting overpowering enemy armor.

Volume 2 of this series of books is an illustrative result of years of teamwork. While the responsibility for the contents lies with the author, Thomas L. Jentz contributed many chapters. The basic data for this project was researched in his thorough and precise manner from all available sources of original documents. Production numbers, chassis numbers and modifications to the vehicles, after-action reports, unit strengths and organizations are his strong points.

Hilary Doyle is impressive in his attention to detail in creating exacting representations of each vehicle in the scale drawings.

The gentlemen Michel Aubry, Kurt Sarrazin, Hans Ström and Karl-Heinz Münch have earned great thanks for contributing valuable records and photographs.

And so a truly well-earned literary monument is established in honor of the Sturmgeschütz, which no longer exists in the German Bundeswehr.

Suggestions and additions from our readers will be gratefully accepted.

Walter J. Spielberger

Introduction

The development of self-propelled vehicles, initiated under contract by the Reichswehrministerium in 1927, also provided for the design of a 77mm cannon on a maneuverable fully-tracked vehicle. The vehicle and weapon were partly protected by armor. With this, the concept of an infantry escort gun was rekindled from the World War I.

The commercial vehicles were enhanced through improvements by the Army. These attempts were halted in 1932 because other plans for motorizing the Army seemed more pressing.

In a memorandum to the Chief of the General Staff and the Commander in Chief of the Army from von Manstein (later promoted to *Generalfeldmarschall*) suggested in 1935 to revive the concept of the Infanterie-Begleitbatterien (infantry escort batteries) of the first World War. He envisaged the solution to his proposal in the form of an armored self-propelled gun for the direct support of the infantry. He coined the phrase Sturmartillerie (assault artillery). His proposal further advocated the creation of an Abteilung (a battalion-sized unit) with three batteries, each with six guns. The suggestion met with initial resistance from many and varied positions of the OKH (Oberkommando des Heeres) but finally was approved by the General Staff and the Commander in Chief of the Heer, *Generaloberst* von Fritsch.

The philosophy put forth in this proposal was outlined to the Chief of Staff in document H.Mot.890/36 g.Kdos of June 8th, 1936:

Abt.Nr. 890/36 g. Kdos.

To: The Chief of Staff of the Army

After the concept of the creation of armored Sturmartillerie (assault artillery) met with the approval of the Chief of the General Staff, it seemed necessary to submit a tactical paper concerning the application of the Assault Artillery in addition to the technical development of the corresponding gun. Otherwise the situation will arise when we have the new weapon without knowing how to utilize it.

It can currently be said, not only with us but with other States as well, that the lines of thought about the utilization of armored vehicles and the assault gun are not clear-cut, and are many times quite blurred. On the one hand, it is said that the armored vehicle should take full advantage of its speed to break through enemy infantry in order to cripple the enemy Artillery and higher headquarters, and also to get to the enemy reserves. On the other hand it is said that we cannot lose contact with the Infantry, without which the very success of an armored victory is in question. The Armor branch is of the opinion that it is the job of the Infantry to maintain contact, which means constant running for the Infantry, and therefore an impossibility. The Infantry wants to keep at least one wave of armor in their area, which in turn effects the speed of armor, and this is the very thing it uses for protection against enemy artillery.

Contrary to this it can be determined that although armor and assault artillery can be viewed as very similar weapons in a technical sense, they must be considered completely different branches in regards to tactics. No one in earlier times thought of attaching the Infantry to the

Cavalry on the attack. And just as few people would have asked a Cavalryman to attack in cadence with the Infantry. It can therefore be easily concluded:

I. Armor units are combined arms units, whose composition makes it possible for them to fight independently to achieve their individual combat missions. If this quality is also applied to the armored vehicle branch, then during the attack they will have motorized artillery for the exploitation of the successes of the Infantry, and moreover the necessary technical troops at their disposal.

Tank units will be utilized in an independent role for strikes, most likely against the flanks and rear areas of the enemy, or at least on a fluid flank. They can also be given the mission of making a breakthrough of the enemy front. In any case, in order to attack independently they must have the necessary tools to fulfill the mission. As soon as one tries to couple them with other units, they lose the value of their very nature.

However, the possibilities for employing tank units is, as was in earlier times with the Cavalry, considerably limited by terrain. Among other types, this includes forested areas, mountains, riverbeds and swamps which limit or rule out their utilization.

It will also be virtually impossible for them to achieve a breakthrough of a well-prepared enemy front. In such an instance, their effect will be decisive if striking the enemy in a sensitive area and surprising him, in the case the enemy is not combat ready, or if they strike an already shattered enemy. To be utilized to their greatest advantage, they should be placed in a light Army (light Division, light Armored Division, light Motorized Infantry Division).

II. Tank Brigades, that is, true Armor Brigades as one would better call them, are weapons of the main thrust of the attack. During the normal course of the attack, they should achieve a quick local victory within the framework of a Corps or Army. For this purpose, they should be coupled with the Infantry Division attacking in the main thrust, as opposed to tank units, which operate independently.

III. The Sturmartillerie, (equally true if they are in the form of armored vehicles or consist of armored and motorized field guns) is, on the other hand, a weapon to assist the normal Infantry Division. Their use during the attack corresponds to the Escort Artillery of the last war, that is, the elite of the light Artillery. In order to make them useful for other purposes, such as on the defensive, there is an additional requirement for the ability to be used as part of the normal Artillery — in an indirect fire role, into the main battle area (usually a distance of 7 kilometers) at a minimum. Finally, they would be a superb offensive anti-tank weapon, and could replace the divisional anti-tank element in this role.

Assault Artillery fights as Escort Artillery within the framework of the Infantry. It does not attack like the tank, does not break through, but carries the attack of the infantry forward by quickly eliminating the most dangerous objectives through direct fire. It does not fight in large numbers like the tank units, but is normally employed at platoon strength.

The platoon, or even the individual gun, makes a surprise appearance in and then quickly vanishes before it can become a target for the enemy artillery.

Being armored and motorized permits the assault artillery to fight in among the Infantry, which means the immediate combined effect at the right moment against decisive targets in a quantity which the artillery in rear positions is not able to deliver.

The gun must be able to take enemy machine gun emplacements out of action with a few rounds. It must also be able to knock out enemy tanks, in comparison to them it has inferior armor, but a superior ability to observe and shoot first.

Each infantry division should have at its disposal at least one battalion-sized unit of Sturmartillerie consisting of three batteries of six guns each. One might even consider dropping either the normal artillery battalion or the divisional anti-tank battalion.

Above all, it can be seen that the Sturmartillerie should not be utilized in the sphere of armor units, but rather in that of the normal Infantry Division. A clean separation of the two branches is necessary if the two do not want to operate according to the improper doctrines.

The Sturmartillerie is to be trained as Artillery units and will have to learn their mission as escort batteries in the environment of the Infanterie.

And finally in order to establish the tactical doctrinal basis, a battery from the Lehr-Inf.Btl. will be temporarily assigned to, and under the command of the Sturmartillerie for trial purposes in the interest of saving time. Initially, it should be sufficient for the battery to consist of six guns, on light tank chassis with temporarily mounted wooden guns, because the first order of business is the establishment and development of the tactics of this Sturmartillerie.

signed,
von Manstein

The 8th (Techn) Abteilung of the General Staff of the Army under the leadership of then *Oberst* i.G. Model was charged with working out the corresponding developmental requirements for the Heereswaffenamt and to hurry the overall development along. Responsibility for this mission lay with the 8th Abteilung premier expert, then *Hauptmann* Hans Röttiger. The official orders for the "Begleitartillerie" (armored artillery for anti-tank and infantry) came from orders AHA (Insp.4) to the Heereswaffenamt contained in document 449/36 gKdos of June 15th, 1936.

This order led to the design development of the Sturmgeschütz with a 75mm gun mounted in an armored superstructure on the Zugführerwagen (Z.W.) chassis, which at this time was still under development. Due to limited funds and technical problems, only five of the 0-Series chassis were produced by the end of 1939. The first vehicles of the first production series did not roll out of the factory before January of 1940.

The General Staff decided to form the Sturmartillerie Abteilungen as part of the Heeresartillerie. In 1936, the Jüterbog Artillery Training Regiment was tasked with developing standards of deployment and utilization for an armored assault gun within the guidelines set forth by In 4 (Inspektion der Artillerie beim OKH).

On December 15th, the General Staff of the Army reported that the first tests with the Pak (Sfl) had been encouraging. At the same time, these previously required armored combat support vehicles were to be addressed:

a) the armored reconnaissance vehicle
b) the armored ammunition vehicle.

In 1937, the test battery of the 7./ALR (mot.) was created. In early 1938, the first field tests were conducted involving two tank chassis with temporary guns mounted.

During the period from the end of 1938 to the beginning of 1939, the first unit-sized training was conducted with the Infantry Training Regiment in Döberitz.

In the beginning of 1940 weapons demonstrations had taken place. The first Sturmgeschütz battery (no. 640) was created on November 1st, 1939 under the 10th/A.L.R. Jüterbog, which was followed quickly by three additional batteries (numbered 659, 660 and 665). These four batteries conducted the first deployments in the French Campaign in 1940. Three additional independent batteries were created in 1940, including one for the Waffen SS, prior to the forming of the Sturmartillerie into Abteilungen of three batteries of six guns each. From the end of 1940 and into the beginning of 1941, the Sturmartillerie grew quite considerably. Up to the beginning of the Russian Campaign on June 22 1941, there were a total of 11 on-line Abteilungen and nine batteries (including those of the Waffen SS). Starting in 1944, the Sturmgeschütz Abteilungen were reorganized into Sturmgeschütz Brigades in order to differentiate between them and those Sturmgeschütz units in the Panzerjäger-Abteilungen organic to the Infanterie-, Jäger-, Grenadier- and Gebirgs-Divisions. By the end of the Second World War, there were more than 75 Sturmartillerie Batterien, Abteilungen and Brigaden which had been deployed in all combat theaters.

In the beginning, the development of the Sturmgeschütz was done under the cover name sPak. On February 7, 1940 the Heereswaffenamt changed the designation to 75mm Kanone (Pz.Sfl.).

It was not until March 28, 1940 that the name was changed to the now familiar Sturmgeschütz. This name was a shortening of the official Heereswaffenamt designation of Gepanzerte Selbstfahrlafette für Sturmgeschütz 75mm Kanone (Sd.Kfz.142), (armored self-propelled vehicle for the 75mm assault gun). Other designations, which appeared either during or after the War, were either unofficial or abbreviations of the official designation. After the introduction of the Sturmgeschütz on the Panzerkampfwagen IV tank chassis, the official designation was changed once again, in order to allow for the differentiation between the following models:

— Sturmgeschütz III für 7.5cm Sturmkanone 40 (L/48) (Sd.Kfz.142/1)
— Sturmgeschütz III für 10.5cm Sturmhaubitze 42 (Sd.Kfz.142/2) and
— Sturmgeschütz IV für 7.5cm Sturmkanone 40 (L/48)

Development by the Reichswehr

The idea of taking horse-drawn Infanterie Begleitgeschütze (Infantry Escort Guns), motorizing, armoring them and making them more mobile had arisen early in the Reichswehr, although due to the political and economic atmosphere at the time, the development and delivery of such vehicles was practically impossible.

Permission for the construction of fully-tracked vehicles for use by the German commercial industry, which until this time was forbidden by the Versailles Treaty, was not granted by the International Control Commission until 1923. In 1926/27, the 3rd Company of the Kraftfahrabteilung 7, the 3rd Company of Kraftfahrabteilung 6 as well as other Kraftfahrabteilungen received fully-tracked tractor vehicles for use in cross-country training.

The commercial W.D.-Schlepper (tractor) with a 50 hp engine as a self-propelled gun carriage for the 77mm Feldkanone 96/16 (field gun).

Commercial Solution

A portion of those allocated vehicles were the W.D. (NOTE: W.D.= Wendeler und Dohm) caterpillar tractors of the Deutsche Kraftpflug-Gesellschaft firm in Berlin, the production of which was the responsibility of the Hannover Maschinenbau-AG (HANOMAG) in Hannover. The heavy model (Typ Z 50) (NOTE: beginning with tractor nr. 50001) of the firm's tractor could produce 50 horsepower (—cylinder, bore/stroke 135 x 155mm, 8876cc displacement) and was fitted with a Graetzin heavy oil carburetor for operation with petroleum. Its external measurements were 4400 x 1900 x 2300mm, the track width was near 400mm, its empty weight about 6800kg. In third gear it could achieve a top speed of 6 kilometers per hour.

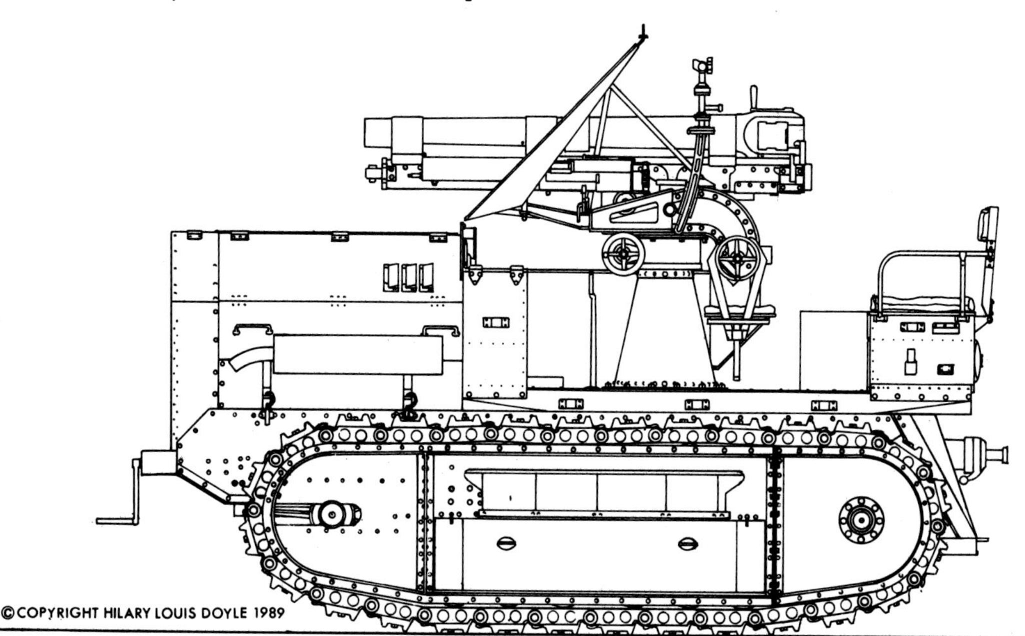

Hanomag W.D.-Schlepper with 77mm FK 96/16

The Reichswehr used one of these vehicles to begin testing for use as a self-propelled track in 1927. It was fitted with a 77mm gun of the Rheinmetall-Borsig firm, which had a limited traverse capability and was turned over to the army for testing. It was planned to mount a machine gun with armor protection next to the main gun. The cannon had a muzzle velocity of 465 meters per second, the projectile weight was 6.85kg. On the vehicle, the gun had an elevation range of -7 to +15 degrees, and all-round firing capability was desired. Due to its limited mobility on streets and cross-country, some resistance was given to the idea of obtaining other commercial tracked vehicles for this purpose.

Battlefield Support Vehicles

By as early as 1930, the German Reichswehr had expressed interest in a cross-country capable combat support vehicle for supplying motorized units. By using the Maffei artillery towing vehicle MSZ 210, of which a small series was produced, a proposal was brought forth suggesting a fuel tank truck with a trailer which could each carry 2,000 liters of fuel. The April 5th 1930 plan did not go into development because the half-tracked towing vehicle did not correspond to the needs of the troops.

Army Solution

Other attempts to create a self-propelled vehicle for the 75mm gun were initiated beginning in 1928 by the Fried.Krupp AG firm in Essen. The leichte Selbstfahrlafette (LSK) (light self-propelled vehicle) was available in 1930 as a prototype designed to carry light infantry guns.

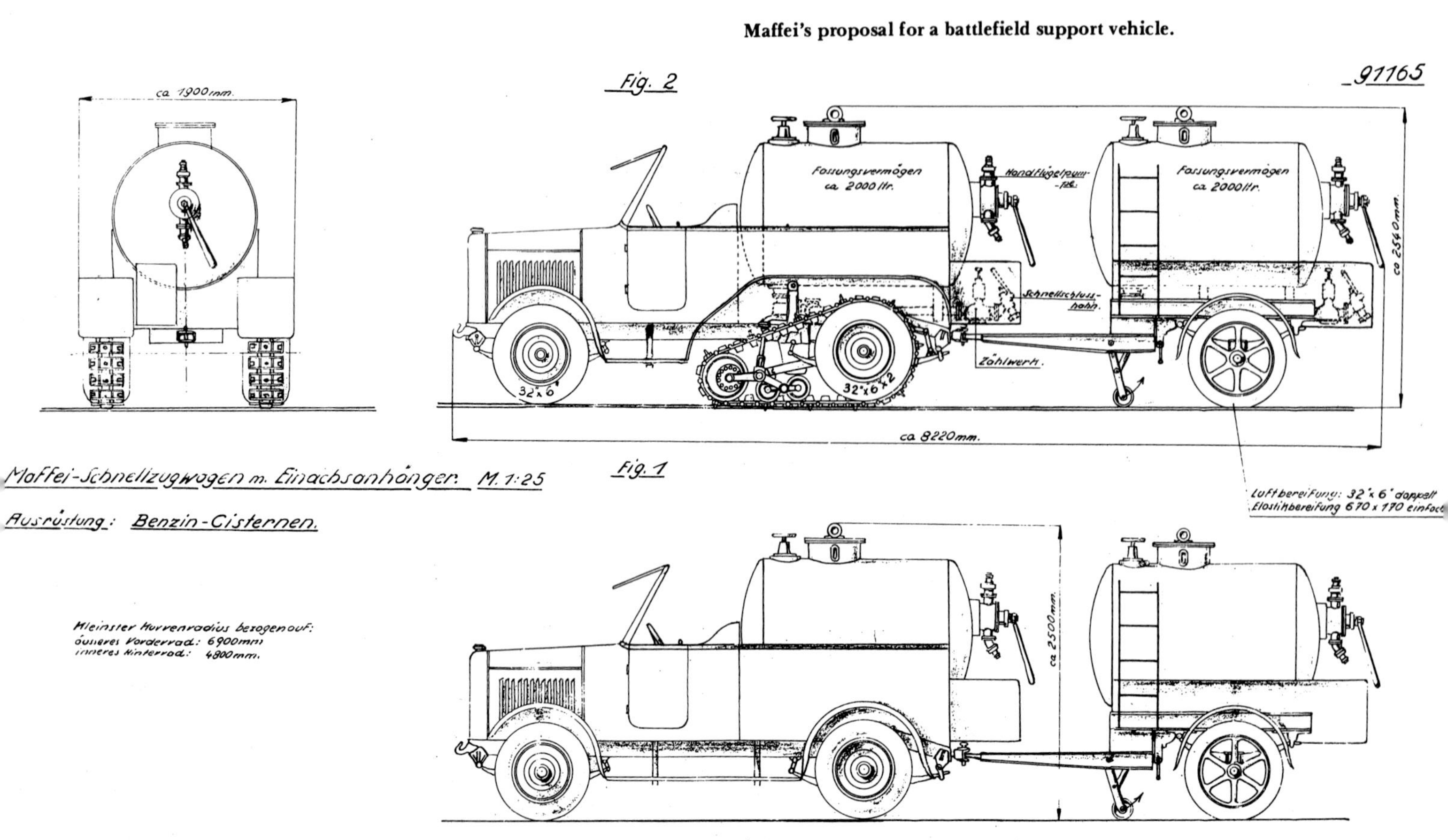

Maffei's proposal for a battlefield support vehicle.

Krupp's light self-propelled vehicle, the LSK.

The vehicle was at the Krupp facility for the purpose of modification from November of 1930 to November of 1931. It was test-driven for 379 kilometers in 1931, for 1007 kilometers in 1932 and 589 kilometers in 1933. A 100hp Daimler truck engine was installed into this fully-tracked vehicle. In 1932 Krupp initiated the installation of a long clutch. Its use by means of Bowden cables indicated good characteristics. An opening was made in the rear wall of the vehicle to permit access to the gearbox. This opening was closed with a bolted cover. This meant the towing pintle had to be moved.

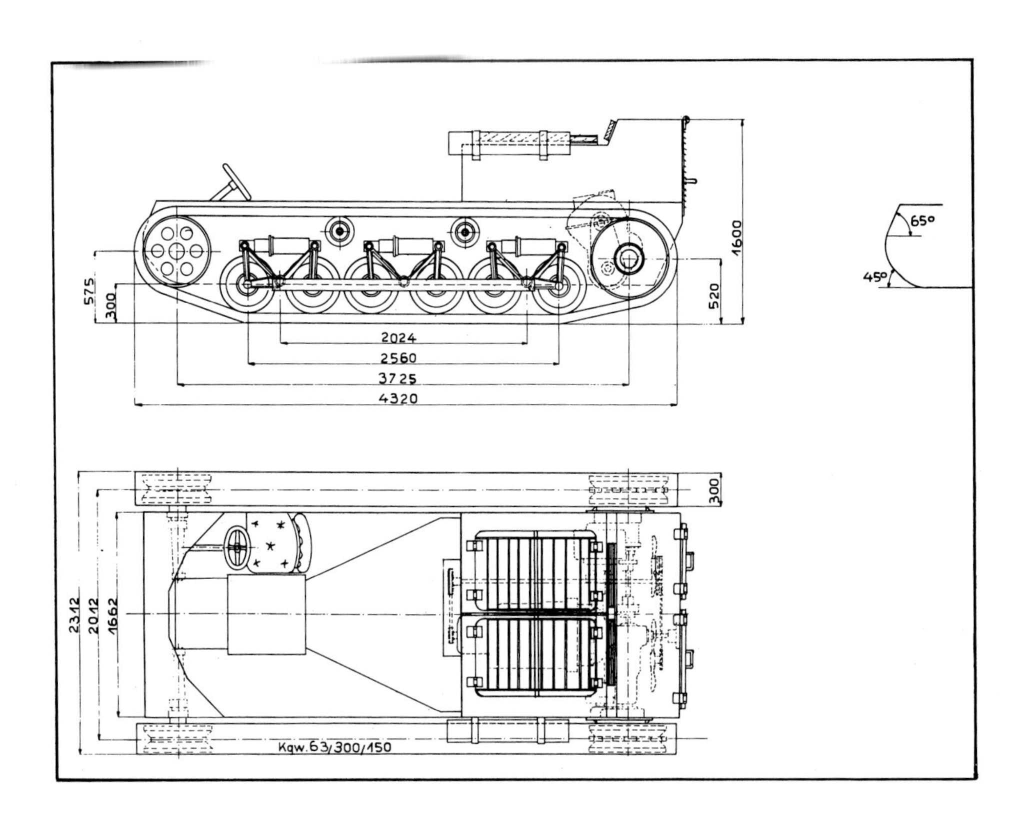

Views of the LSK from above and the side.

The Krupp leichte Selbstfahrlafette, LSK.

The tracked running gear consisted of six road wheels and two support rollers on each side. The engine lay to the rear. Road contact length of the track was 2560mm. Each track consisted of 86 115mm sections. The track used for testing was given the designation Kgs 61/280/89. The weight of the vehicle was 8910kg. One prototype was produced.

R.K.Schlepper

Parallel to the development of fast-moving fully-tracked vehicles for the German Reichswehr, the Maffei firm in Munich was testing their normal half-tracked vehicles, as well as a vehicle with changeable drive. While moving cross-country, this vehicle could use its lowerable track drive, while during normal road use it could raise the tracks to travel on four wheels. The use of wheels during road travel permitted a reduction in track wear and an increase in speed. The resulting Maffei R.K.-Schlepper (Räder-Ketten Schlepper (wheeled/tracked tractor)) appears to have had standard frame dimensions (according to blueprint ZM 116864 of October 6 1930) on which the front axle was supported by suspension arms with leaf springs, and the rear truck axle, delivered by Henschel, was supported by cantilever springs. The running gear situated between the axles provided a long wheelbase of 4351mm.

Maffei RK-Schlepper with the 75mm Kw.G.14.

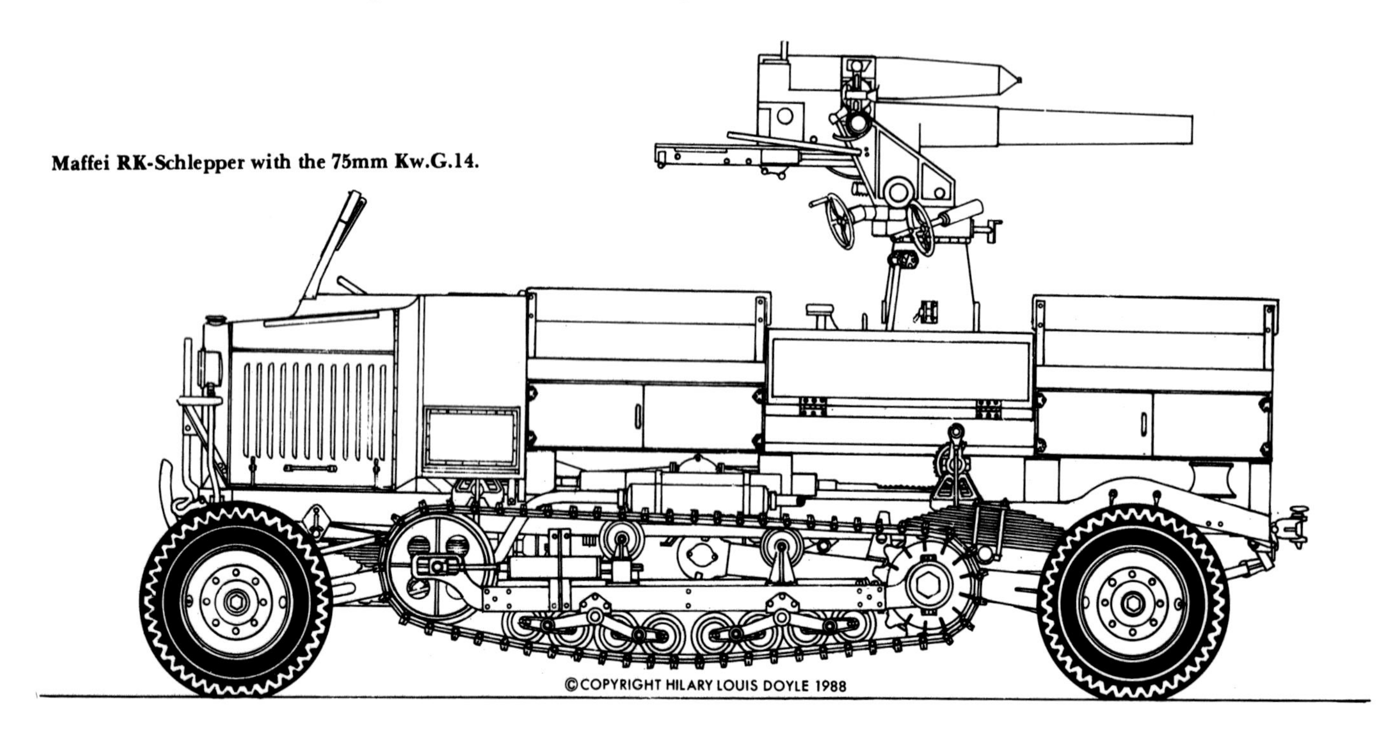

Left side view of the Maffei RK-Schlepper.

These two photos show the RK-Schlepper cross-country trials.

Overhead view of the RK-Schlepper chassis.

The forward-mounted Magirus V100 four cylinder normally-carbureted in-line engine could produce 60hp and 29.6mkg of torque at 1800rpm. To the right of the engine was the driver compartment. The gear ratio of the 4-speed gearbox was 7.1: the flow of power was carried from the engine via an intermediate gearbox to either the rear axle with a reduction ratio of 5.3 or via the side drive with a reduction ratio of 3 to the drive wheels for the tracks. The tracks were driven in the rear and were supplemented by eight pairs of suspended road wheels as well as an idler wheel. Two support rollers guided the returning track. The road wheels were supported by coil springs, which, with a spring thickness of 18mm, provided 100mm of spring deflection, and with a thickness of 16mm provided 128mm of spring deflection. The lifting device for the tracks was driven from the engine via a reverse idler gear and a conical clutch. A winch was operated from the same drive. It was installed on the end of the frame and could pull 3000kg. The lifting clutch for the tracks could deliver 3.42 horse power, which raised or lowered the tracks a distance of 330mm in 2.25 minutes. Normally, the rear axle was fitted with dual tires. For this system, the vehicle utilized tire size 34X7, and had a top speed of 57.6 kilometers per hour.

The vehicle was thought of as a replacement for the Reichswehr's Kw 19, Sd.Kfz.1, which was left over from World War I. The special open construction was suitable for the installation of a 75mm Flak (Kw.G.14) on a pivoting mount together with the needed crew and ammunition.

But although this solution was a promising one, both as a wheeled and a tracked vehicle, its further development was halted.

The Sturmgeschütz Course of Development

The official contract for the creation of the Sturmgeschütz was, as already mentioned, published on June 15th, 1936.

The following developmental requirements were put forth:

— the armament was to be a gun of at least 75mm caliber

— the gun was to be able to traverse more than 30 degrees without movement of the vehicle itself

— the gun elevation was to be sufficient for a minimum

firing range of 6000 meters
— shells fired from the gun were to be able to penetrate all types of armor known at the time at distances up to 500 meters
— all round protective armor was required for the vehicle itself, which would be of open construction and lack a turret. The frontal armor was to be capable of resisting hits 20mm anti-tank rounds and be designed with an incident angle of 60 degrees, while the sides were to provide protection from armor-piercing small arms fire
— the overall height should not exceed that of a man standing erect
— the remaining dimensions were resultant upon the armored chassis which was to be used
— other designs were to take into account such things as ammunition, communications gear and crew strength.

Lengthy discussions were needed in order to implement these requirements. In addition, the above mentioned open construction was eventually closed in order to better protect the crew in close combat.

The basic vehicle finally selected as the chassis was from the Panzerkampfwagen III. The first prototypes for the Panzerkampfwagen III, which was designed under the cover name of Zugführerwagen (Z.W.) (platoon leader vehicle) became available at the end of 1935. On the basis of their experience and due to the urgency of the project, Daimler-Benz AG Werk 40 in Berlin-Marienfelde was put in charge of production of this series, and at the same time made responsible for the development of the chassis.

Between 1937 and 1939 the following Panzerkampfwagen III's were produced:
Series 1./Z.W.: (chassis no. 60101-60110)
Series 2./Z.W.: (chassis no. 60201-60215)
Series 3a/Z.W.: (chassis no. 60301-60315)
Series 3b/Z.W.: (chassis no. 60316-60340)
Series 4./Z.W.: (chassis no. 60401-60441 and 60442-60496)

The five 0-Serie Sturmgeschütz with soft steel superstructures were assembled using 5 chassis from the second series (Typ 2./Z.W.) originally intended for Panzerkampfwagen III (chassis serial numbers 60201-60225).

Precise production records were not kept by the Heereswaffenamt in the period prior to December 1938. It is therefore difficult to determine just when these five prototype vehicles were manufactured. Nevertheless, there are still clues in the files of those firms involved in the development (Daimler-Benz for the chassis and superstructure, Krupp for the 75mm gun) which indicate that Daimler-Benz had already produced three of the ZW series vehicles by December of 1937, and the last two ZW series chassis for Panzerkampfwagen III were also produced and entered service with Panzer-Regiment 1 in Erfurt on December 6, 1937. It can therefore be assumed that the five ZW chassis for the 0-Serie Sturmgeschütz experimental vehicles had been produced by the end of 1937.

Front view of the 0-series on the 2./Z.W. chassis. The round maintenance access hatches on the lower hull front plate are typical.

0-Series vehicle as seen from the front right.

A 0-Series vehicle in Jüterbog, Fall 1940.

Unfortunately, there have been no further documents found up to this time which indicate when Daimler-Benz delivered the completed vehicles with superstructure. An original document, reporting the status of tank production dated September 30, 1936, revealed: Four Panzerkampfwagen III chassis with wooden superstructures are to be used for trials of the Sturmgeschütz in April and May 1937. Further, a model as a Sturmgeschütz is to be completed in July 1937.

The 0-Series Sturmgeschütz differentiated themselves from the later production series primarily in the running gear: on each side of the vehicle this consisted of the drive wheel, eight road wheels, an idler wheel and three return rollers, surrounded by the track.

Sturmgeschütz, 0-Series.

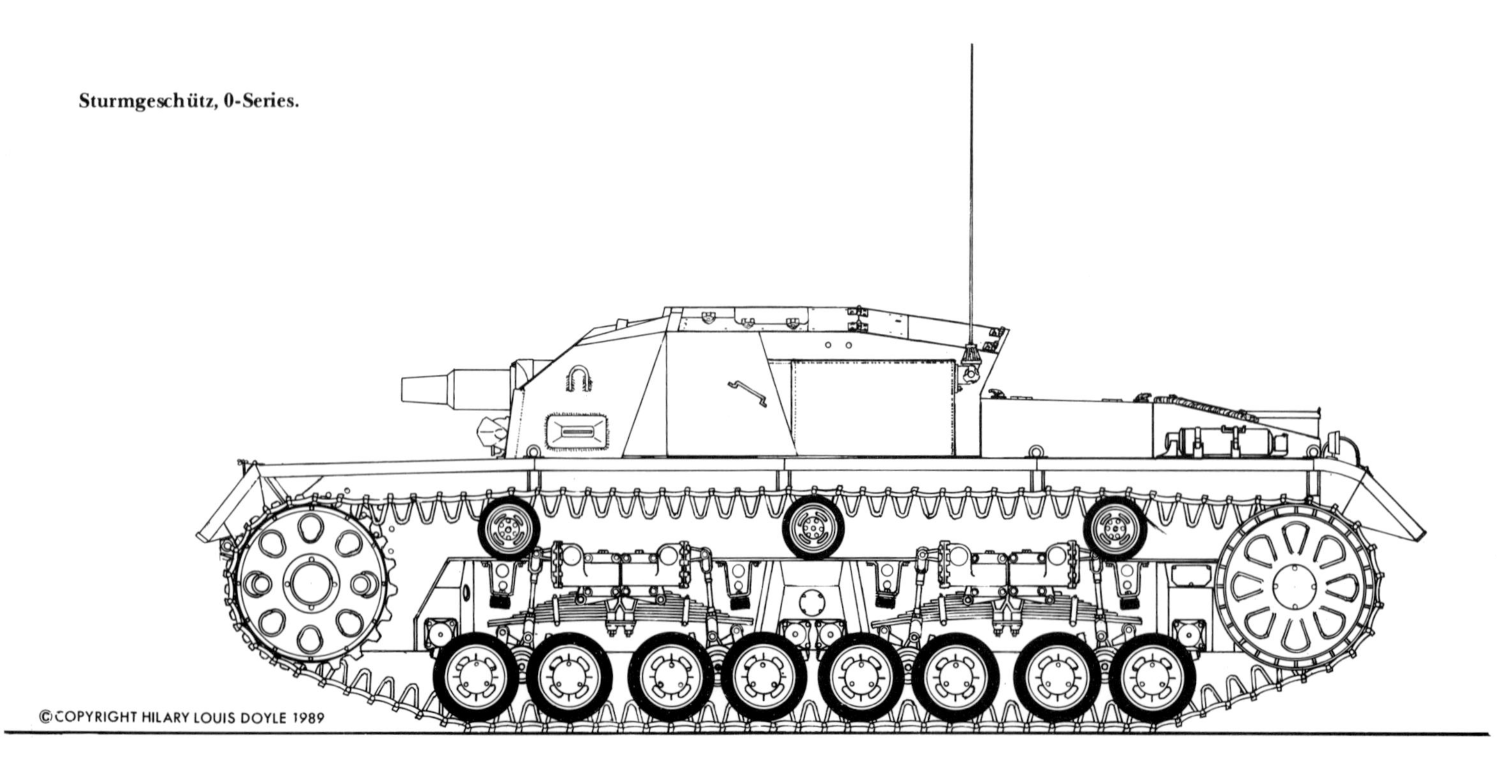

Rear view, loading ammunition at the ALR.

The unlubricated track used in testing had a width of 360mm (track pin length of 380mm) with 121mm sections. Each pair of roadwheels were mounted on a double-swing arm, pivoting on the free end of a suspension arm. The other end of the suspension arm pivoted in a mounting bracket on the hull side. A common set of leaf springs were mounted for each pair of suspension (two sets of leaf springs, for the four suspension arms). Upward movement of the roadwheels on the double-swing arms was limited by rubber bump stops. To suppress swaying induced by driving over uneven surfaces, a shock absorber was connected to the double-swing arm for each pair of road wheels. The Fichtel & Sachs shock absorbers were one-way shocks — their damping effect was only against the upward movement of the road wheels.

The Maybach HL 108 powerplant installed in the rear was a 12-cylinder 60 degree V, gasoline carbureted engine with a two-part cast crankcase. The upper and lower blocks were bolted together. The crankshaft was centrally mounted. The lower portion of the crankcase formed an oil pan. The cast-iron cylinder sleeves were replaceable. Each side of the engine had a cylinder head, manifolds for intake and exhaust channels, spherical combustion chambers with inclined valves. There was one intake and exhaust valve per cylinder, operated by the camshaft mounted in the cylinder heads via rocker arms. Performance was rated at 230 metric hp at 2300 rpm.

A balanced drive shaft carried the torque to the main clutch. The clutch, transmission and steering mechanism were mounted together as an integral unit. The housing for the steering mechanism was flangged to the front end of the transmission. The five-gear synchromesh Aphon "SFG 75" was designed and produced by Zahnradfabrik Friedrichshafen (ZF).

The five 0-Serie Sturmgeschütze were listed on the Army inventory from September of 1939. They were only used in training and never in combat, since the superstructures had been constructed from soft steel. These five experimental vehicles were handed over to the Sturmartillerie Schule Jüterbog, where they were in use for training crews as late as 1941.

Gepanzerter Munitionstransportwagen (armored ammunition transport vehicle) at ALR.

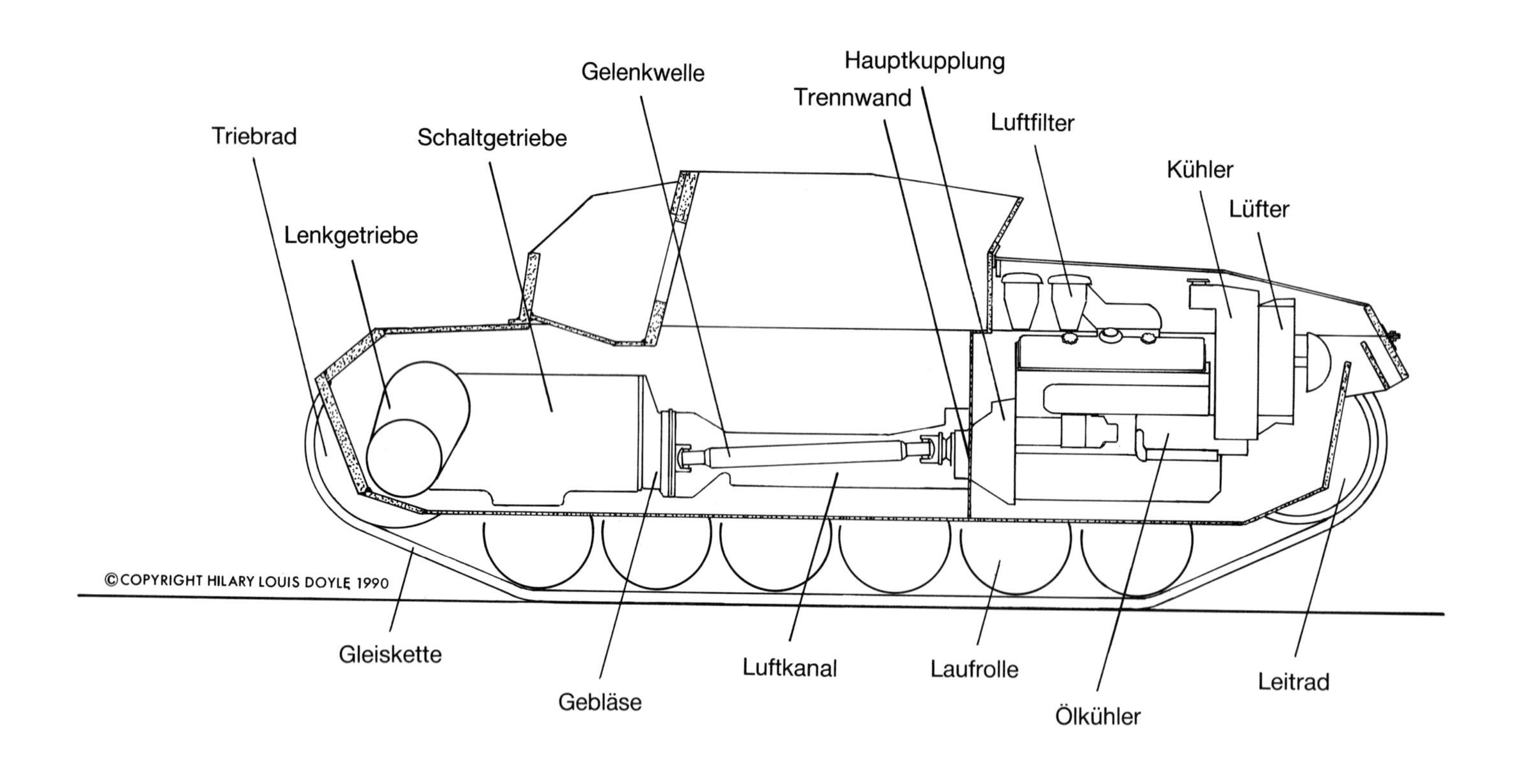
Gelenkwelle
Hauptkupplung
Trennwand
Luftfilter
Triebrad
Schaltgetriebe
Kühler
Lüfter
Lenkgetriebe
©COPYRIGHT HILARY LOUIS DOYLE 1990
Gleiskette
Luftkanal
Laufrolle
Leitrad
Gebläse
Ölkühler

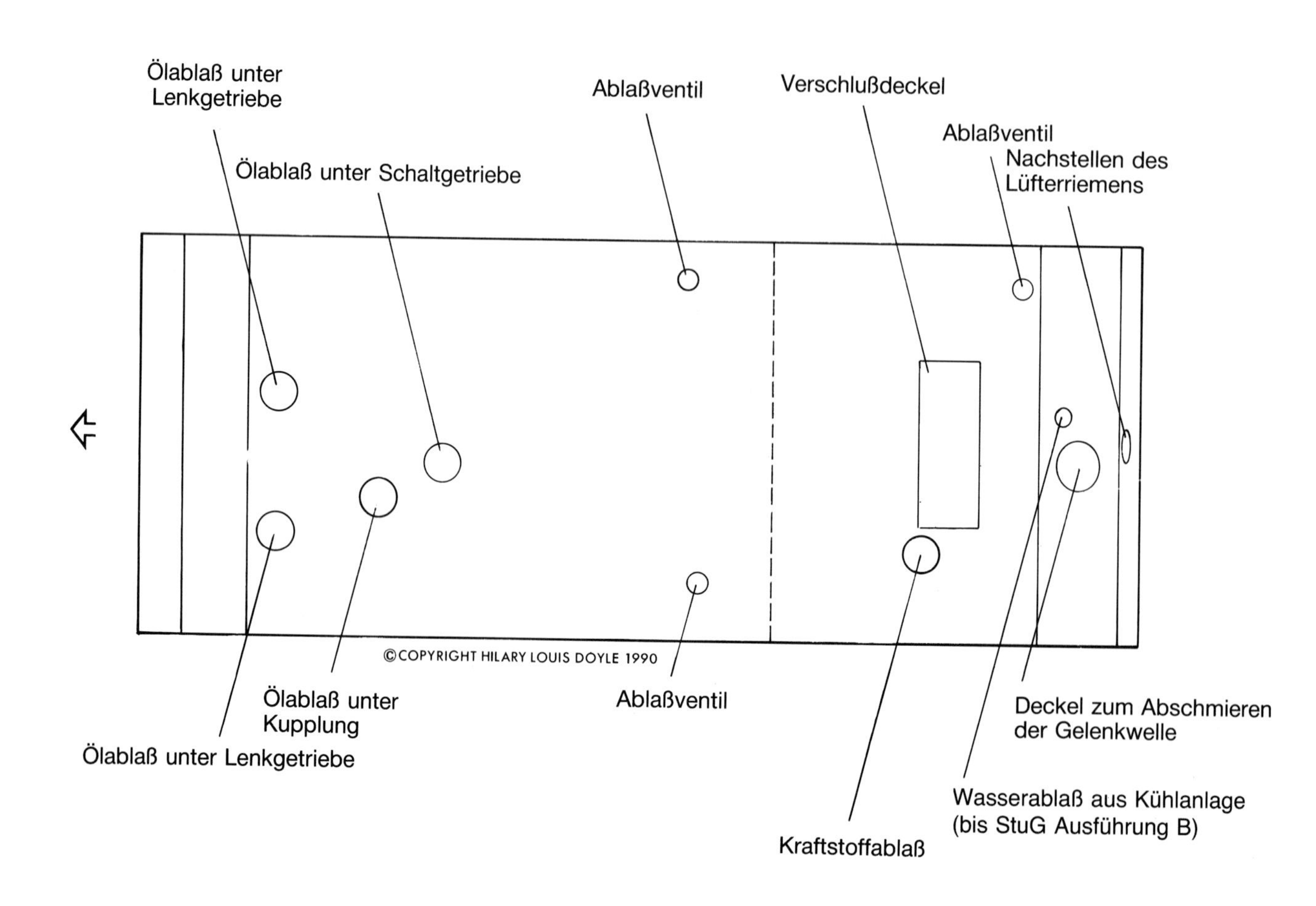
Ölablaß unter
Lenkgetriebe
Ablaßventil
Verschlußdeckel
Ablaßventil
Nachstellen des
Lüfterriemens
Ölablaß unter Schaltgetriebe
©COPYRIGHT HILARY LOUIS DOYLE 1990
Ölablaß unter
Kupplung
Ablaßventil
Ölablaß unter Lenkgetriebe
Deckel zum Abschmieren
der Gelenkwelle
Wasserablaß aus Kühlanlage
(bis StuG Ausführung B)
Kraftstoffablaß

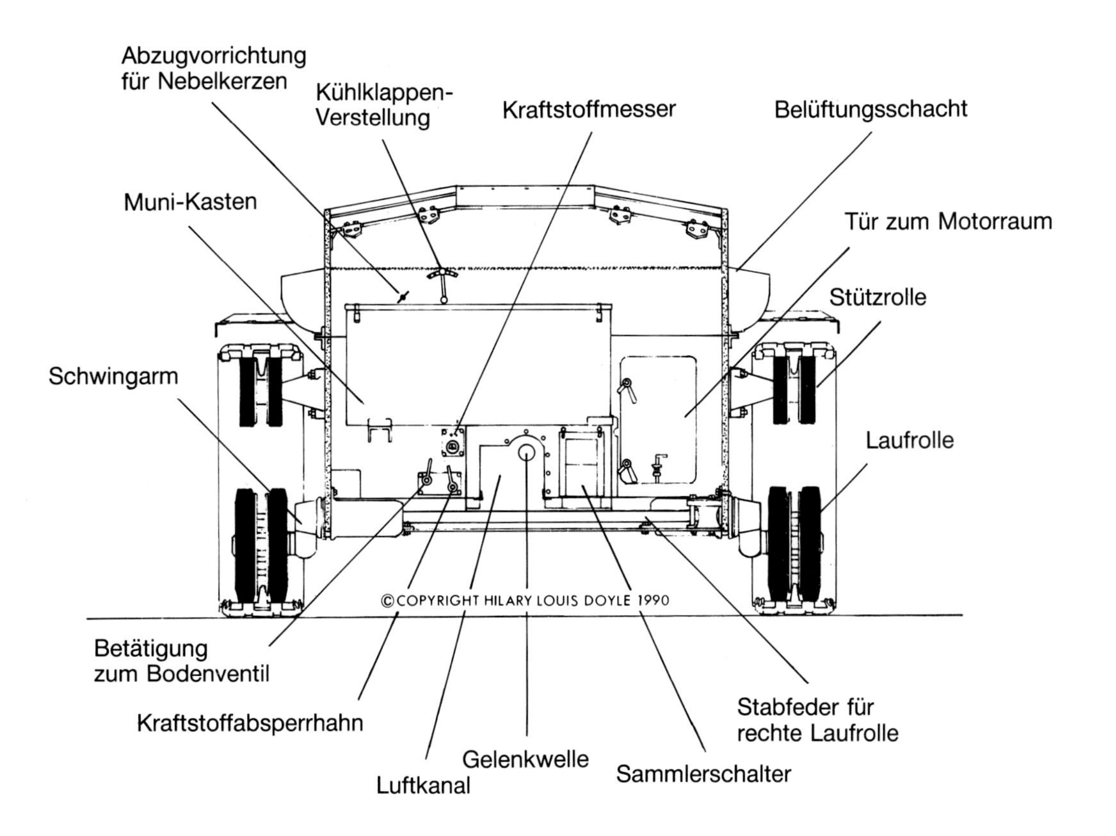
Abzugvorrichtung
für Nebelkerzen
Kühlklappen-
Verstellung
Kraftstoffmesser
Belüftungsschacht
Muni-Kasten
Tür zum Motorraum
Stützrolle
Schwingarm
Laufrolle
©COPYRIGHT HILARY LOUIS DOYLE 1990
Betätigung
zum Bodenventil
Kraftstoffabsperrhahn
Stabfeder für
rechte Laufrolle
Gelenkwelle
Sammlerschalter
Luftkanal

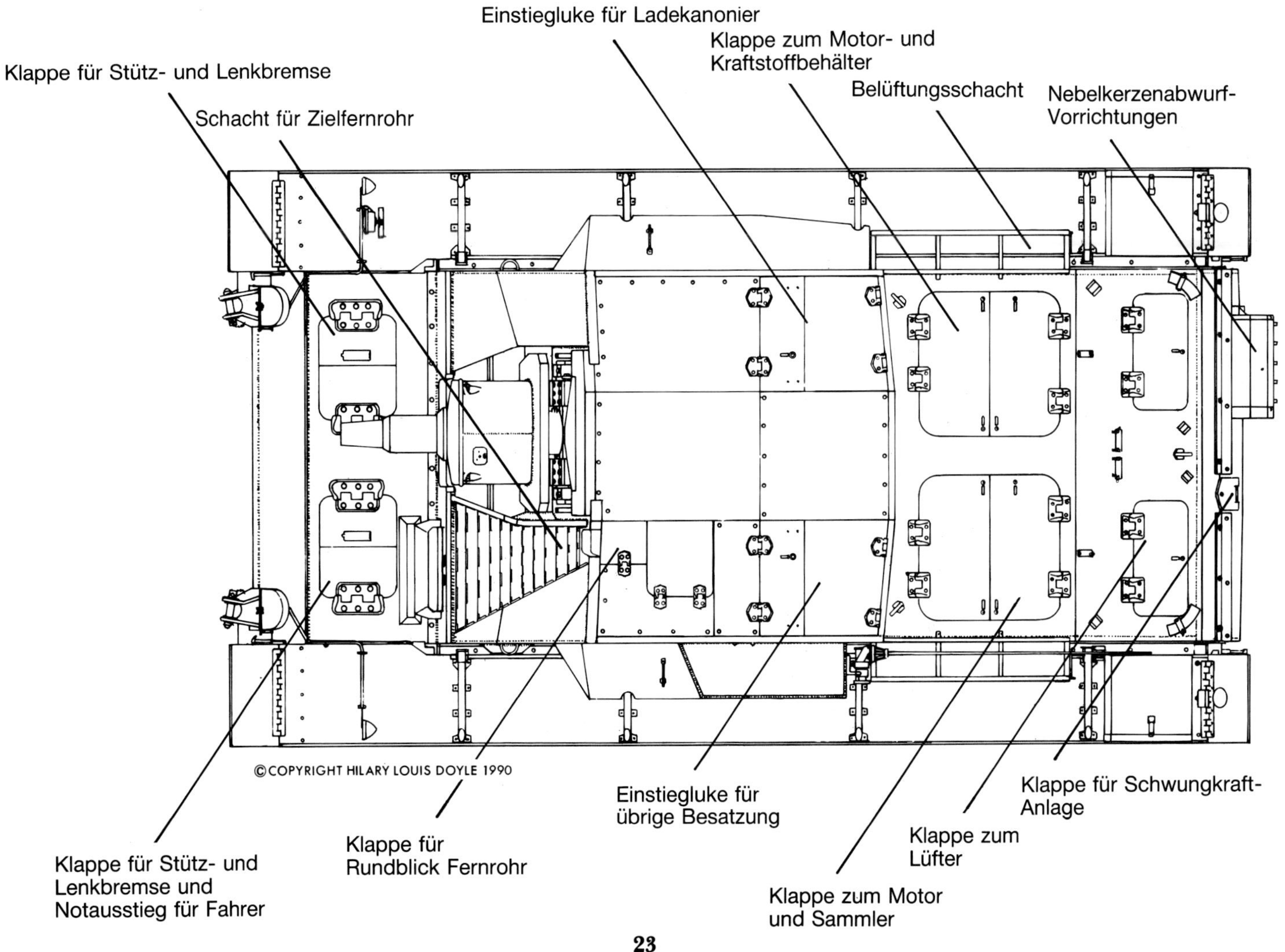
Einstiegluke für Ladekanonier
Klappe zum Motor- und
Kraftstoffbehälter
Klappe für Stütz- und Lenkbremse
Belüftungsschacht
Nebelkerzenabwurf-
Vorrichtungen
Schacht für Zielfernrohr
©COPYRIGHT HILARY LOUIS DOYLE 1990
Einstiegluke für
übrige Besatzung
Klappe für Schwungkraft-
Anlage
Klappe für
Rundblick Fernrohr
Klappe zum
Lüfter
Klappe für Stütz- und
Lenkbremse und
Notausstieg für Fahrer
Klappe zum Motor
und Sammler

Sturmgeschütz, Ausf.A

The Heereswaffenamt contracted Daimler-Benz to produce 30 sPak chassis. The chassis series numbers for these first 30 Sturmgeschütz Ausf.A ran from 90001-90030.

On May 30, 1939, Fried.Krupp AG recorded that the final design for the sPak chassis would utilize the running gear and transmission of the Z.W.38. Krupp wanted to be included in the road testing in Berka and Kummersdorf. The decision in favor of the 5./Z.W. (Panzerkampfwagen III chassis) for the Sturmgeschütz was accompanied by long and sometimes fierce arguments between the designers of the running gear (mostly Krupp) and the Heereswaffenamt WaPrüf 6 (Kniepkamp).

During a conversation with Kniepkamp and Krupp representatives on the 23rd of May 1939, there was urgent query from General Becker as to how long the manufacture of the transmissions for the Z.W. (Panzerkampfwagen III) was going to flounder. Kniepkamp indicated that the manufacturing difficulties had been rectified and suggested that in order to continue production of the tank chassis, that the transmissions be fitted with a "Hochtrieber" device. This device was also known as an accelerating gear. This "Hochtrieber" was designed to accelerate the rotating portions of the transmission to the high rpm rate required for shifting gears, which could be higher than the maximum rpm rate of the engine. Once fitted with a transmission modified with a "Hochtrieber", the Panzerkampfwagen III were to be tested for acceptance, then the superstructures were dismounted, the transmissions were to be removed, and the superstructures mounted again. Afterwards, these same modified transmissions were to be installed into five additional Panzerkampfwagen III for acceptance tests, removed, reinstalled in new vehicles and so on. This ideawas dropped based on consideration of the extra work created by assembly, disassembly, mounting and removal of the transmissions and superstructures.

In contrast to the statements by Kniepkamp, other representatives from Daimler-Benz were of the opinion that the transmissions still did not properly function without breaking down. In addition, the shock absorbers that were absolutely necessary for the torsion bar suspension design had not yet been accepted. The new shock absorbers designed by Boge were not expected to be ready until July of 1939.

The Sturmgeschütz, Ausf.A, left side.

Right side view of the Sturmgeschütz, Ausf.A.

The Sturmgeschütz, Ausf.A, view of the left front.

The Sturmgeschütz, Ausf.A, view of the left rear.

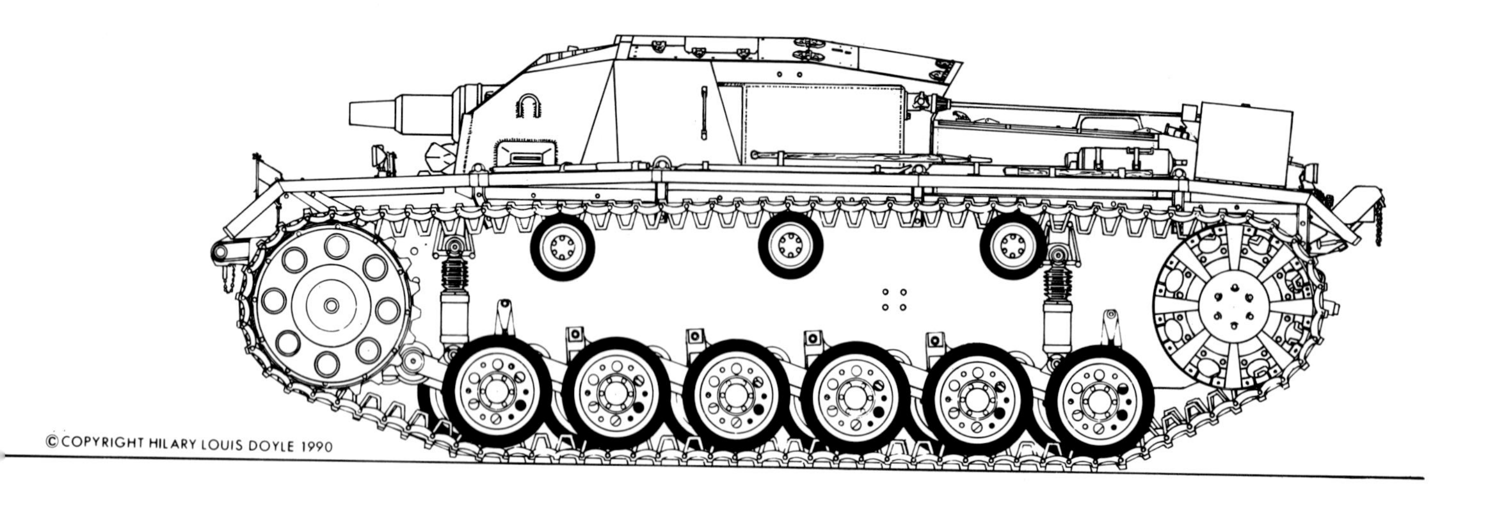

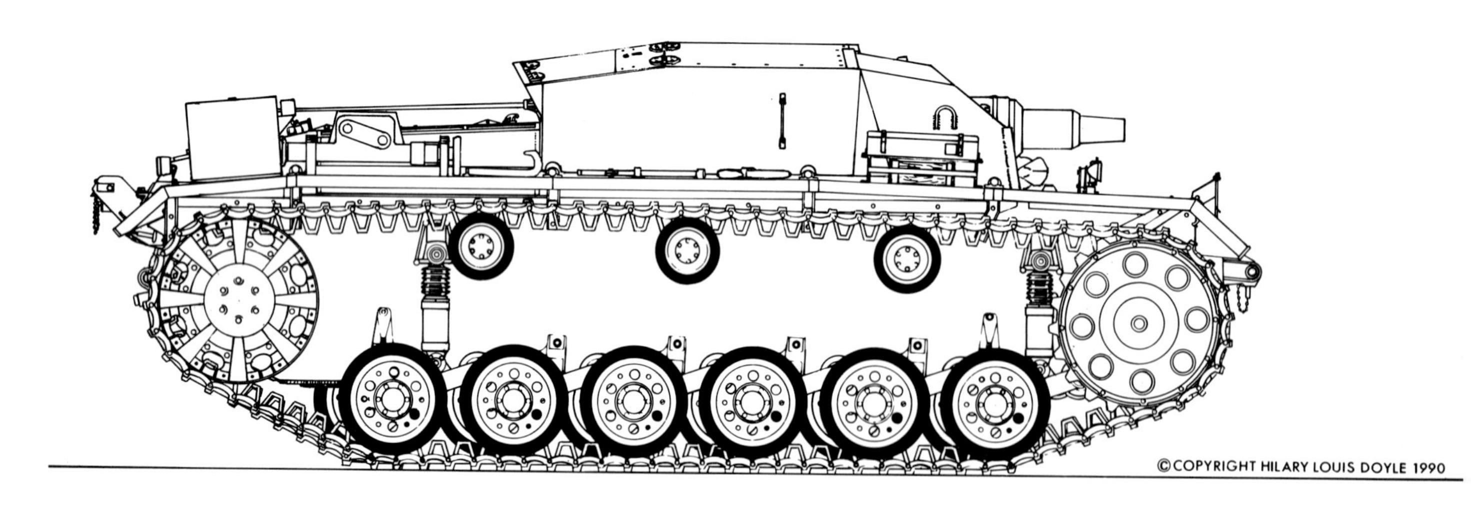

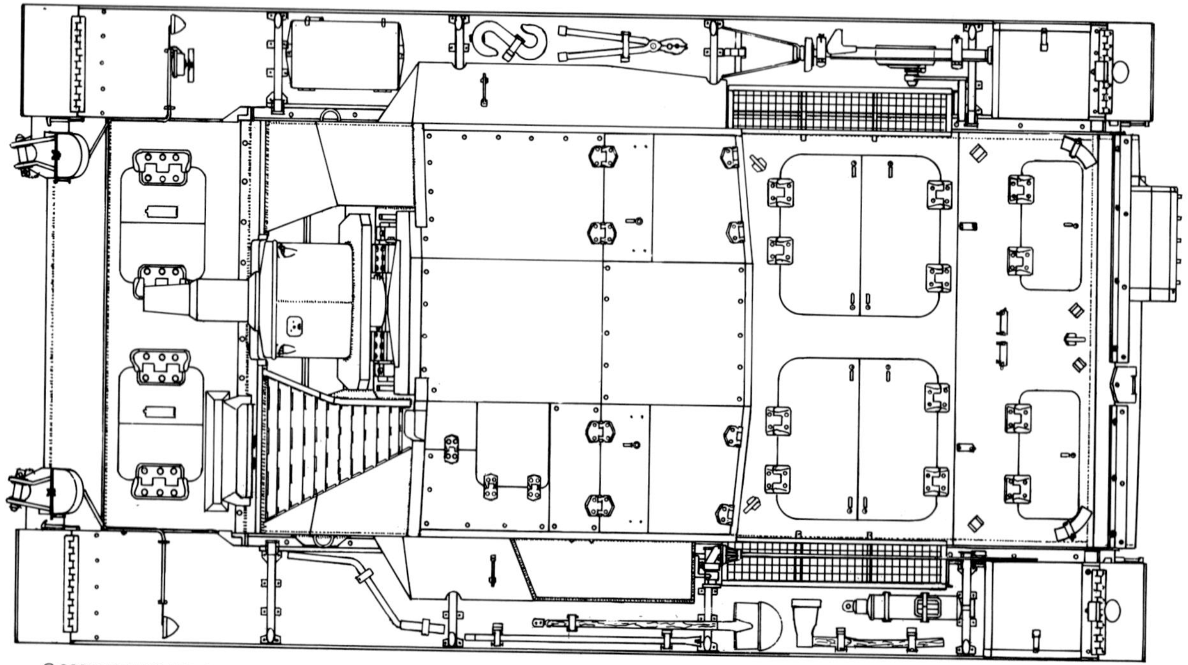

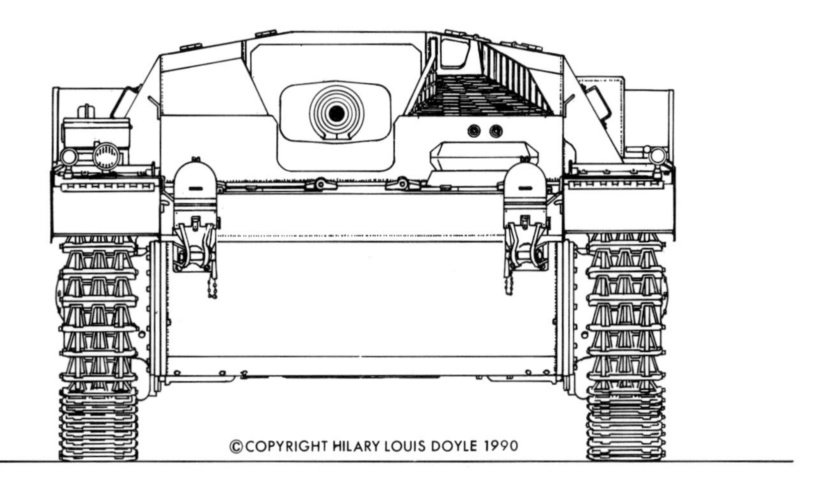
©COPYRIGHT HILARY LOUIS DOYLE 1990

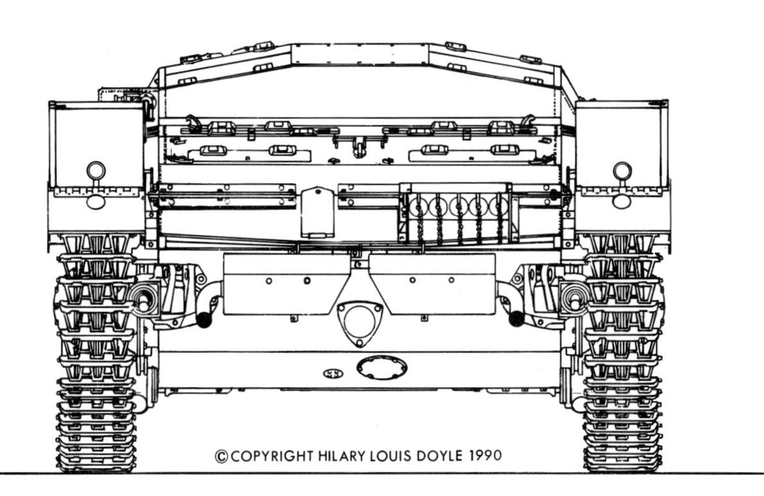
©COPYRIGHT HILARY LOUIS DOYLE 1990

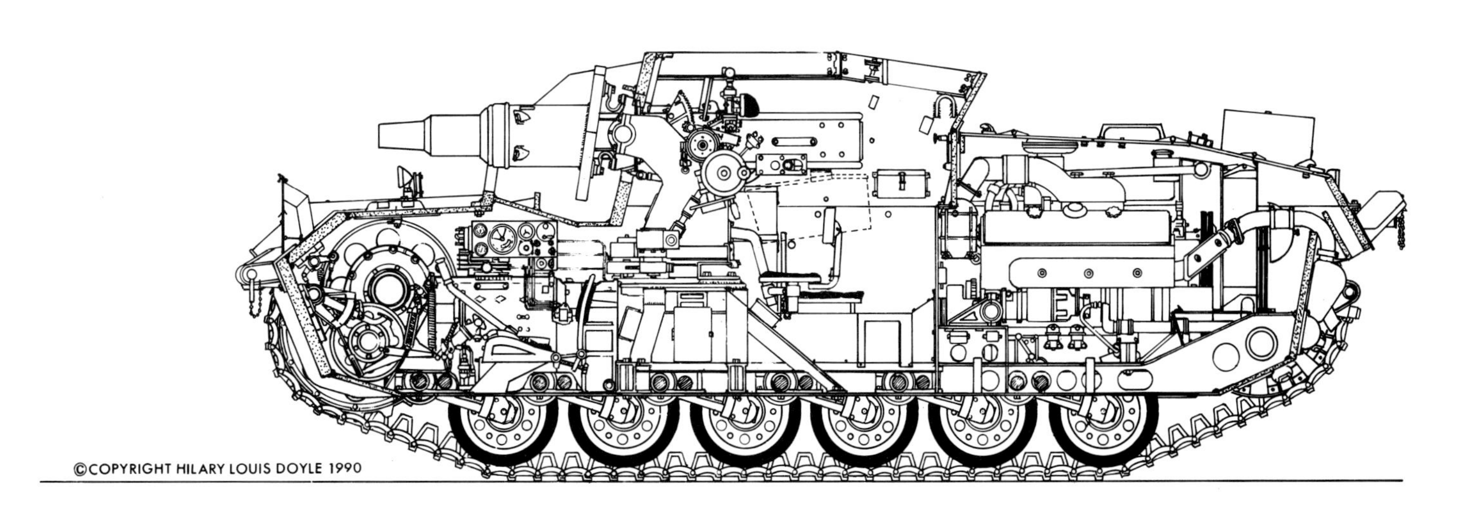
©COPYRIGHT HILARY LOUIS DOYLE 1990

A memorandum dated October 13, 1939, concerning the "Pz.Sfl.III (sPak)" (the official name for the Sturmgeschütz until March 1940) indicated the following situation:

1. Development of the Pz.Sfl.III (sPak) has been concluded. The introductory phase has begun.
2. On hand are 5 Pz.Sfl.III (sPak) with fully-functional weapons but with soft steel superstructures.
3. Output of the first production series consisting of 30 Pz.Sfl.III (sPak) is to occur from December 1939 to April 1 1940. Output of the second series of 250 is to start in April 1940 with 20 vehicles per month, barring any complications.
4. Further development of the Pz.Sfl.III is projected to occur with an upgraded gun, the 7.5cm Kanone L/41, muzzle velocity 685 meters per second. The first fully-functional test vehicle with sheet steel superstructure is planned for May 1940.

On December 12th, 1939, destructive test firings against sPak armored components were conducted at Kummersdorf. A complete sPak armored hull with superstructure and mantle were tested. The "test standard" used was the 37mm anti-tank gun firing a shell weighing 0.695kg with an initial muzzle velocity 760mps, at a distance of 100 meters.

The results:

After the first shot on the center shield above the armor jacket of the gun mantle, the upper weld between the shield and the jacket had a 300mm fissure in it. The two upper mounting plates between the armor and the gun mount had been bent together, about 2mm, otherwise no distinguishable peculiarities.

Two additional rounds hit the upper right corner of the shield and one round hit the crown. The effect of these shots was such that the weld between the jacket and the mantle was completely cracked and the bolts of the upper right mounting plates were sprung. In an earlier firing test of a Krupp-Appa III gun mantle, the welds between the jacket and gun mantle were also sprung.

The WaPrüf was to inform Krupp if there was to be any modification to the design of the gun mantle.

Daimler-Benz was not only the developing firm for the design, but was also responsible for the assembly of the first series of "Panzer-Selbstfahrlafette III."

Normally, the assembly plant only conducted their work utilizing components and systems delivered by other firms. These were delivered to the assembly plant ready to install by the so-called subcontractors. The assembly was carried out either in groups of vehicles all completed together or on an assembly line. Suppliers up to this time for the Sturmgeschütz construction program were:

Brandenburger Eisenwerke GmbH
Production of the hull, superstructure, gun mantle armor

Fried.Krupp AG
Design of the 75mm gun and delivery of 14 of these weapons

Wittenauer Machinenfabrik AG
Production under license of the 75mm gun

Maybach Motorenbau GmbH
Design and production of the HL 120 TRM engine and design of the SRG 32 8 145 gearbox

Nordbau
Production under license of the HL 120 engine

Zahnradfabrik Friedrichshafen
Production under license of the SRG 32 8 145 transmission

Hans Windhoff
Design and production of the radiators

Leitz
Production of the periscopic gun sights for self-propelled guns

The failure of one of these suppliers could severely influence the ability of the assembly firm to meet their production goals.

Daimler-Benz Werk 40 in Berlin-Marienfelde was to produce the initial series of 30 sPak (Series 1, Pz.Sfl.III) to meet the following schedule:One in December 1939, four in January of 1940, eleven in February, seven in March, and seven in April. This production schedule was not met. Although Daimler-Benz succeeded in completing the first chassis in December of 1939, it was not, as requested, a complete Sturmgeschütz with superstructure.

In early January 1940, a memorandum explained the reasons for delays in the delivery of this first production series of the 7.5cm Kanone (Pz.Sfl) (another semi-official name for the Sturmgeschütz).

The end of the production of the first 30 examples of the sPak, originally planned for the 1st of April, is changed to ten in the month of April and four in the month of May for

Sturmgeschütz, Ausf.A with Infanterie-Regiment "Grossdeutschland", France 1940.

the following reasons:

1. Due to the requirement for repairing many Panzerkampfwagen III damaged in Poland, the machines utilized for the production of individual parts were taxed to the extent that they have caused a delay in sPak production.
2. The many changes in design. For example: The superstructure was originally planned to be an open design. In view of tactical considerations the vehicle was planned to have a roof, and later the roof design was changed upon request of the troops to facilitate indirect fire. A further displacement of the protective armor side walls for reasons of observation meant multiple changes to the blueprints. The armor manufacturer received the final plans quite late.
3. The normal ZW-transmission (Maybach SRG 32 8 145) with accelerator gear was to be used in the sPak. After the first installation it became clear that a change to the transmission housing was necessary due to interference with the gun mount.
4. The Brandenburger Eisenwerke firm, at this time, the sole producer of the armor plate for the sPak, experienced an unexpected setback due to the failure of the armor material to meet specifications.
5. Bringing in foreign parts was especially difficult for the Daimler-Benz assembly plant; particularly detrimental was the embargo in place since the cooling of international relations. The sub-contractors were no longer in the position to deliver their completed or raw goods to the assembly firm.

France 1940., Sturmgeschütz, Ausf.A with "SS-Leibstandarte."

Sturmgeschütz, Ausf.A. The bullet deflectors can be seen along the sidewalls of the sight aperture.

After minor delays, Daimler-Benz at its Werk 40 in Berlin Marienfelde completed the chassis and assembled the first series of Sturmgeschütz Ausf.A as follows:

	Chassis	Sturm-geschütz
December 1939	1	0
January 1940	4	1
February 1940	11	3
March 1940	7	6
April 1940	7	10
May 1940	0	10

Description

The armored hull also served as the frame for the chassis. Situated within it was the powertrain with the engine in the rear of the hull. Radiators and cooling fans were mounted on both sides of the engine. A fuel tank, installed on the right side of the vehicle, was enclosed to protect it from the heat of the engine.

A firewall separated the engine from the fighting compartment. A door in the firewall permitted access to the engine compartment. The drive shaft (in a tunnel in the fighting compartment) connected the engine to the clutch mounted on the transmission. In front of the transmission, the connecting bevel gears were affixed to the steering gears. Two connectors led to the drive brakes and the final drives on either side. The driver's seat was located to the left of the transmission.

Idler wheels were mounted on adjustable axles at the rear of the chassis. Between the drive wheels at the front and the idler wheels, on each side of the vehicle, there were six road wheels suspended via torsion bars. There were three return rollers mounted on each side of the hull above the road wheels. A Kgs.6111/380/120 track was driven by the drive wheel sprockets and covered on top by the fenders.

One of the two pairs of hatches in the glacis plate could be used by the driver as an escape hatch.

The engine, a Maybach HL 120TR, with gasoline carburetors, was an improved version of the HL 108, which had been used in the Z.W.-Series prototypes. Due to an increase in the bore of the engine from 100 to 105mm, the swept volume was increased from 10.8 to 12 liters. The crank shaft mounted on seven roller bearings had an eighth radial bearing set. The connecting rods mounted on sliding bearings were mounted next to one another. Power output was rated at 300 metric hp at 3000 rpm.

The running gear layout with torsion bars for the 5./Z.W. (Z.W.38).

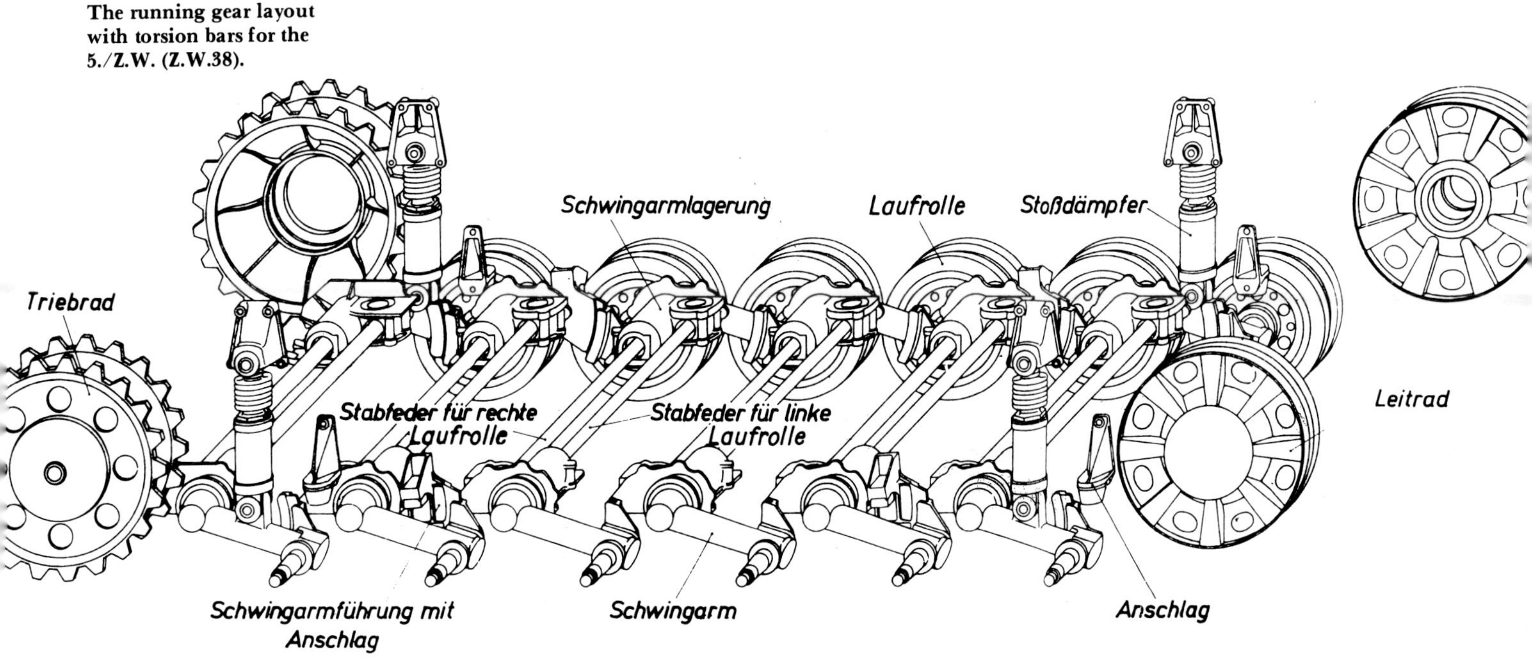

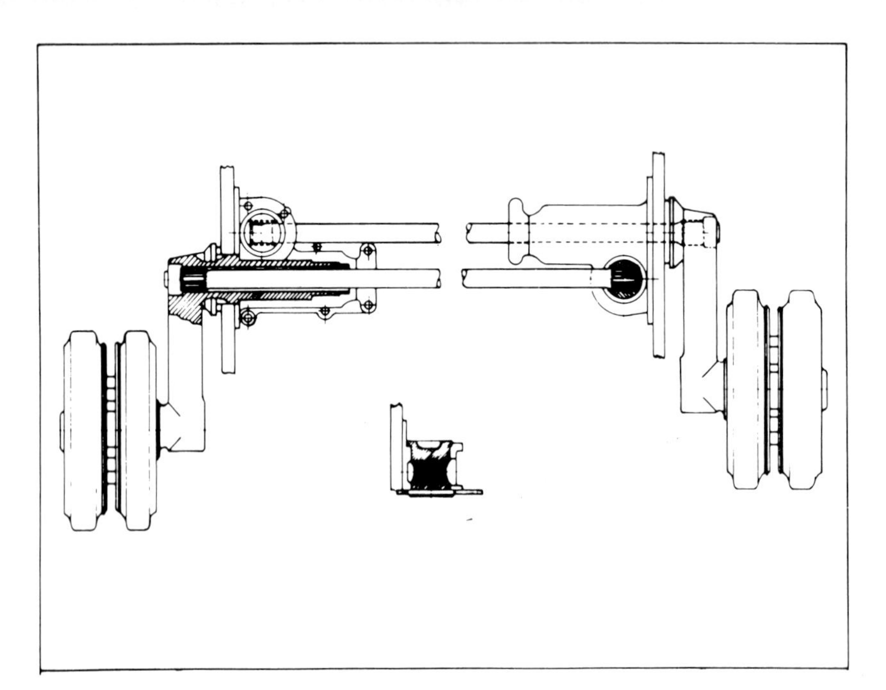

Roadwheel suspension arm assembly.

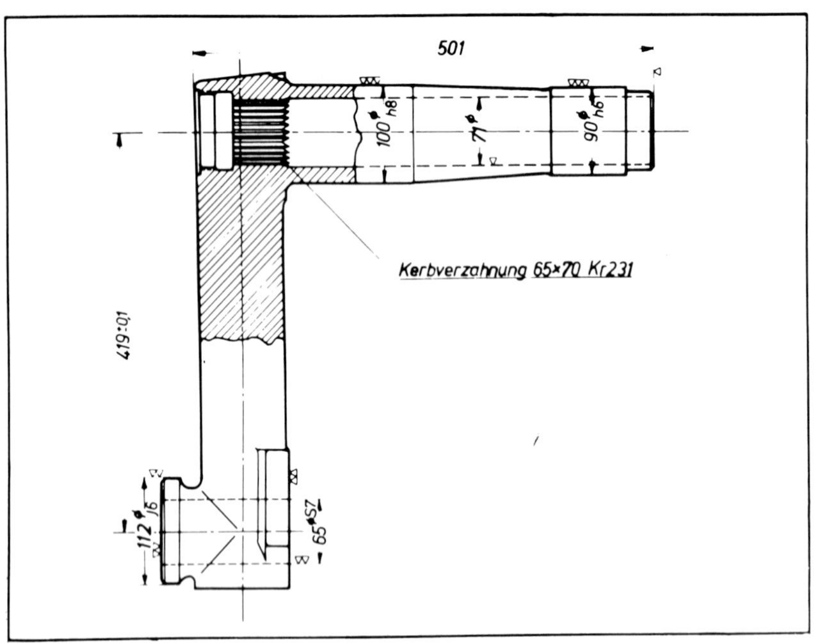

Dimensions for the suspension arm.

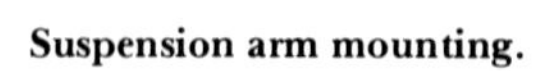

Suspension arm mounting.

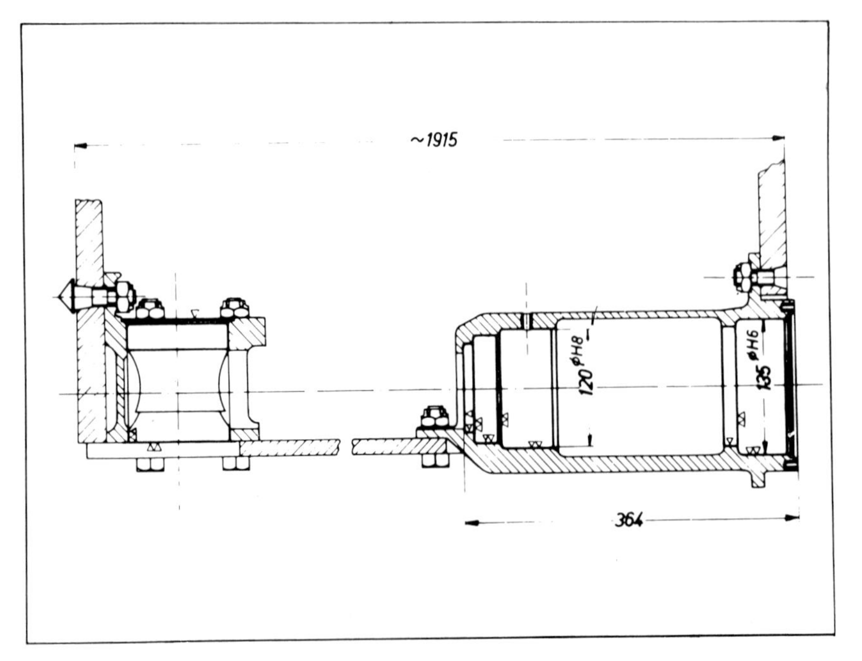

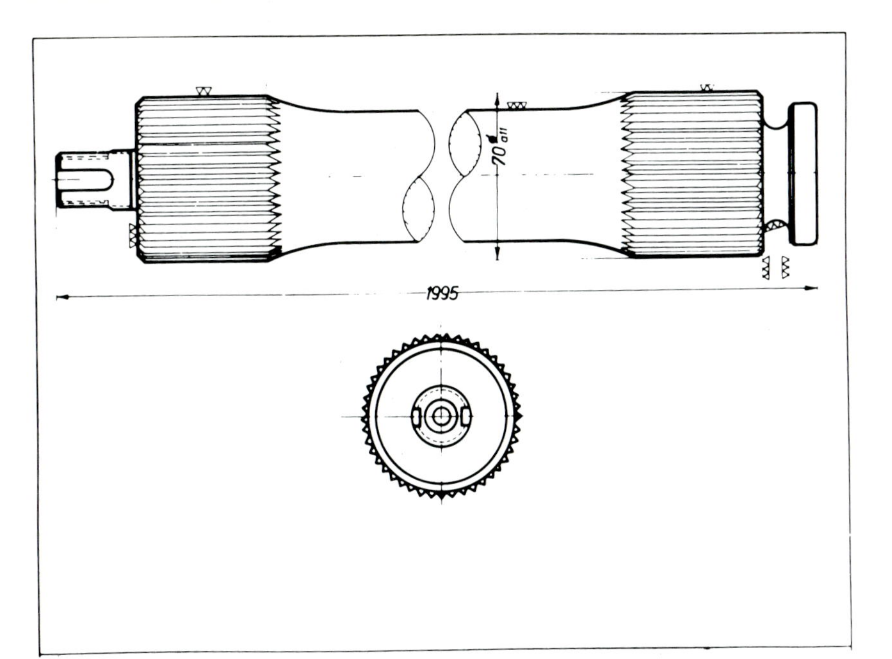

Details of the torsion bar.

Drive sprocket.

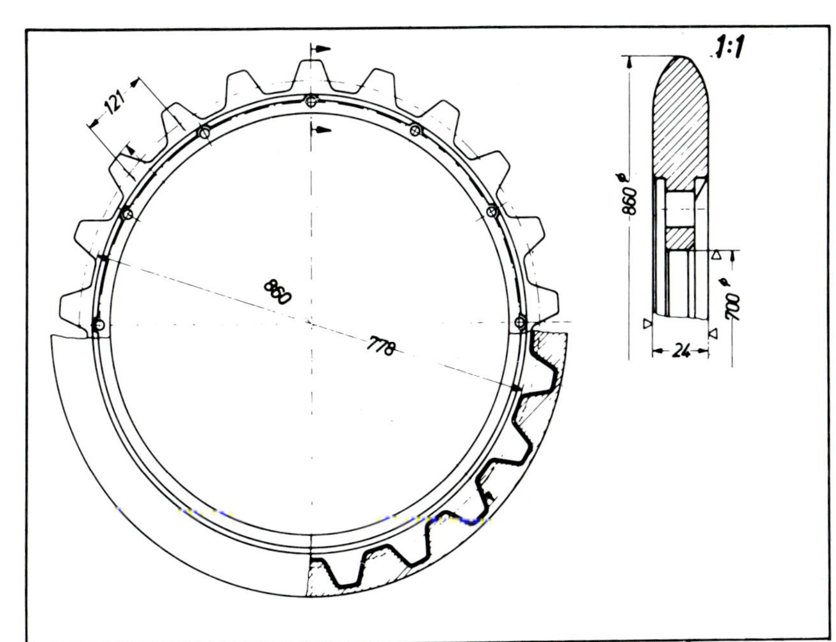

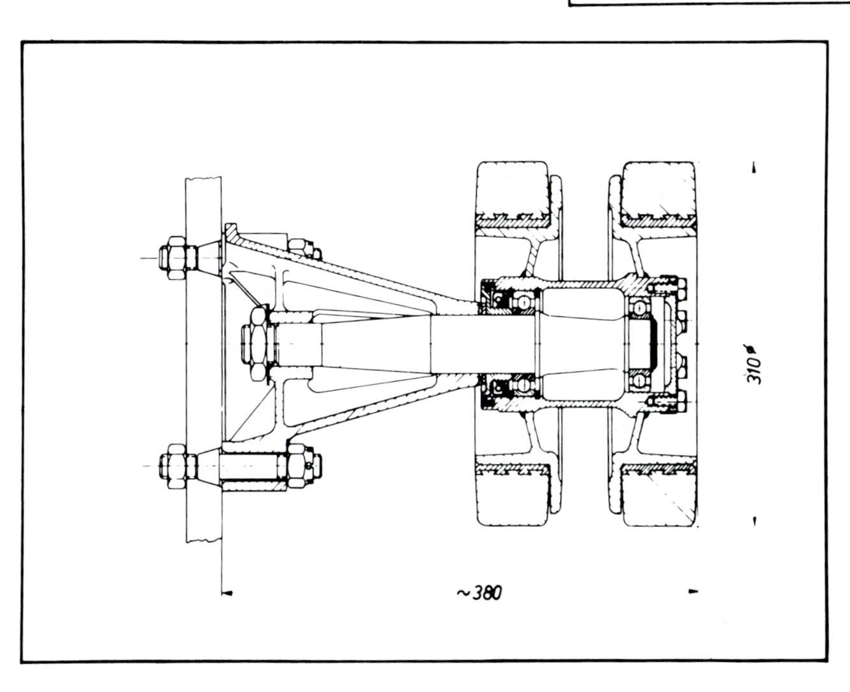

Return roller assembly.

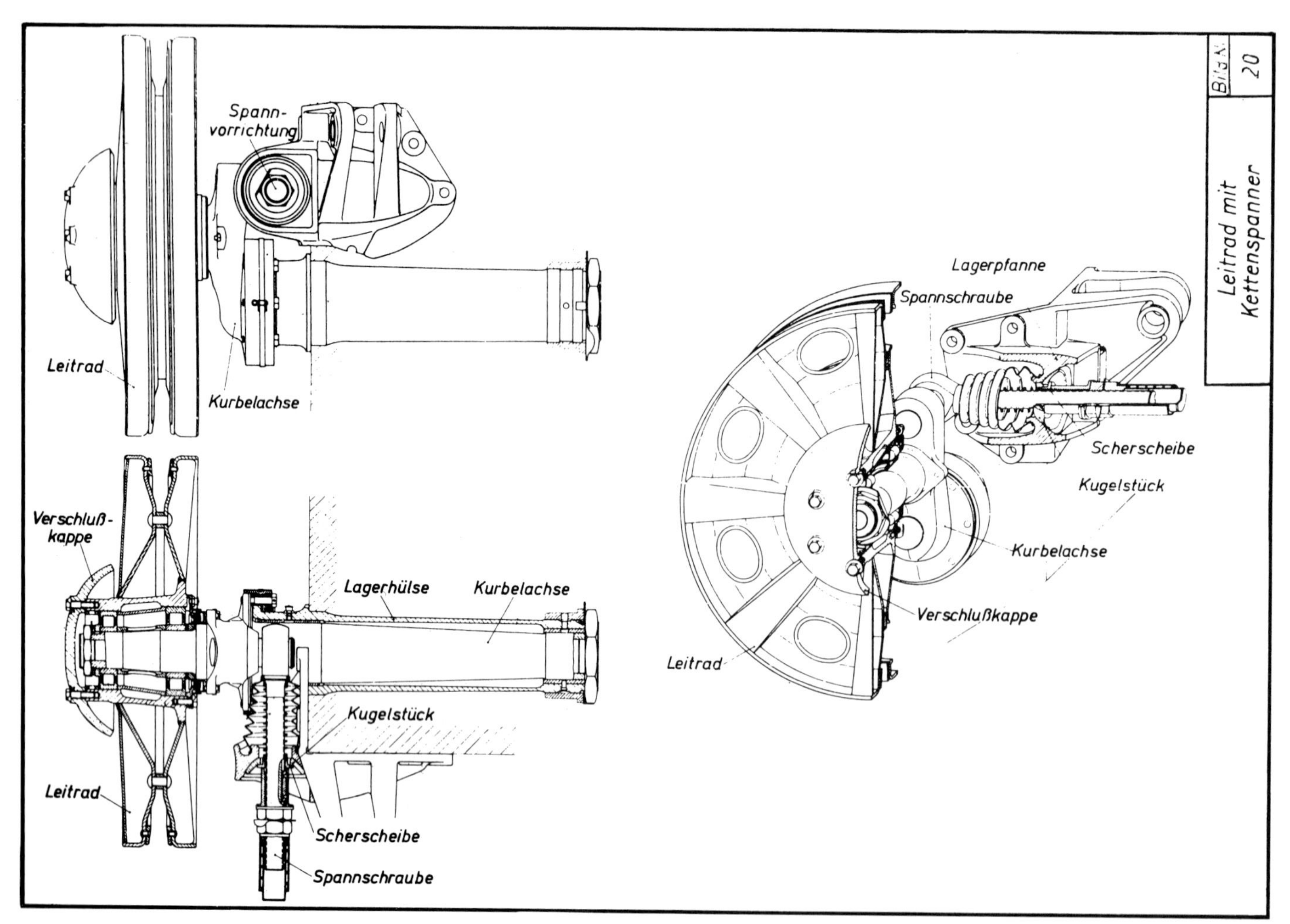

Idler wheel and track tensioner.

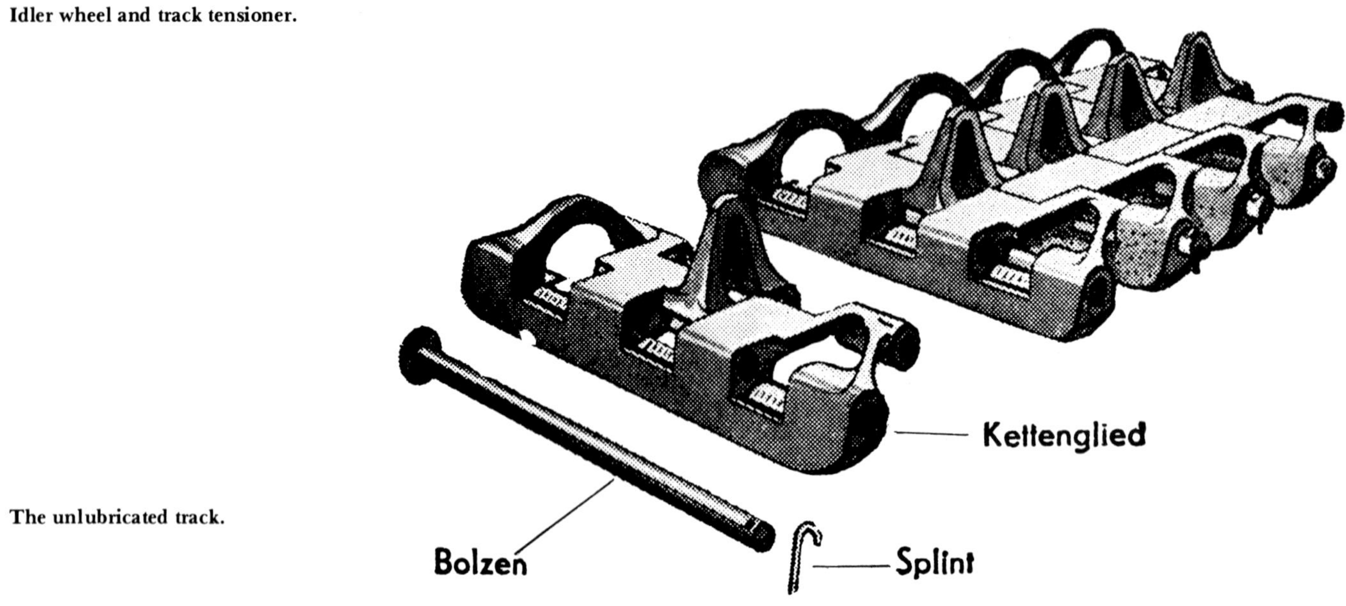

The unlubricated track.

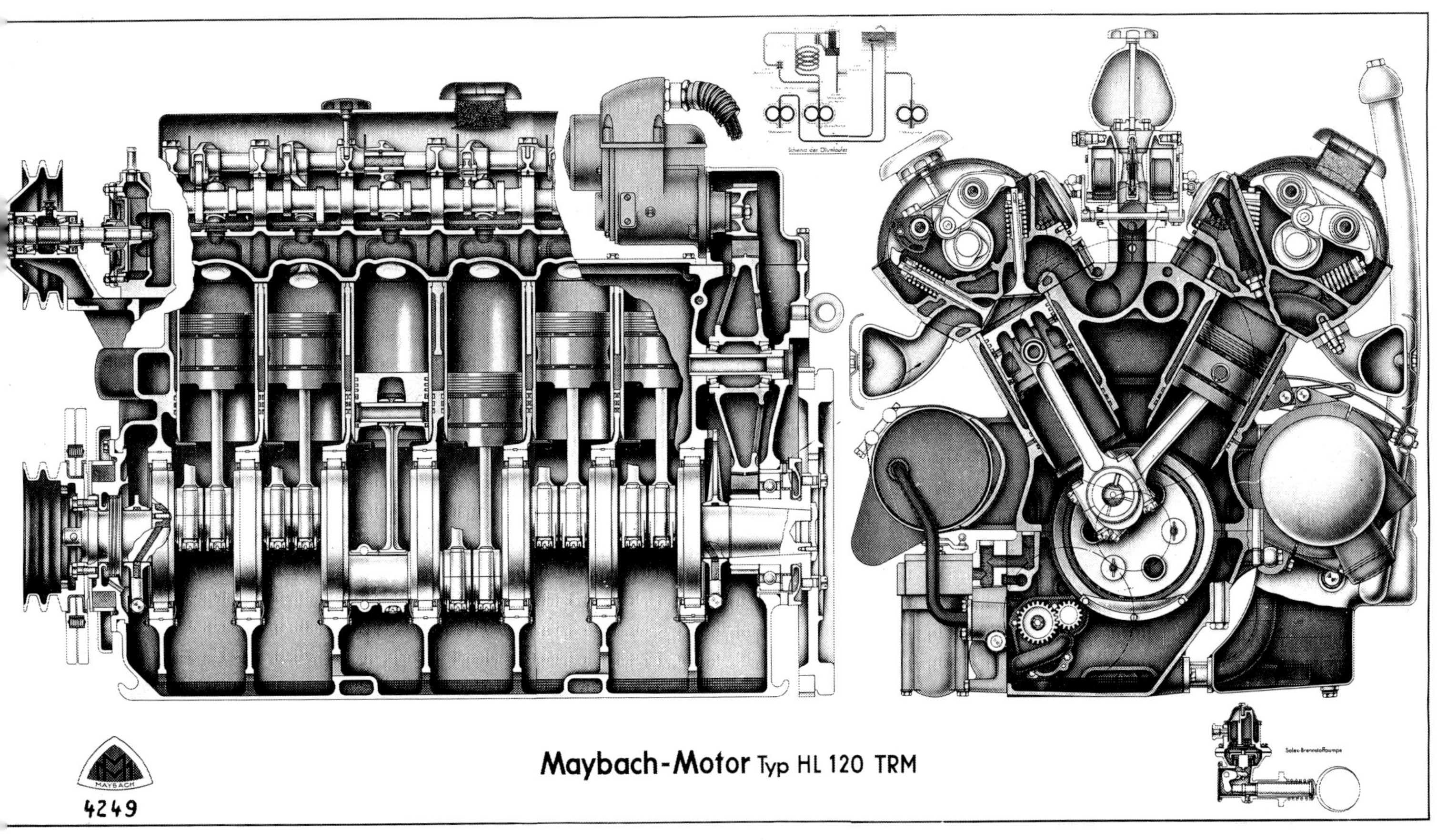

Cross-section and profile plan of the high-performance HL 120 Maybach engine.

The single fuel tank held 310 liters. Two mechanical fuel pumps delivered the fuel to the carburetors. An electric fuel pump aided starting. In the Ausf. A, the drive shaft transferred the torque from the engine via the main clutch to the transmission. The transmission was a Maybach SRG 32 8 145 adopted for a forward mounted clutch. This transmission had 10 forward gears as well as one reverse gear. Gear changes operated semi-automatically. Control of the transmission was through a vacuum operated selector and gear changing system coupled with a hydraulically operated gear synchronizing system (the Hochtrieber) and hydraulically operated clutch. When changing gears, the desired gear was manually selected using the preselector lever which set cams to a predetermined combination in a vacuum distribution box. To complete the gear change, the clutch pedal was depressed. After depressing the clutch pedal, the change became entirely automatic, since it initiated the series of events in the regulator that physically changed the gears.

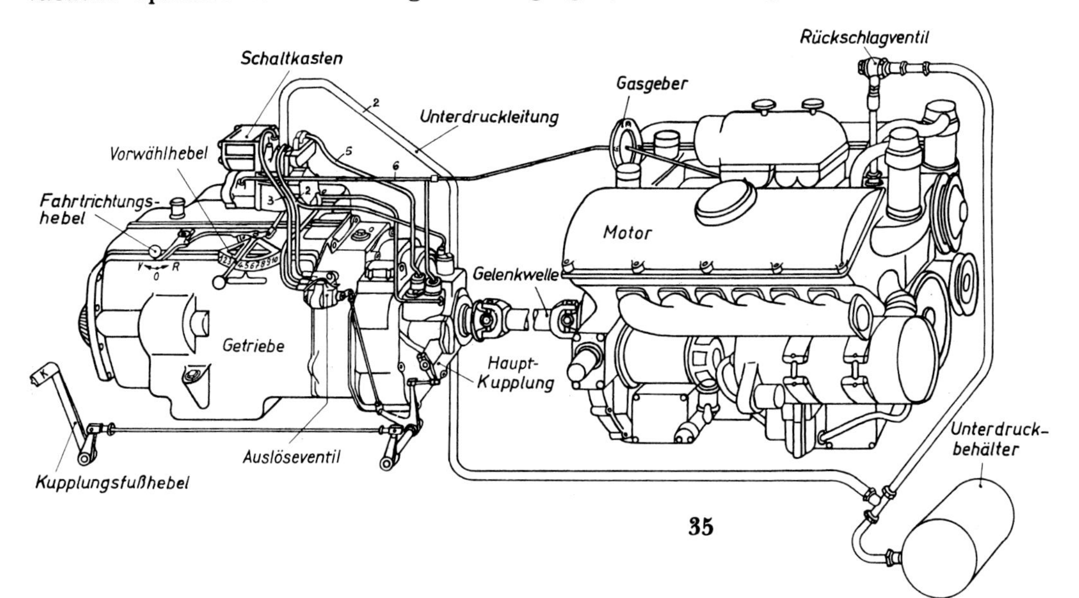

Engine and transmission for the Sturmgeschütz, Ausf.A.

Comparison of the configuration for the drive train for the Ausf.A (above) and the drive train introduced with the Ausf.B (below).

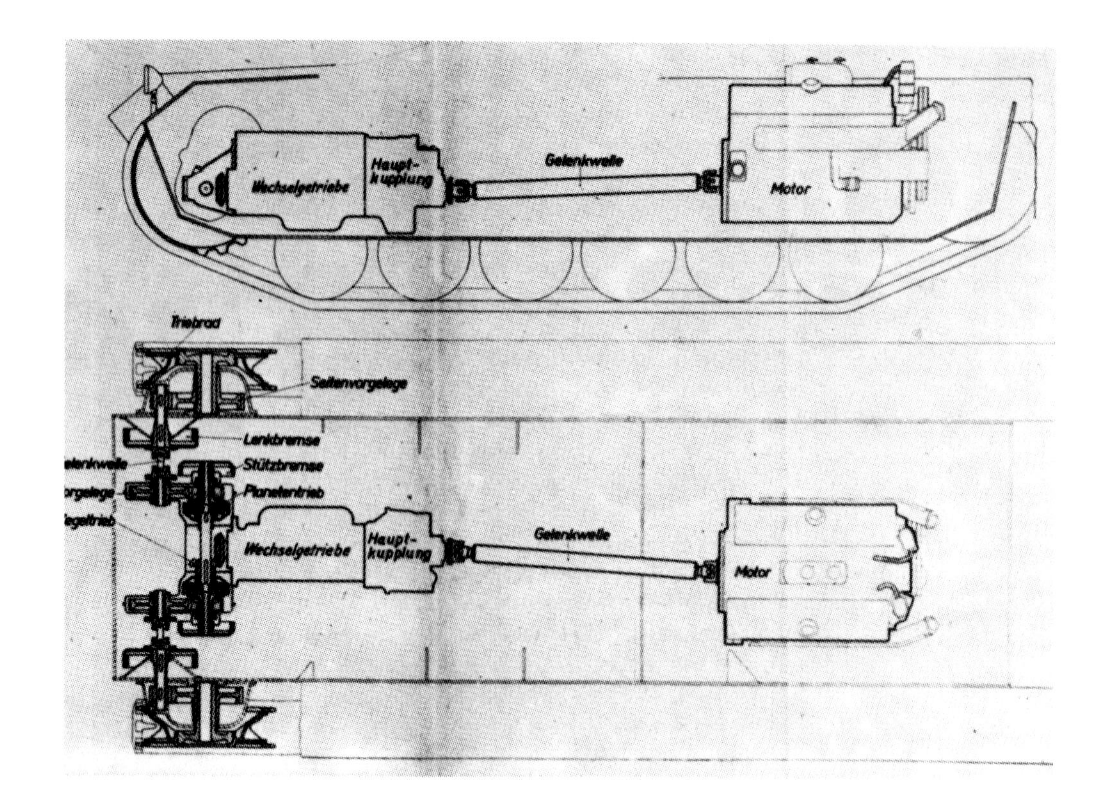

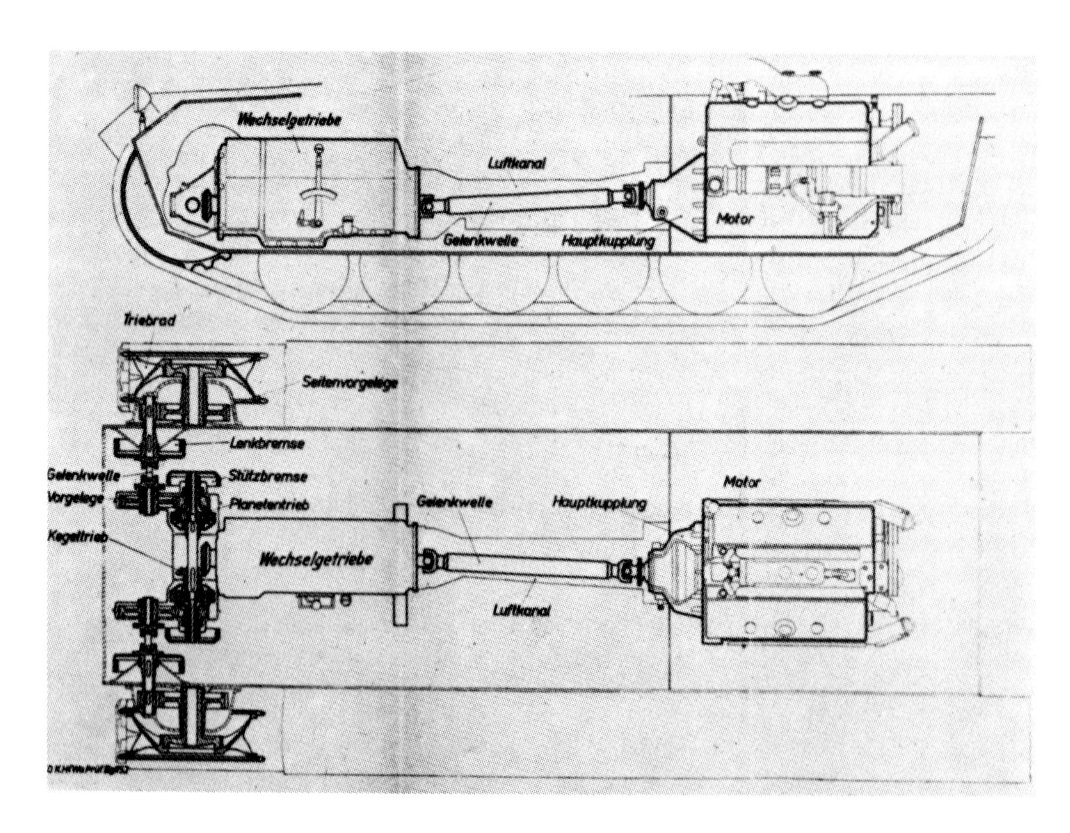

With the engine running at 2800rpm, the speed attained in each gear was:

1st gear	1:8	4.55 km/hr
2nd gear	1:5.88	6.27 km/hr
3rd gear	1:4.39	8.39 km/hr
4th gear	1:9	11.17 km/hr
5th gear	1:43	15.116 km/hr
6th gear	1:81	30.36 km/hr
7th gear	1:1.33	27.71 km/hr
8th gear	1:1	36.85 km/hr
9th gear	1.34:1	40 km/hr
10th gear	1.82:1	40 km/hr

Max. allowable speed: Reverse 1:7.9

The overall gear reduction ratio was 1:14.5. Ninth and tenth gears were overdrive gears — driver's were warned not to exceed a maximum speed of 40 kilometers per hour.

By the 21st of June 1939, Maybach had already delivered the documents needed to produce transmissions to Altmärkische Kettenwerke GmbH (Alkett). Also in 1939, the transmission known up until that time as the Schaltregelgetriebe (SRG) was renamed **VARIOREX**. All firms producing this transmission were to mark it with "Variorex transmission — under license of Maybach." Instead of the marking "SRG" which had been used up until this point, each new type transmission was marked with "VG" and an additional 3-element number:

1. The hundreds and tens values of the metric horsepower.
2. The hundreds and tens values of the torque (mkg).
3. The overall ratio of the transmission.

The Sturmgeschütz was fitted with a Bosch 12V lighting and starter system. Two Varta-Sammler 12 volt, 105 amp-hour batteries were installed. To prevent interference to radio reception and transmissions, the electrical system was suppressed in accordance with specification class M1.

The armor superstructure which had been closed on top to protect the crew, also protected sensitive internal equipment. Tools, spare parts and equipment were stored on the interior walls of the superstructure as well as under the deck plates. At the rear of the superstructure roof were two separate hatches; the right for the loader and the left for the commander. The gunner's hatch was near the front left corner of the superstructure roof. In front of this another hatch was installed for the head of the indirect periscopic gun sight.

The pedestal gun mount consisted of two box-shaped supports, connected on top, which supported the base plate for the gun. The rear support was braced by two struts on the left and right of the drive shaft tunnel. The supports were bolted to the hull side walls and to the floor, with spacers for initial alignment installed between the bolted surfaces.

Anti-skid floor plates rested above the torsion bars.

The ammunition was stored in metal bins with closeable covers. In the space to the right of the transmission, an ammunition rack held four smaller boxes, two with three rounds in each and two with two rounds in each. A total of 44 7.5cm rounds could be properly stored within the vehicle.

On the rear wall of the fighting compartment there were mountings to hold two sub-machine guns.

The 7.5cm Kanone L/24, with a barrel length of 1766.5mm, was installed as the main weapon. This same gun tube was also installed in the Panzerkampfwagen IV, Ausf.A through F. As mounted in the Sturmgeschütz, the gun had a maximum range of 6000m. The types of ammunition normally fired were:

— 7.5cm K.Gr.rot Pz (Vo 385 m/s) (capped armor piercing shell with high explosive filler)
— 7.5cm Gr.34 (Vo 420 m/s) (high explosive shell)
— 7.5cm Gr.38 Hl (Vo 450 m/s) (HEAT round)
— 7,5cm Nebel-Gr. (smoke shell)

The gun's weight was 490 kilograms, its price, RM 9150-. As mounted, the gun could be traversed through 24 degrees and moved through an elevation arc from -10 to +20 degrees.

The Rundblickfernrohr 32 (panoramic periscope) with 4X magnification and a 10 degree field of view served as an indirect fire gun sight. The head of the periscope protruded through the foremost hatch in the superstructure roof.

This first series of Sturmgeschütz only had "UKW-Empfänger h" (VHF radio sets) to receive but not transmit. The crew communicated amongst themselves via speaking tubes.

Production of the first series of 30 Sturmgeschütz, Ausf.A was completed in May 1940.

For reasons that have yet to be discovered from original documents, during the period from June through September 1940, 20 Sturmgeschütz were produced using normal Panzerkampfwagen III chassis. Their superstructures had the new hatch configuration above the gunner that has been previously associated with a design change for the Sturmgeschütz, Ausf.B. During this period, the frontal armor plates on Panzerkampfwagen III hulls were still only 30mm thick. Therefore, these converted chassis were reinforced with an additional 20mm armor plate bolted to the front nose plate.

This series of Sturmgeschütz retained the standard Panzerkampfwagen III chassis characteristics, including the escape hatches on the hull sides, the two maintenance hatches on the glacis plate (hinged to the front and rear) and the two vent cowlings on the upper nose plate to provide cooling air to the steering brakes. A hinge for the left maintenance hatch interferred with mounting a shot deflector on the glacis plate for the driver's visor. The running gear was still configured for the narrower 360mm wide track links. As revealed in the official parts manual for the "Fahrgestell Sturmgeschütz (7,5 cm) (Sd.Kfz.142)" dated February 1944, these 20 Sturmgeschütz were designated as Ausf.A and assigned chassis numbers in the 90401-90500 range. The Ausf.A designation was correctly applied because these converted Panzerkampfwagen III chassis still possessed the 10 speed Maybach SRG 32 8 145 drive train. The Ausf.B drivetrain was redesigned based on the reliable 6 speed, ZF SSG 77 transmission.

The first Sturmgeschütz produced by Alkett were completed in June 1940. The Alkett Sturmgeschütz production program for the first six months was planned and completed as follows:

	Planned	Produced	Accepted
June 1940	8	12	12
July 1940	22	22	12
August 1940	32	20	10
September 1940	24	29	29
October 1940	30	35	35
November 1940	36	35	35

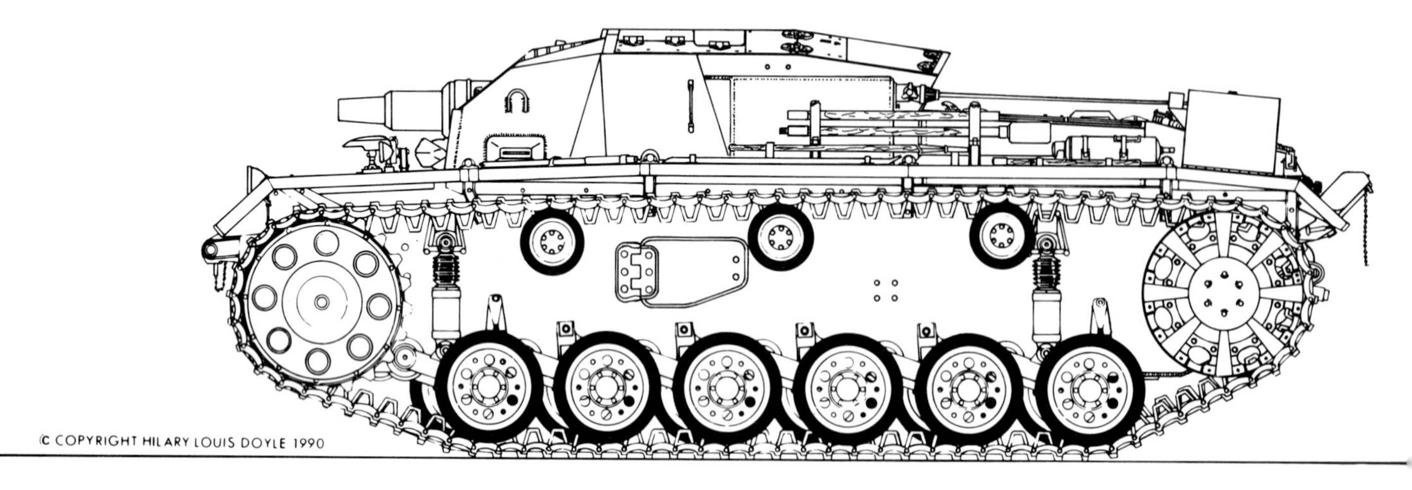

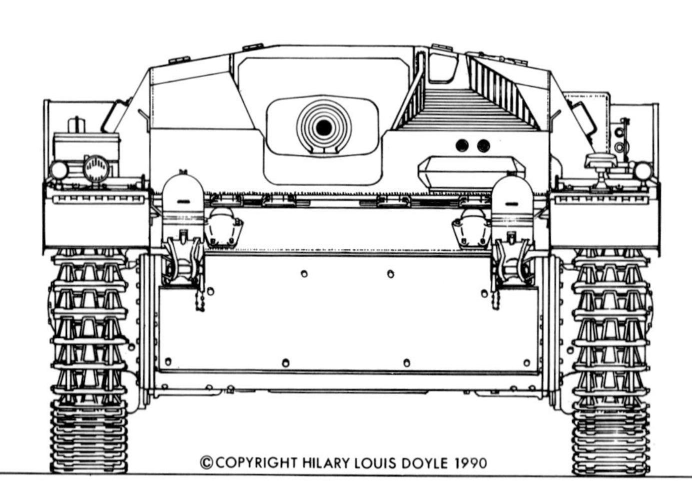

Sturmgeschütz, Ausf.A (with Panzerkampfwagen III chassis).

Sturmgeschütz, Ausf.A, produced on a Panzerkampfwagen III chassis.

Front view of the vehicle, of which twenty were produced.

The escape hatches on the sides of the Panzerkampfwagen III chassis were retained. ALR Jüterbog, 1940.

Sturmgeschütz, Ausf.A, with Panzerkampfwagen III chassis, produced with original running gear for the 360mm-wide track. A unique feature is the armored cover for the inertia-crank starter.

Sturmgeschütz, Ausf.B

The initial contract for the Ausf.B called for completion of 250 units with chassis number series 90101-90400. This was followed by a contract extension for 50 additional units with chassis number series 90501-90550.

Beginning with the Ausf.B, there were several important changes introduced in the drive train. The Maybach HL 120 TR engine was modified to become the Maybach HL 120 TRM (an engine with a dry sump lubrication system and Schnapper Magneto). Based on continuing problems with the Maybach VARIOREX transmission, a six-speed Aphon transmission designed and produced by Zahnradfabrik Friedrichshafen was installed. It was the SSG 77 six-speed synchromesh transmission rated for 77 mkg of torque. The main triple-plate clutch (LA 120 HD) was now flanged directly to the engine.

The hatches on the superstructure roof above the gunner were redesigned for the Ausf.B.

During the course of Ausf.B production, the following modifications were also introduced:

— the first return roller was moved farther forward, closer to the drive wheel, in order to dampen vertical surges in the track and to lessen the likelihood of thrown tracks

Sturmgeschütz, Ausf.B, chassis number 90111, produced by Alkett in July 1940. ALR in Jüterbog.

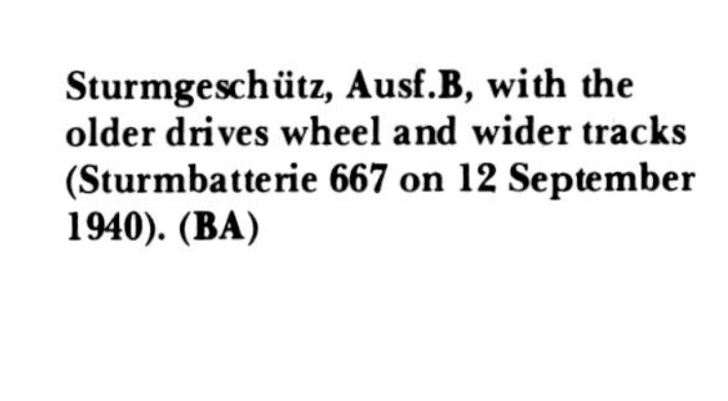

Sturmgeschütz, Ausf.B, with the older drives wheel and wider tracks (Sturmbatterie 667 on 12 September 1940). (BA)

Sturmgeschütz, Ausf.B, with the older drive wheels modified for the 400mm tracks. 2m antenna. An armored guard was introduced to protect the smoke candle rack. (BA)

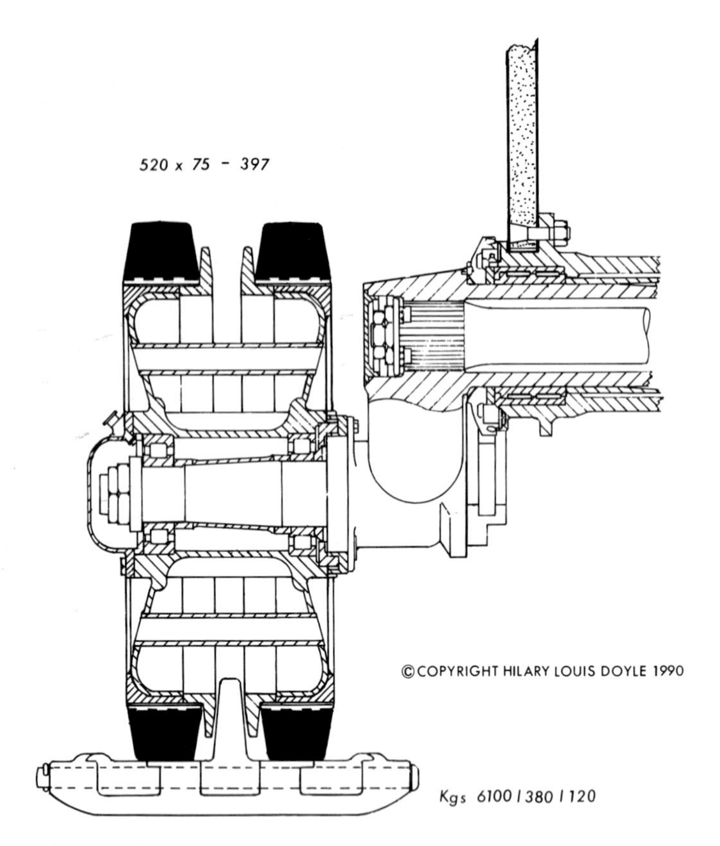

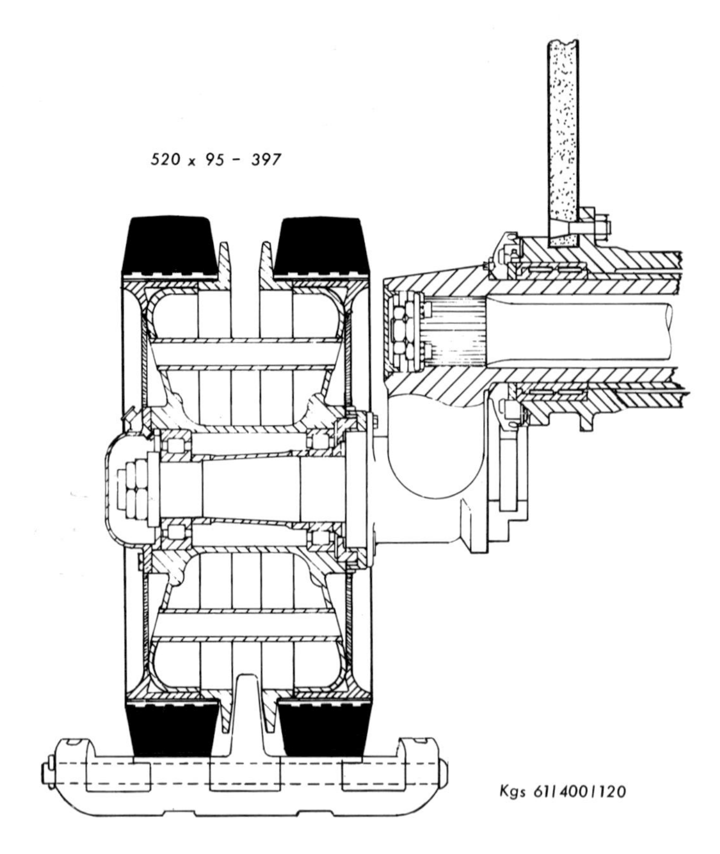

A comparison between the narrower and wider tracks and roadwheels.

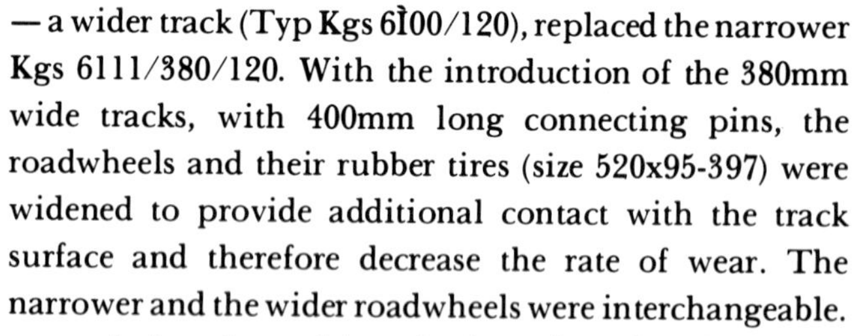

— a wider track (Typ Kgs 6100/120), replaced the narrower Kgs 6111/380/120. With the introduction of the 380mm wide tracks, with 400mm long connecting pins, the roadwheels and their rubber tires (size 520x95-397) were widened to provide additional contact with the track surface and therefore decrease the rate of wear. The narrower and the wider roadwheels were interchangeable.
— a redesigned cast drive wheel was introduced
— a removable guard over the clutch housing at the firewall was installed (for chassis numbers from 90321 to 90400 and from 90501 to 90550)

A published memorandum from the Army ordered that for the Sturmgeschütz (Sd.Kfz.142) the remaining armored vehicles were to have the following changes: On Sturmgeschütz (Sd.Kfz.142) chassis numbers 90101 through 90320, replace the guard over the clutch housing that is welded to the firewall with a removable guard. O.K.H. (Ch H Rüst u. BdE), 20. 9. 1941 — 3348/41 — AHA/Ag K/In 6 (III d)

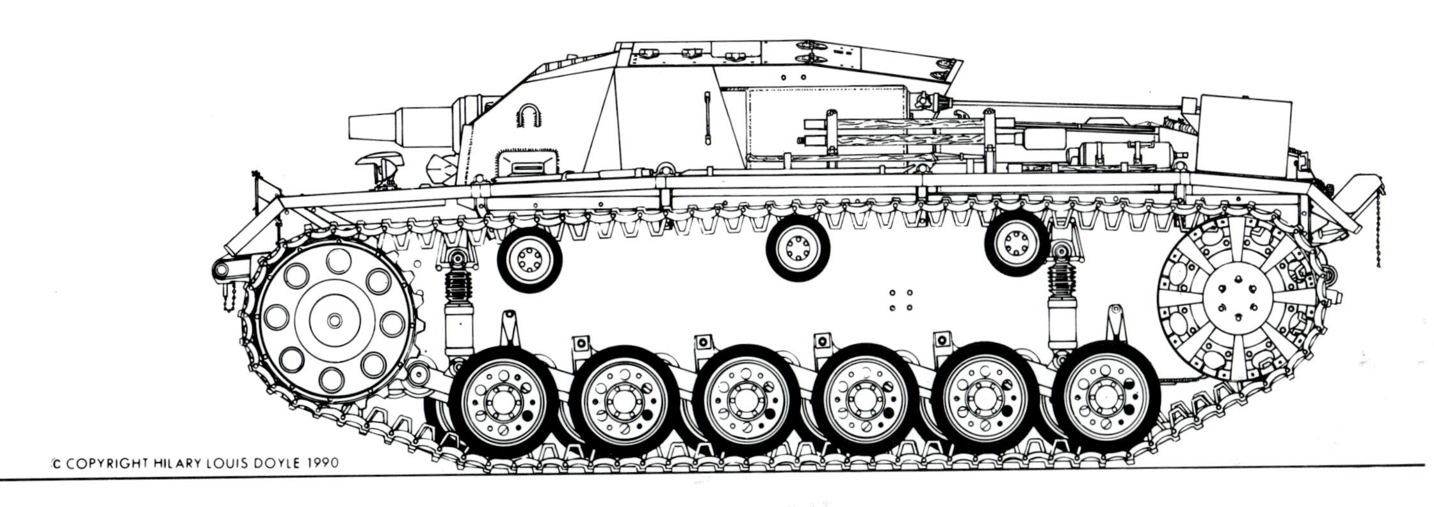

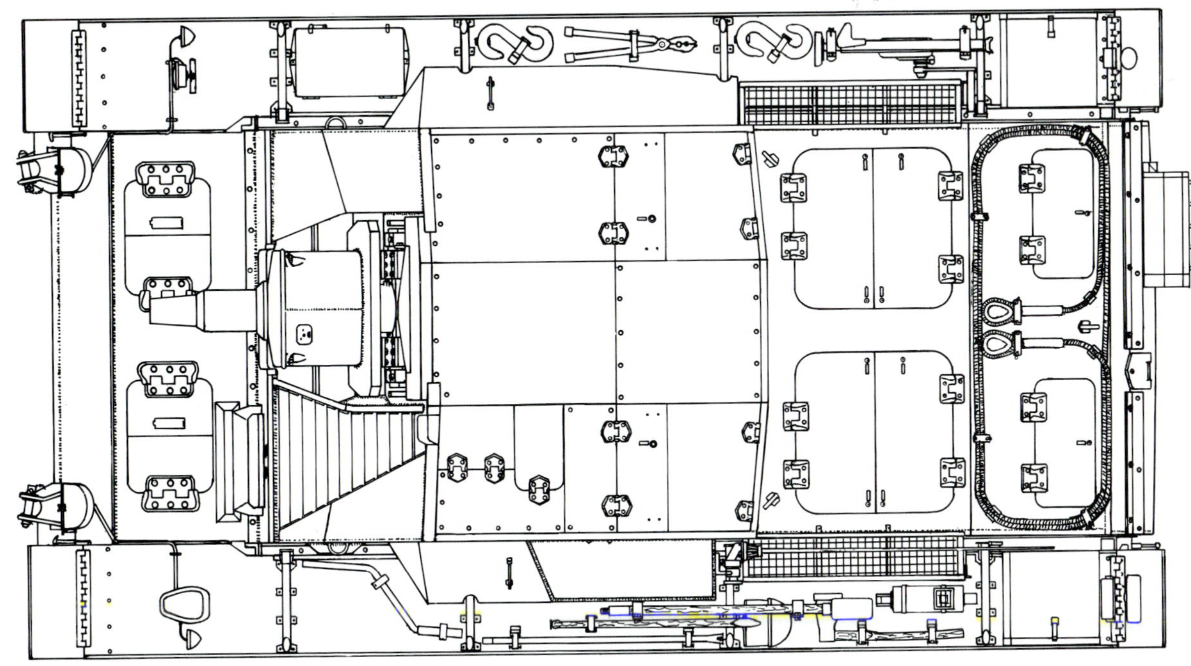

Above and left: Sturmgeschütz, Ausf.B (with drive wheels for the 380mm wide tracks).

Sturmgeschütz, Ausf.B (with drive wheels originally designed for 400mm wide tracks).

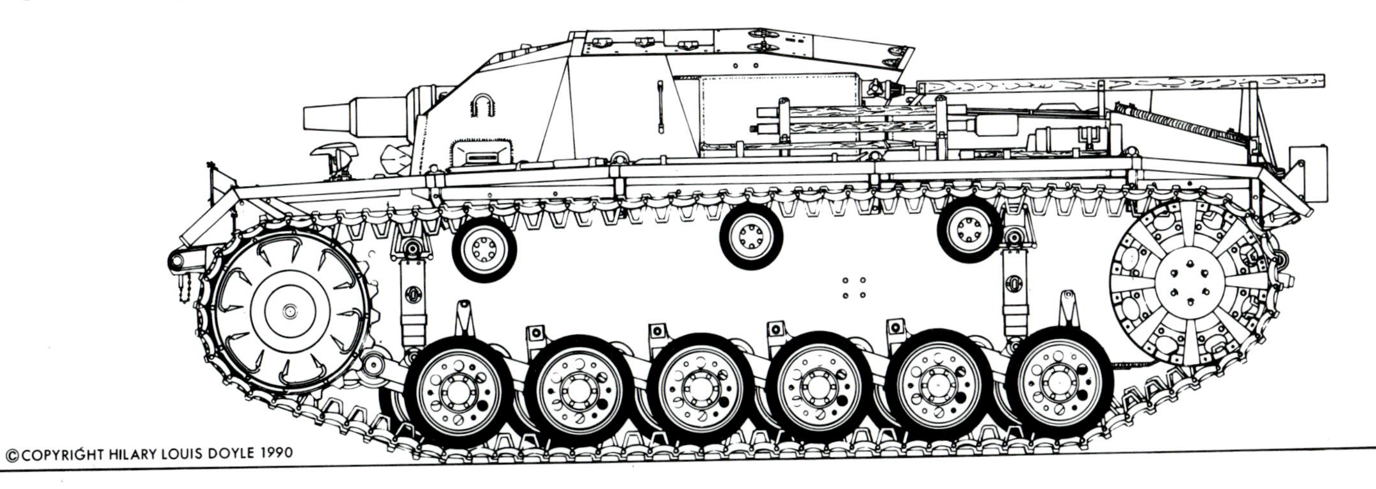

Above left: Sturmgeschütz, Ausf.B, as view from above the front of the vehicle.

Above right: Sturmgeschütz, Ausf.B, with all the superstructure hatches open.

Sturmgeschütz, Ausf.B, with the port for the sight aperture closed.

Left: A Sturmgeschütz, Ausf.B. Changing the drive sprocket on the drive wheel. The sight aperture port has been closed. (BA)

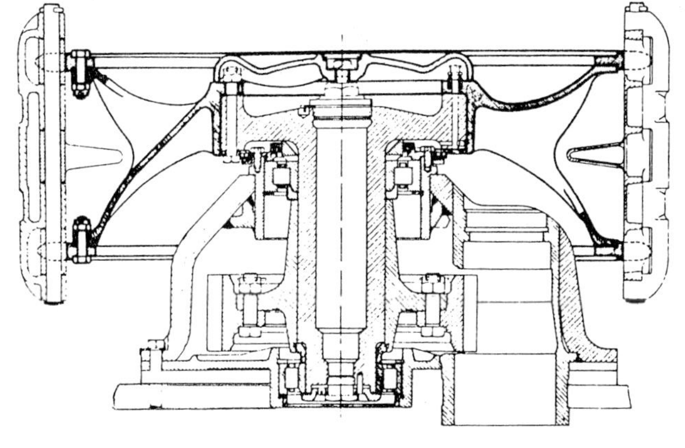

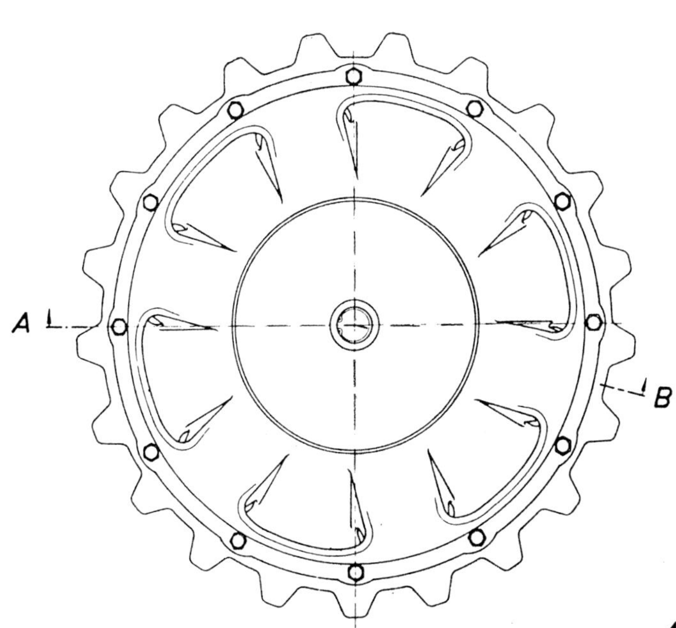

Below: The new drive wheel design for the Kgs 61/400/120 track.

Sturmgeschütz, Ausf.C

Fifty Sturmgeschütz, Ausf.C were produced beginning in April of 1941 (with chassis numbers 90551 through 90600).

On September 13, 1940 it was agreed in a meeting with Daimler-Benz that the Sturmgeschütze of the third production series would be equipped with a redesigned gun sight for self-propelled guns. The weak point in the armor protection, the gun sight aperture on the left front of the superstructure, was to be eliminated. For this reason it became necessary to alter the periscope mounting. Krupp was contracted to complete this undertaking. On September 27 1940, Krupp inquired as to when the new longer sight was to arrive at Daimler-Benz. October 1 was set as the date. Krupp wrote to the Daimler-Benz Werk 40 on September 30th 1940:

We have examined the installation of the gun sight in the Sturmgeschütz and have determined the appropriate opening in the superstructure roof. An insert will be used to raise the sight mounting by 80 mm and the mount will be moved forward by 60 mm. This gun sight configuration is only acceptable if a decrease in the field of view is permitted. Further, the side armor must be changed to slope more acutely as shown in draft sketch AKF 21011.

In March of 1941 start-up problems were reported with the Ausf.C. Due to the changes associated with the installation of the new direct fire, periscopic gun sight, the design for the superstructure roof had been changed.

The small two-part hatch above the gunner was replaced by a single piece enlarged hatch. An section was cut out of the right side of the new hatch to accommodate the head of the new direct fire, periscopic gun sight. The forward, sloped roof on both sides of the superstructure was changed. The sight aperture that had been present for the previous direct fire gun sight was eliminated. With the exceptions of redesigned idler wheels and the introduction of new engine air intake filters, the chassis itself remained virtually unchanged from the Ausf.B. During Ausf.C production, the locking mechanism for the maintenance hatches on the glacis plate were changed from those opened with a skeleton key to those opened with a square drift key.

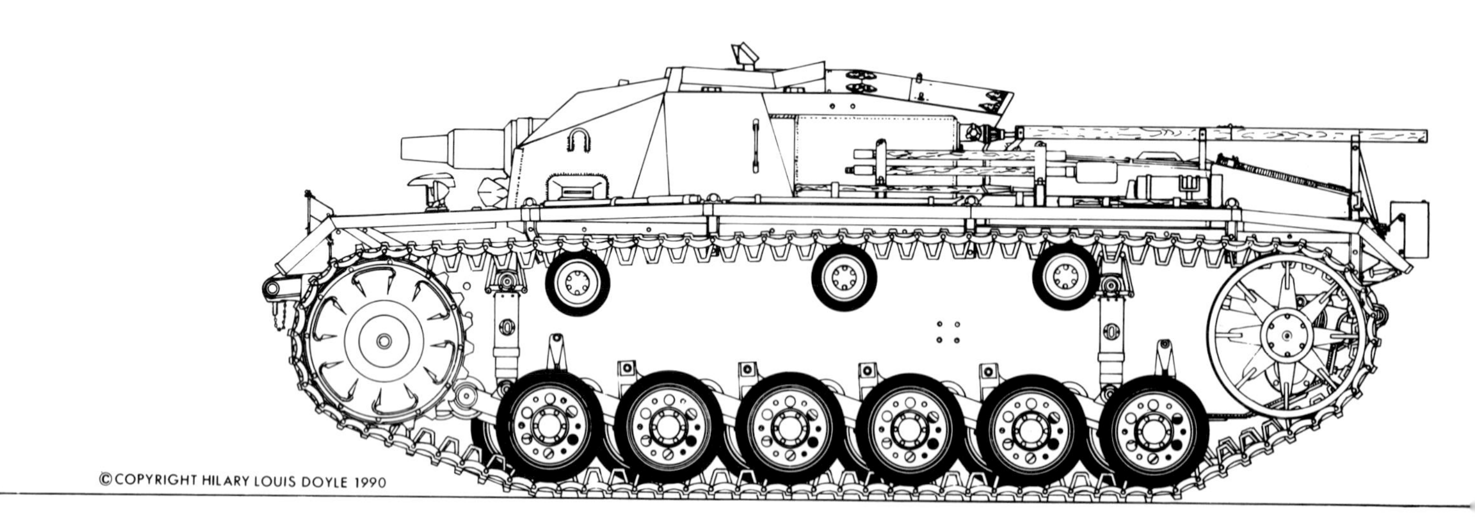

Sturmgeschütz, Ausf.C.

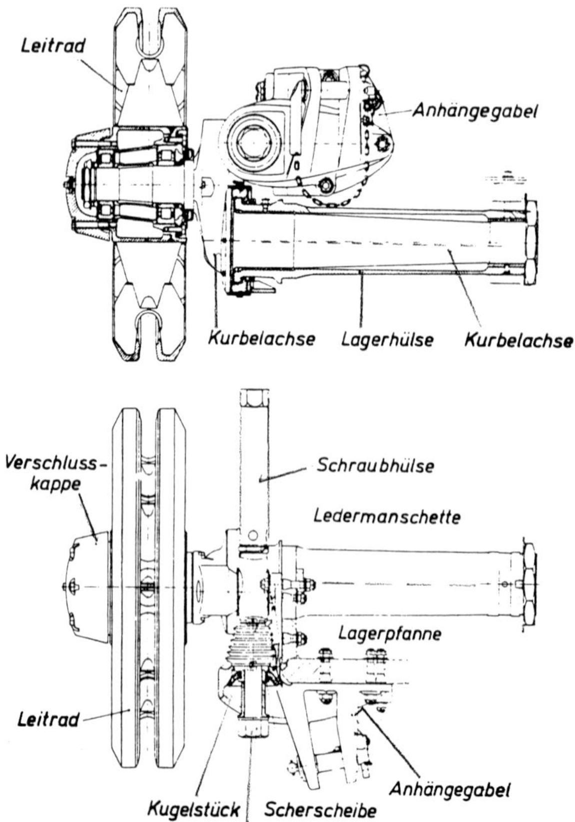

The new idler wheel with track tensioner for the type KgS 61/400/120 track.

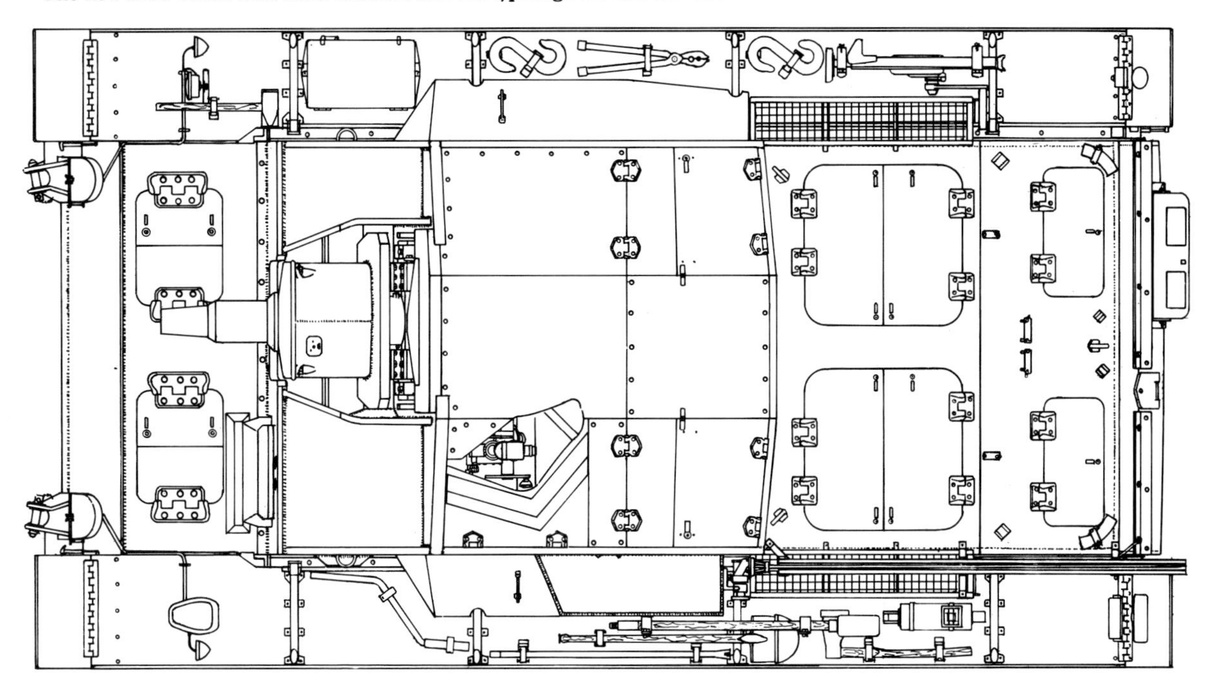

The Sturmgeschütz, Ausf.C, chassis number 90555, produced in April 1941. Photograph taken in Jüterbog. The gun sight protruded through an opening in the superstructure roof.

Sturmgeschütz, Ausf.C. A new idler wheel was introduced. Rear view of chassis number 90555.

Sturmgeschütz, Ausf.D

One hundred fifty of the Sturmgeschütz, Ausf.D were manufactured beginning in May of 1941 (chassis numbers from 90601 to 90750).

In the models Ausf.A through D, a box-shaped armored pannier on the left side of the superstructure was used for radio equipment. However, as a variation from models Ausf.A through C, an on-board intercom system was installed in the Ausf.D.

Otherwise, there were no essential differences between the Ausf.C and Ausf.D. Clearly, the Ausf.D was simply a contract extension placed by the Army for 150 additional Sturmgeschütz. These were needed to outfit newly created Sturmgeschütz units and to replace losses at the front.

Sturmgeschütz, Ausf.D.

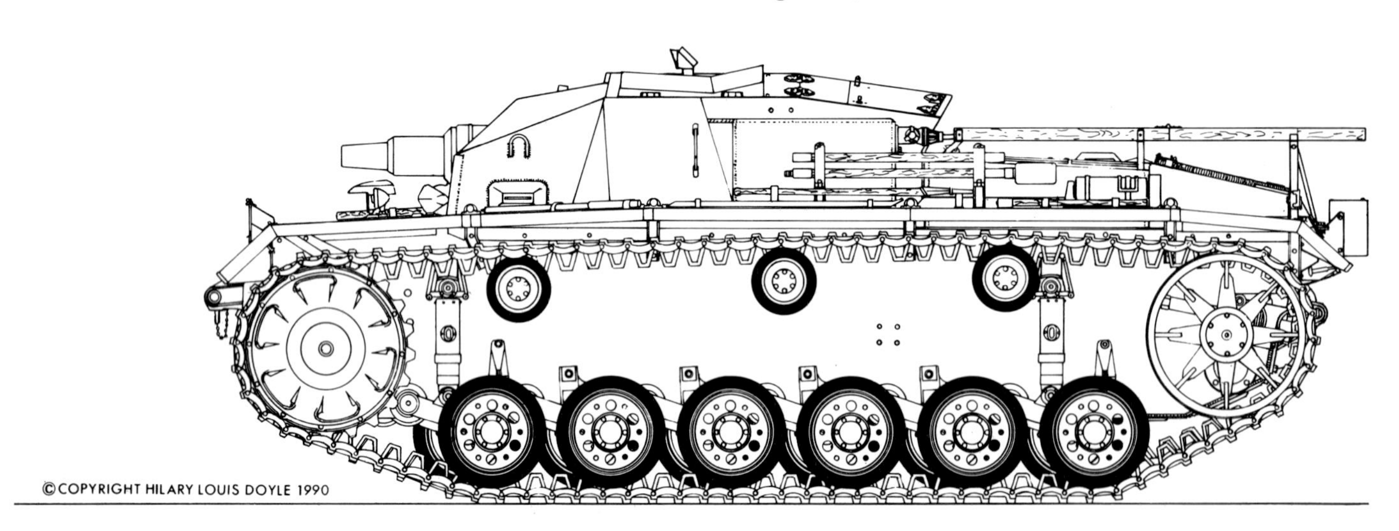

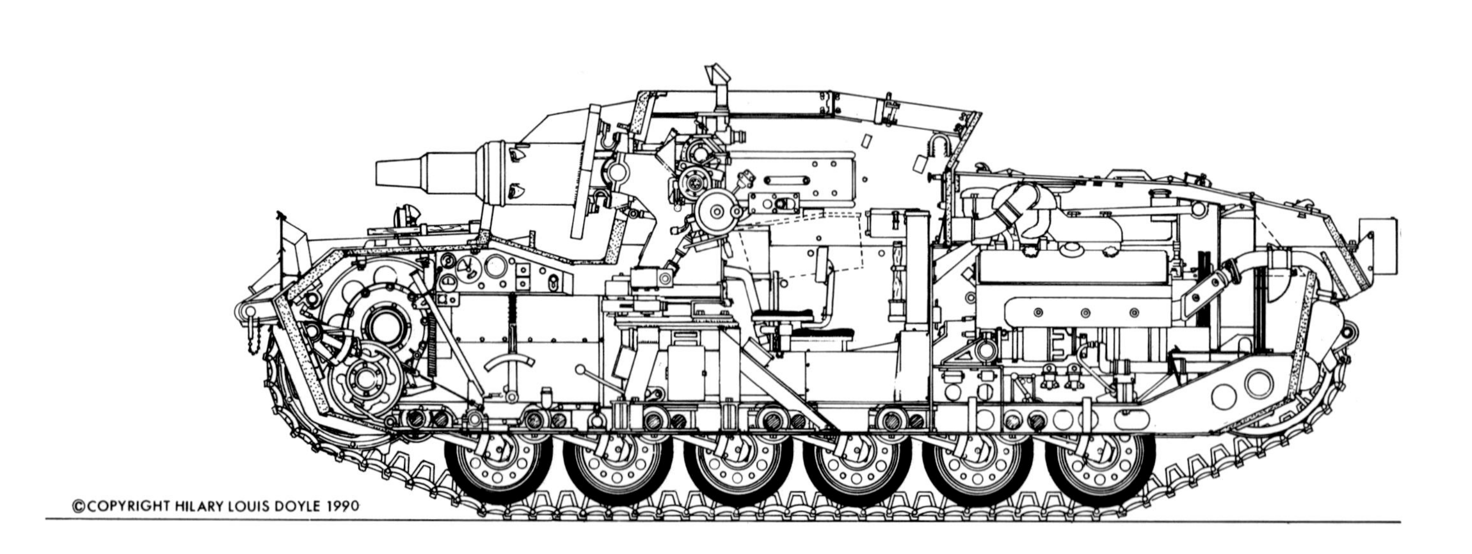

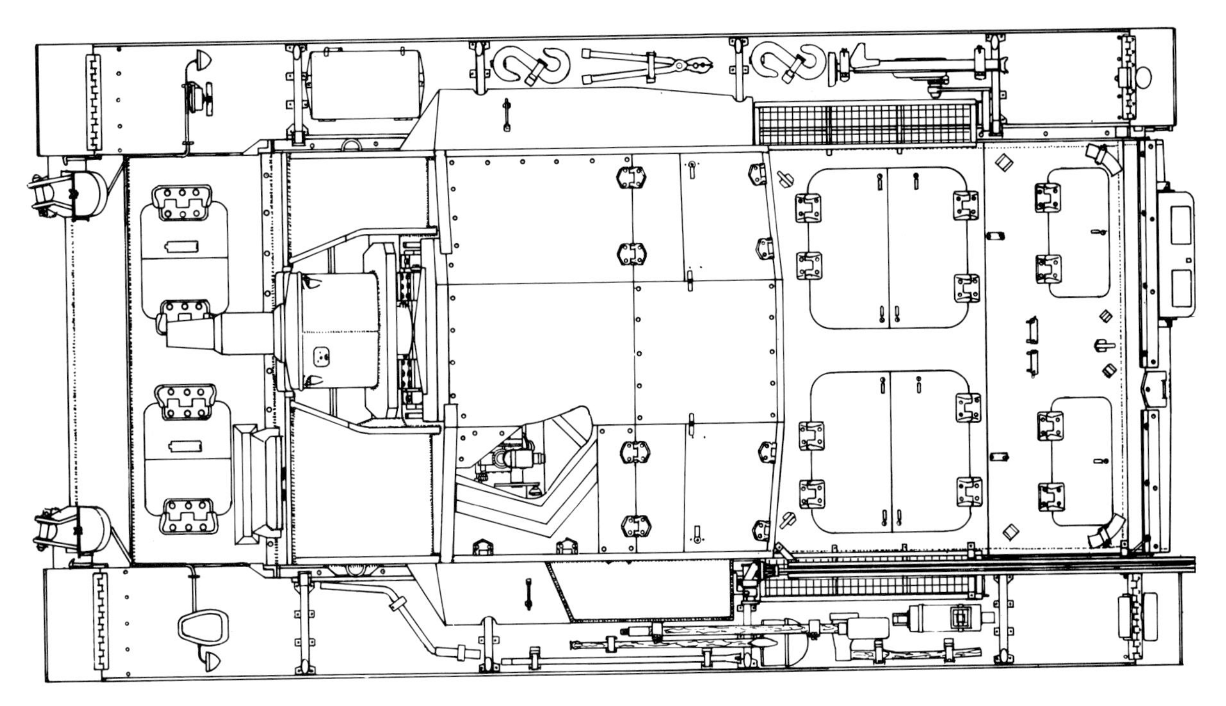

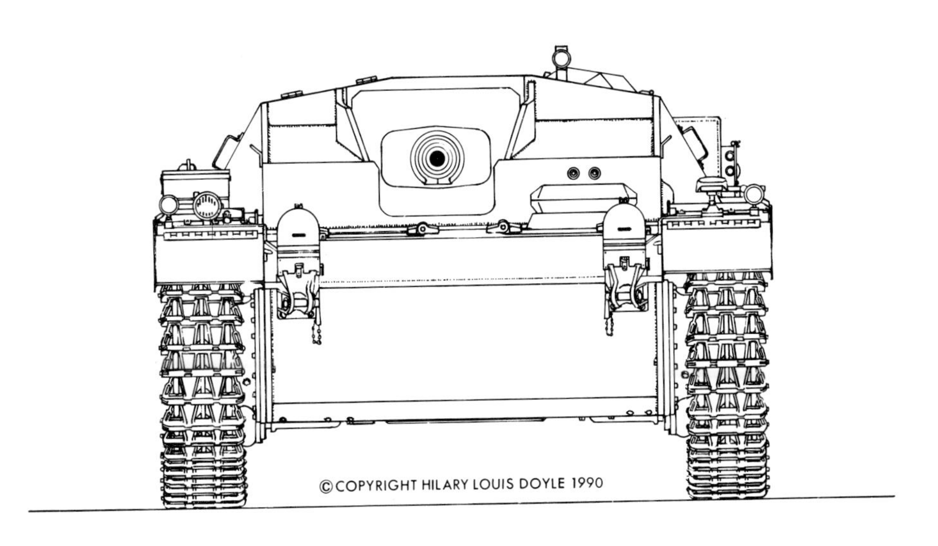

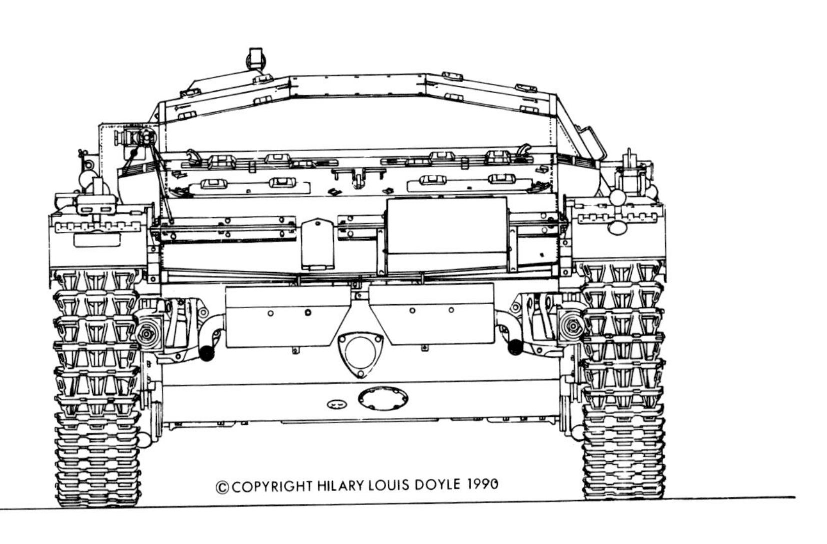

Front view of a Sturmgeschütz, Ausf.D.

Sturmgeschütz, Ausf.D, chassis number 90630, produced in June 1941. (StuG-Abt.189). (BA)

Sturmgeschütz, Ausf.D, chassis number 90661, produced in July 1941. (BA)

Sturmgeschütz, Ausf.D, chassis number 90683, produced in July 1941. In North Africa with Sonderverband z.b.V.288.

Sturmgeschütz, Ausf.D. Rear view of chassis number 90683.

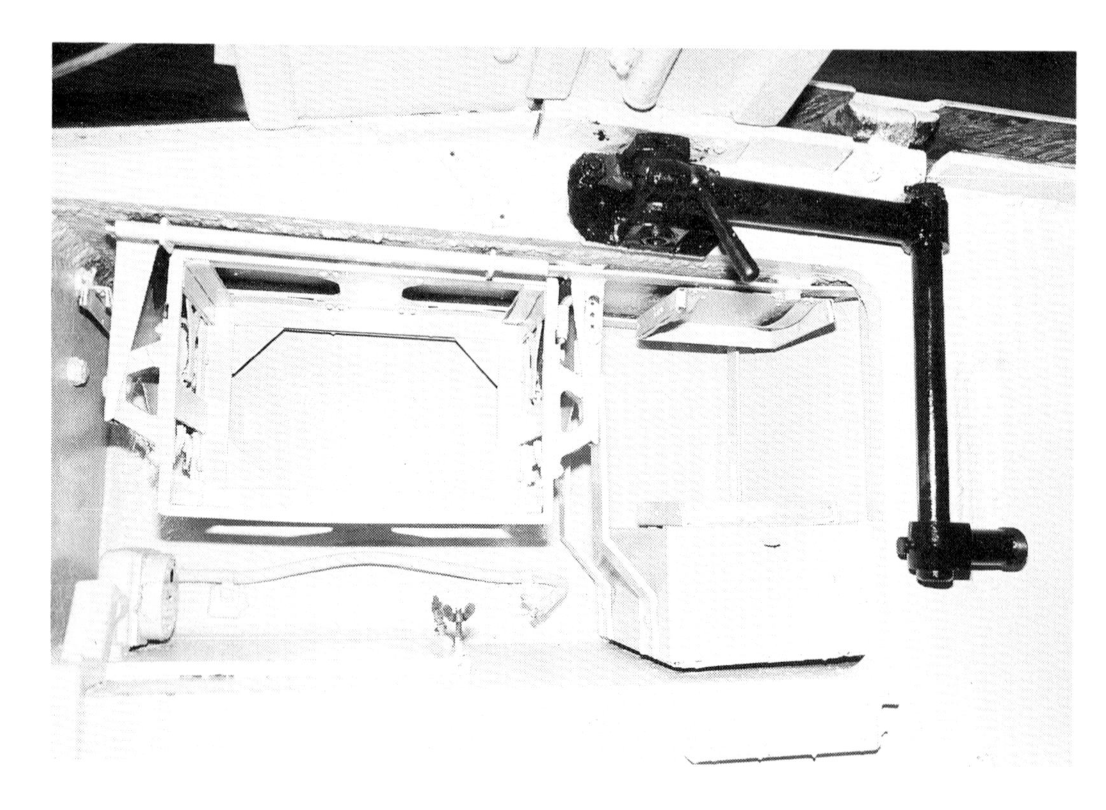

Sturmgeschütz, Ausf.D. Radio racks and arm for mounting the Scherenfernrohr 14 (scissors-periscopes).

Sturmgeschütz, Ausf.E

A total of 284 units of this model were produced beginning in September of 1941 (chassis numbers from 90751 to 91034).

The Ausf.E was to be utilized as a platoon leader's or battery commander's vehicle which necessitated the addition of a pannier on the right side of the superstructure for additional radio equipment. Panniers of a new design were now present on both superstructure sides. With this modification, enough space was gained to store six additional rounds of ammunition. The total stored ammunition load was now fifty 7.5cm rounds. A bracket for the M.G.34 machine gun was added in the right rear corner of the fighting compartment and a container was added on the rear ammunition bin (12 rounds) for seven drum-type magazines. In addition to the Rundblickfernrohr 32 and Sfl.Zielfernrohr 1 periscopic gunner's sights, the commander was provided with "SF14Z" stereoscopic, scissors periscopes.

A Sturmgeschütz, Ausf.E. This is an Ausf.E chassis (number 90773) with an Ausf.D superstructure. Produced in September 1941.

Sturmgeschütz, Ausf.E, platoon leader's vehicle with two whip antennae.

The Sturmgeschütz could now be outfitted for the platoon leader with additional communications equipment, a "10WSh" ten-watt transmitter and two "UKW-Empf.h" VHF receivers as well as a loudspeaker. During the production of the Ausf.E, the following changes were ordered:

— General Army communication Nr.101 dated 20 December 1941, entitled Spare Track Links for the Sturmgeschütz: For the purpose of mounting spare track links (11 track links with pins), a mount is to be welded across the hull front plate on the Sturmgeschütz (Sd.Kfz.142) Ausf.A through Ausf.E. O.K.H. (Ch H Rüst u. BdE, 4. 11. 1941 AHA/Ag K/In 6 (III Ing.)

— General Army communication dated 20 December 1941: Sheet metal cowlings are to be installed on the hull rear to deflect the discharged cooling air away from the ground. Two spare road wheels are to be mounted on the track fenders of the Sturmgeschütz (Sd.Kfz.142) Ausf.A through Ausf.E.

— General Army communication Nr.23 dated 21 February 1942, entitled Torsion Bars for the Sturmgeschütz: There are three forward torsion bars (55mm diameter) and three rear torsion bars (52mm diameter) on each side in the Sturmgeschütz. During installation of the bars, ensure

The left front of the fighting compartment of a Sturmgeschütz, Ausf.E.

The right front of the same compartment.

that the distance between the center of the road wheel and the underside of the armor hull is set at 145mm.
In the event that an outdated torsion bar configuration is encountered, specifically 55mm (forward), 52mm, 44mm, 44mm, 52mm, and 55mm (rear), the torsion bars are to be adjusted as follows: The distance from the center of the road wheel to the underside of the armor hull for the 1st, 2nd, 5th and 6th torsion bars is to be set at 145mm and 193mm for the 3rd and 4th.
In order to simplify the supply of replacement parts, only torsion bars of 52 and 55mm diameter are to be delivered. All Sturmgeschütz are to be converted to the new torsion bar configuration.
O.K.H. (Ch H Rüst u. BdE), 9.2.1942 — 610/42 — AHA/Ag K/In 6 (Ing)
— General Army communication Nr.38 dated 7 April 1942: Warm air heaters are to be installed as a backfit modification in all Sturmgeschütz. O.K.H. (Ch H Rüst u. BdE), 20.3.1942 — 1330/42 — AHA AgK/In 6 (Ing)

Smaller hinges on the sides of the maintenance hatches on the hull glacis plate were introduced with the initial design for the Sturmgeschütz, Ausf.E.

The Ausf.E was the last of the Sturmgeschütz series to be produced with the short-barrelled 7.5cm Kanone L/24.

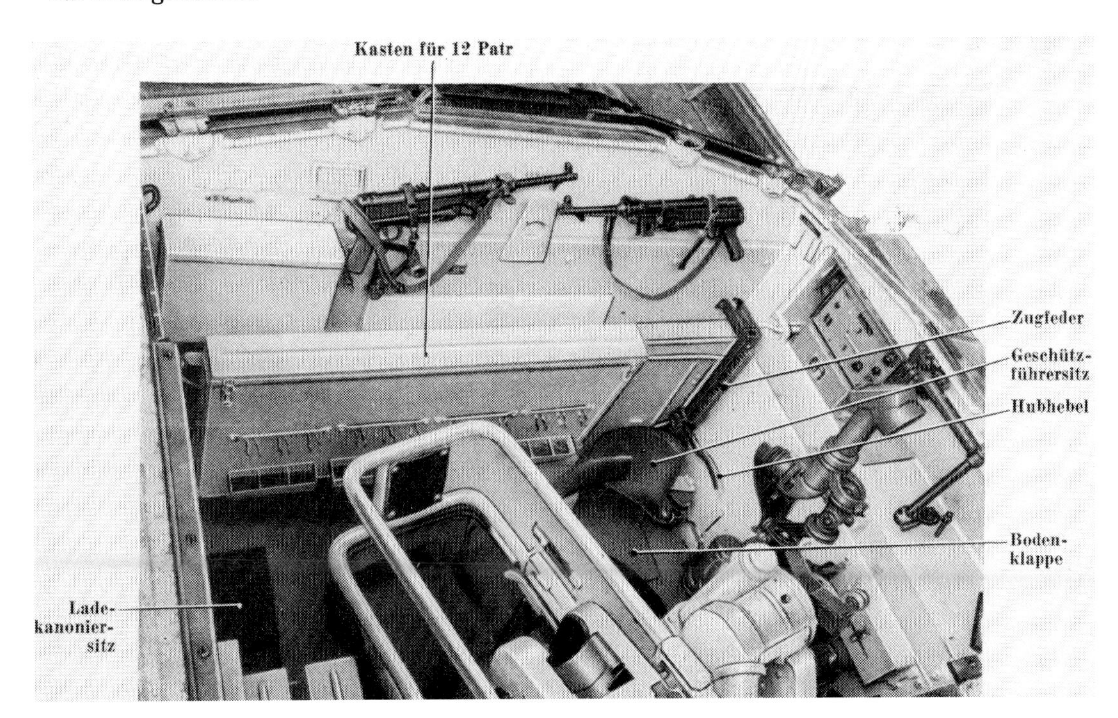

Left rear of the Sturmgeschütz, Ausf.E fighting compartment.

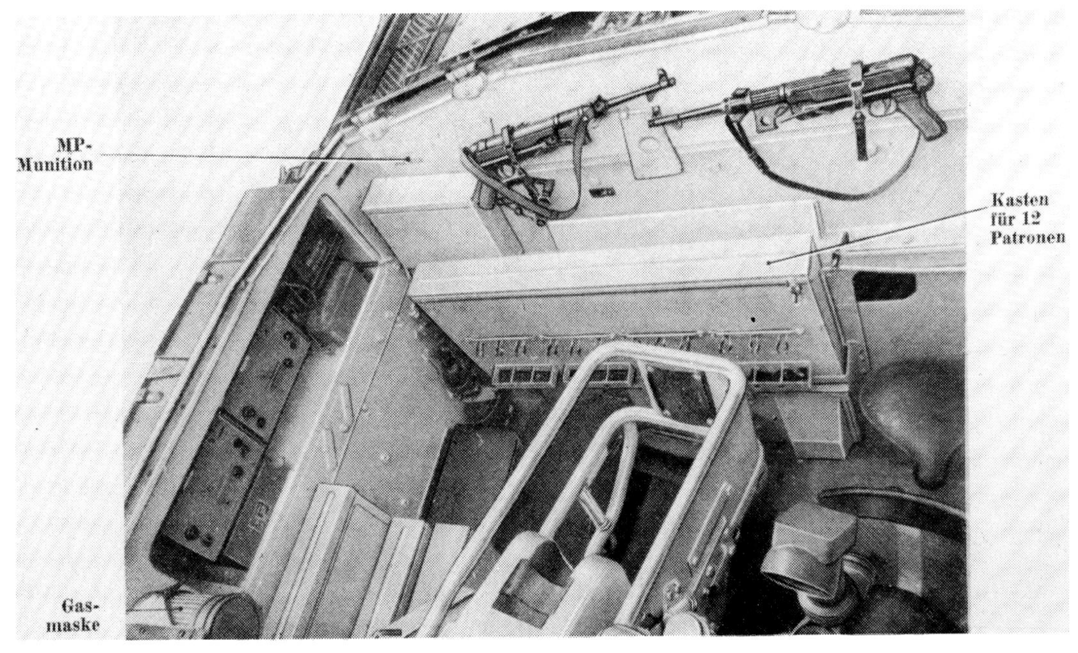

Right rear of the same compartment.

The Sturmgeschütz, Ausf.E without the superstructure, showing the crew in position. Introduced with the Ausf.E, smaller hinges were utilized for the maintenance hatches in the glacis plate.

The Sturmgeschütz, Ausf.E. Driver's visor open (left) and closed (right).

Drawings of the driver's visor and periscopes.

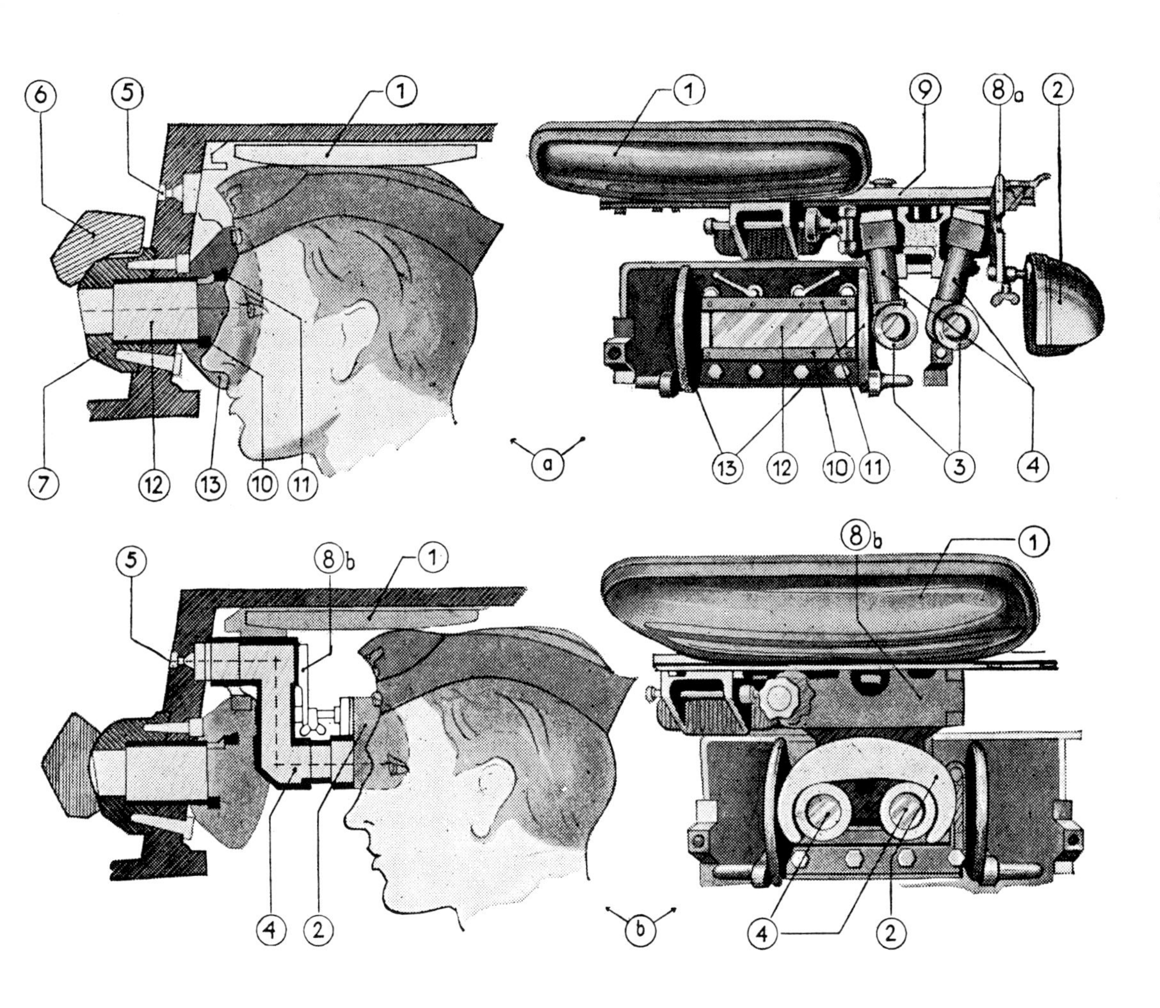

Sturmgeschütz, Ausf.E. Gunner mounting the sight.

Sturmgeschütz, Ausf.E. Loudspeaker for the gunner.

Sturmgeschütz, Ausf.E. Radio racks with radio equipment mounted and the scissors-periscopes in stowed position.

Sturmgeschütz, Ausf.E. Radio equipment operated by the loader.

Sturmgeschütz, Ausf.E. The gun leader installing the whip antennae.

Details of the running gear, front and rear.

Sturmgeschütz, Ausf.E (produced by Alkett starting in September 1941). Overall view of front/left side.

A Sturmgeschütz, Ausf.E as seen from behind.

Sturmgeschütz, Ausf.E, chassis number 90993, tested at the Heereszeugamt on February 8, 1942.

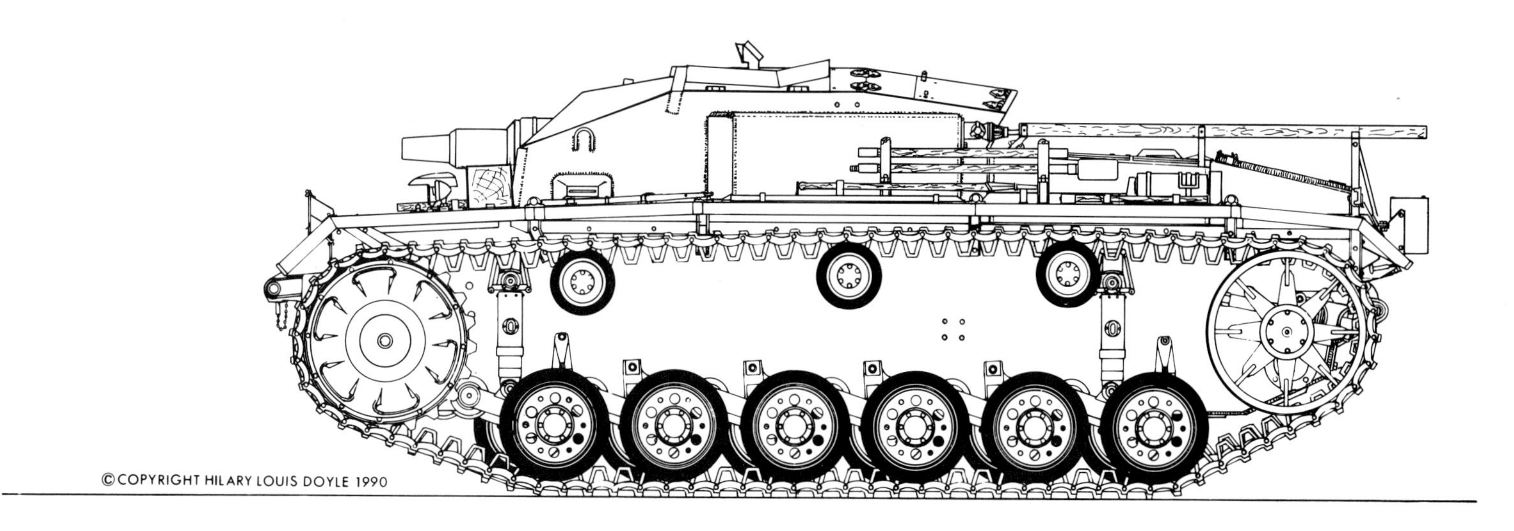

Sturmgeschütz, Ausf.E.

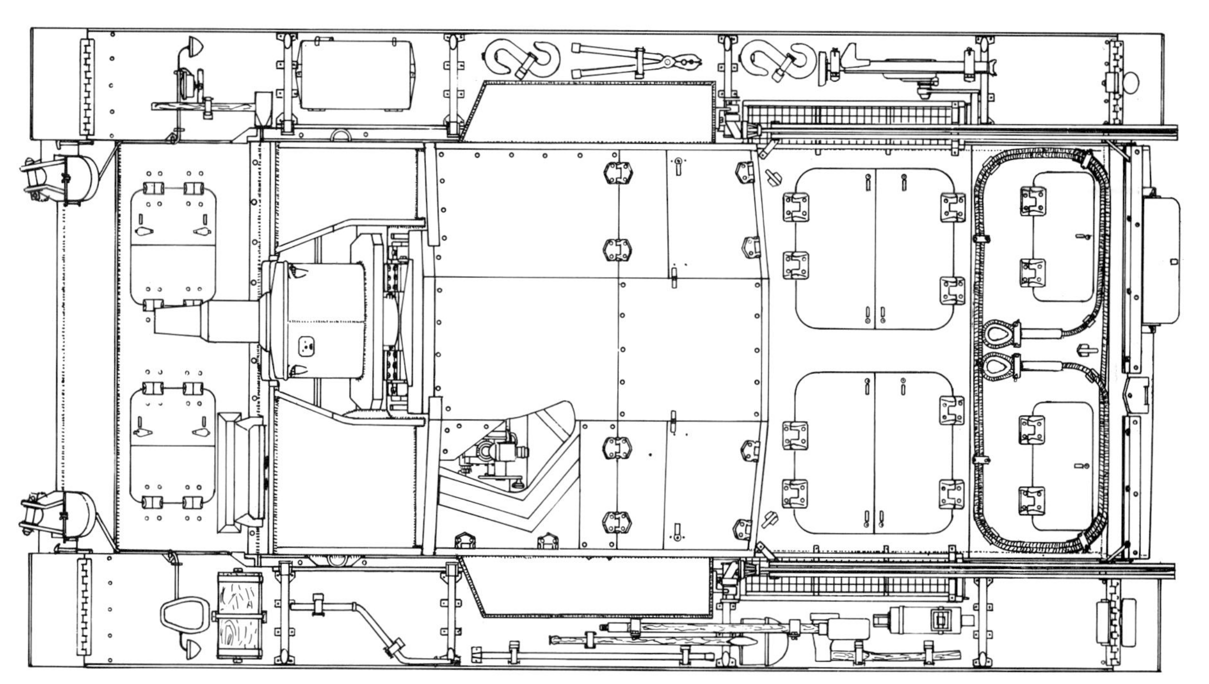

Development of the Long 7.5cm Sturmkanone 40

Shortly after the start of the Sturmgeschütz development, the installation of a more effective main weapon was being considered.

The short 75mm gun satisfied the concept for an infantry support weapon, but its capabilities against an armored opponent could stand improvement.

Krupp took the lead in the development of a long gun for the Sturmgeschütz and constructed a prototype of the weapon which was demonstrated and tested.

On the occasion of a meeting in Berlin on August 1, 1938, Krupp announced that the first conceptual design for a 7.5cm Kanone L/40 for self-propelled guns would be ready by August 20 1938. On January 12, 1939, Krupp received the official contract for the production for an "sPak (verstärkt) L/42."

On November 4 1939, Krupp sent a wire to the OKH WaPrüf 4 informing them that the wooden model for the "lang sPak", ordered by WaPrüf 4 on 21 July 1939, was ready for delivery. The wooden model of the "lang sPak" was delivered to Daimler-Benz on 5 December 1939.

On 22 February 1940 Waprüf 4 ordered 30 rounds of ammunition for the gun, and on 9 March 1940 muzzle brake number "V" was transported from Meppen to Hillersleben. On 26 April 1940, Krupp informed the OKH that the "lang 7.5cm Kanone (Pz.Sfl.)" was finished and could be expected in Meppen on 9 May 1940. A Krupp teletype message of 21 May 1940 informed the Oberkommando des Heeres that the "sPak L/41 Nr.R V1 in gun mount L V1" would be available on June 6, 1940. The results of the test-firing were reported on July 16, 1940. 133 rounds were fired, one cartridge did not eject from the gun. Muzzle velocities for the armor piercing shells were recorded as 676, 675, 671, 673 and 671 meters per second.

Based on these tests, the weapon was accepted on July 16, 1940.

A message from August 13 1941 reported: The delivery note dated 30 July 1940 confirms the transfer of one Sturmkanone with barrel Nr. R V1 by rail to the Daimler-Benz AG Werk 40 in Berlin-Marienfelde.

On October 17, 1940, Krupp received a contract to convert the wooden model "Pz.Sfl.IIIb" to a falling-block breech. On 13 March 1941, Krupp determined the armor penetration capability of the 75mm L/40 gun with a barrel length of 3023mm. With a muzzle velocity of 634 meters per second, at a range of 400 meters, the predicted penetration was 70mm for armor sloped at 30 degrees from the vertical (NOTE: Krupp reported that on 13 March 1941 an experimental barrel for the 75mm L/33 gun had been produced and achieved a muzzle velocity of 580 m/s during comparison testing. At a range of 400 meters, the penetration performance was predicted to be 59mm. The length of the gun was 2470mm, the weight of the armor piercing round was 6.8 kilograms. This weapon was not and never had been intended for installation in the Sturmgeschütz.).

75mm Sturmkanone 40 L/43.

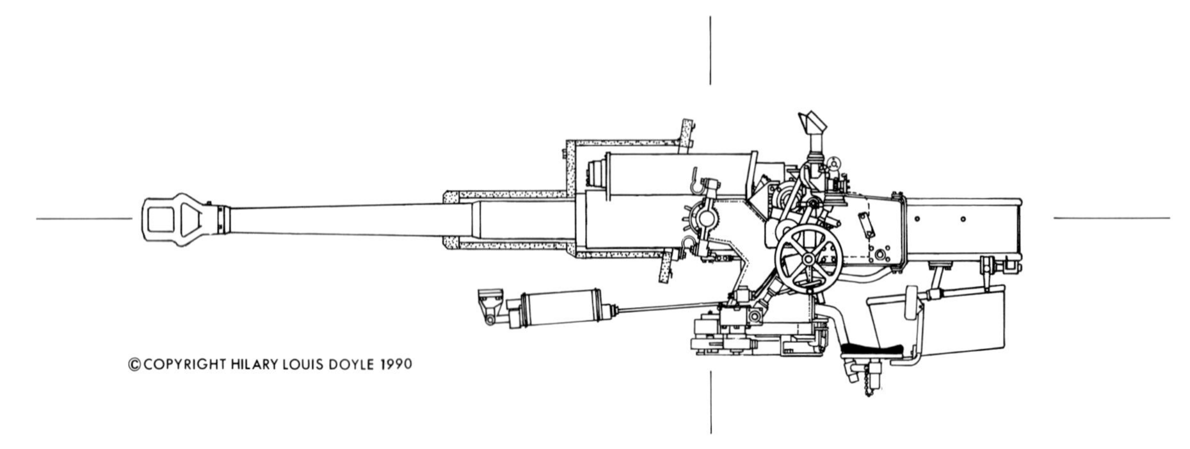

On Wednesday, the 19th of March 1941 at 1100 hours, the presentation of the "lang sPak" in a "Pz.Sfl.III" took place at Daimler-Benz in Berlin-Marienfelde. On the 31st of March 1941, the a demonstration was held for Hitler. The possibility of procuring the Sturmgeschütz (lang) with the 75mm L/41 gun was discussed. Predicated on a favorable outcome to the field tests, production was to begin in the late spring of 1941. On April 15 1941, Krupp provided the following data for their long 75mm gun: 7.5cm Kanone L/40 as shown in Krupp blueprint 5-B-291, projectile weight 6.8kg, muzzle velocity 670 m/s, barrel length (without muzzle brake) 3023mm, weight 1400kg.

On November 5th, 1941, the OKH sent a message to Krupp informing them that the 7.5cm Sturmgeschütz (Sfl.IIIb) was only to be procured by the Wehrmacht and was not intended for export. In their response, Krupp reminded the OKH that "Our responsibility with this piece of equipment is solely for the weapon and armor."

As was often the case in many of the weapons systems developed during the Second World War, there were disappointments, when on 20 November 1941, WaPrüf 4 wrote to Krupp:

It is reported that all construction work on the "lang Sturmgeschütz (Nr.5-291)" is to be halted. The gun has been replaced by a newer development (Kw.K.44 auf Sfl.).

Despite the order to cease further development on the Krupp 75mm L/40 gun, the Heereswaffenamt published Manual Nr.420/152 on January 6, 1942, dealing with the preparation of ammunition for the "Sturmgeschütz lang, 7.5cm Kanone (Krupp)."

In the mean time, further development of a longer gun was activated by the Oberkommando der Wehrmacht communication OKW 002205/41 gKdo of 28 September 1941 to the Oberkommando des Heeres.

The Führer has, due to thorough appreciation of the value of the Sturmgeschütz ordered the following:

1. The armor is to be strengthened on newly produced Sturmgeschütz without regard to the disadvantages of increased weight and the resulting loss of speed.

2. The Sturmgeschütz must receive a long-barreled 75mm gun with a higher muzzle velocity. The advantage over new types of enemy tanks can only be regained in this way.

These demands were fulfilled by the Heereswaffenamt awarding a contract to the firm of Rheinmetall-Borsig for three experimental 75mm Kanone 44 L/46 guns.

The first production plans for the longer gun were published by the Heereswaffenamt on December 2, 1941:

	7.5cm Kan. L/24	7.5cm Kan.44 L/46
December 1941	40	0
January 1942	40	0
February 1942	40	0
March 1942	4	0
April 1942	15	10
May 1942	0	15

This was the first mention of production plans or the resulting schedule for this longer gun for the Sturmgeschütz.

The nomenclature of the longer gun was changed from 7.5cm Kanone 44 L/46 to 7.5cm Sturmkanone 40 L/43 on March 16th 1942.

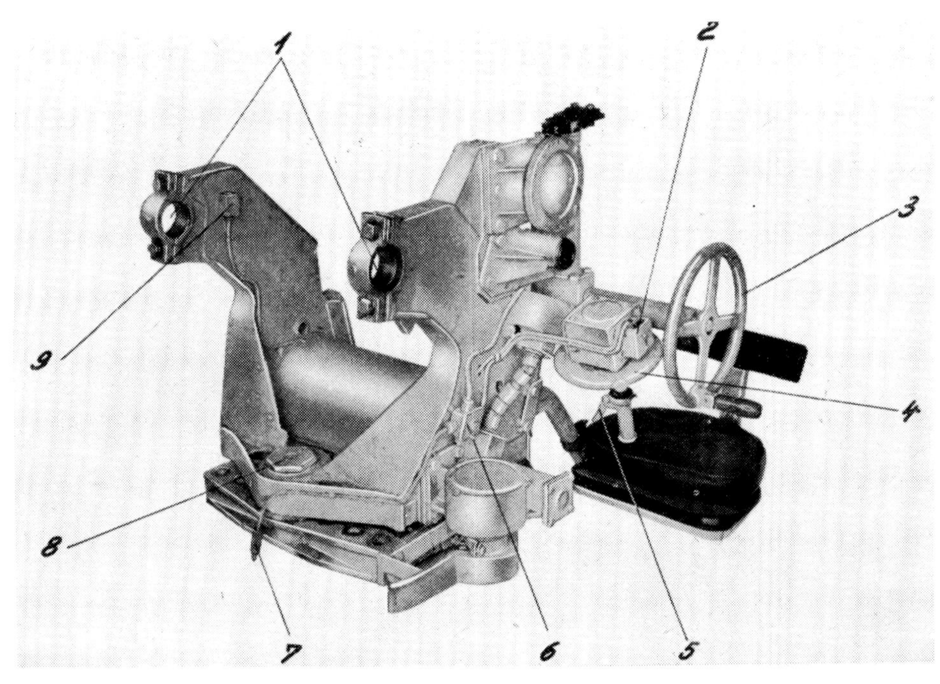

Upper gun mounting with base plate. View from the front left.

1. trunnion mount
2. control light
3. elevation wheel
4. traverse wheel
5. firing trigger
6. junction box
7. connector for electrical supply
8. central pivot
9. junction boxes for electrical firing mechanism

Development and production of these first experimental guns commenced immediately. Two of the guns were ready for installation and testing of the sights on 27 January 1942. Three experimental guns were delivered in February and installed and accepted for delivery as completed Sturmgeschütz in March 1942.

On 14 April 1942, results were reported for the test firing which had taken place in Hillersleben on April 3rd. The first gun tested (R V 1), installed in a Sturmgeschütz, had experienced twice failed to eject spent shell cases out of 87 rounds fired, utilizing shells which had been warmed to 10 and 35 degrees centigrade. On 13 May 1942, Hitler mentioned these difficulties with the new Sturmkanone 40. Failure to eject casings was said to have occurred even with the first shot. An investigation into the matter was to start immediately.

The StuK40 L/43 was produced from March until May of 1942. Further production of this gun was with the longer StuK40 L/48. The following production statistics record the end of production for the StuK L/24 and the start of the StuK 40:

	StuK L/24	StuK 40
February 1942	43	3 experimental guns for WaPrüf
March 1942	41	2 L/43
April 1942	14	51 L/43
May 1942	9	66 L/43
June 1942	11	78 L/48
July 1942	0	88 L/48

The production goal for September 1942 was not achieved due to a bottleneck in muzzle brake manufacturing. Several Sturmgeschütz, Ausf.F and F/8, therefore, appeared with the ball-shaped single-chamber muzzle brakes.

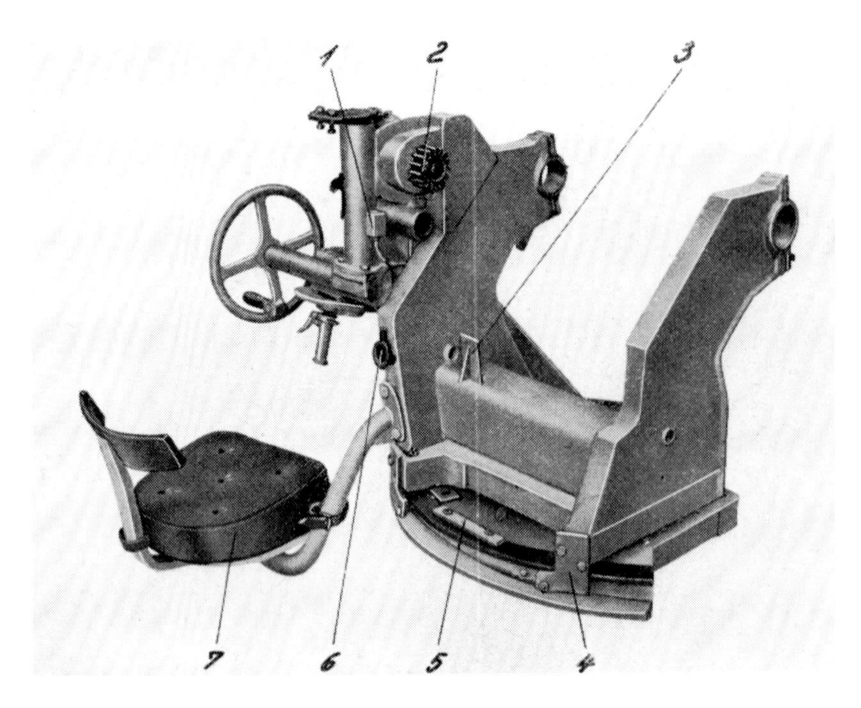

Upper gun mounting with base plate, from the right rear.

1. junction boxes for nighttime illumination of sights
2. pinion (for elevation gear)
3. elevation limiter
4. guide
5. traverse limiter
6. mount for emergency firing device
7. gunner's seat

Complete gun. View from the left rear.

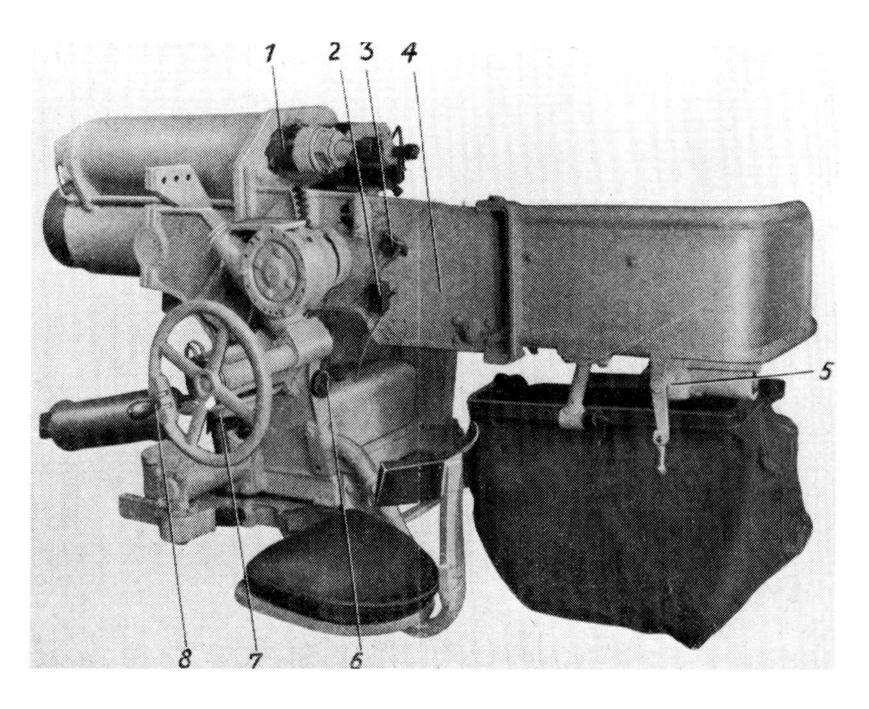

Gun mount as seen from the left rear.

1. recoil cylinder
2. switch
3. signal light
4. deflector
5. travel lock
6. emergency firing device
7. traverse gear
8. elevation wheel

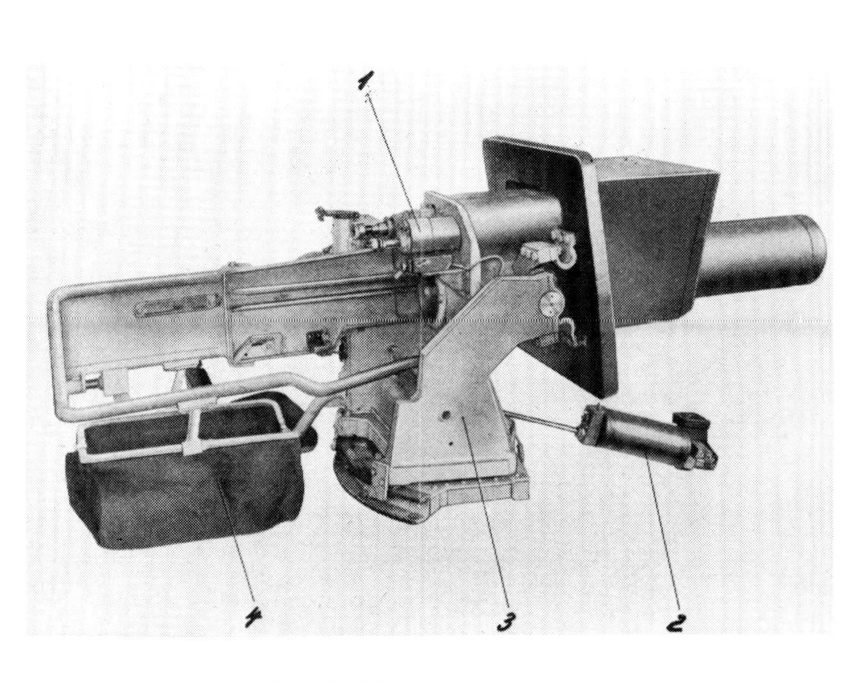

Gun mount, as seen from the right.

1. recuperator
2. counterbalance
3. upper gun mount
4. spent shell sack

1. Sfl.ZF1a gun sight
2. traverse adjustment
3. level
4. range indicator
5. adjustable indicator
6. elevation adjustment
7. square drive gear
8. coarse adjustment of the elevation angle
9. locking screw

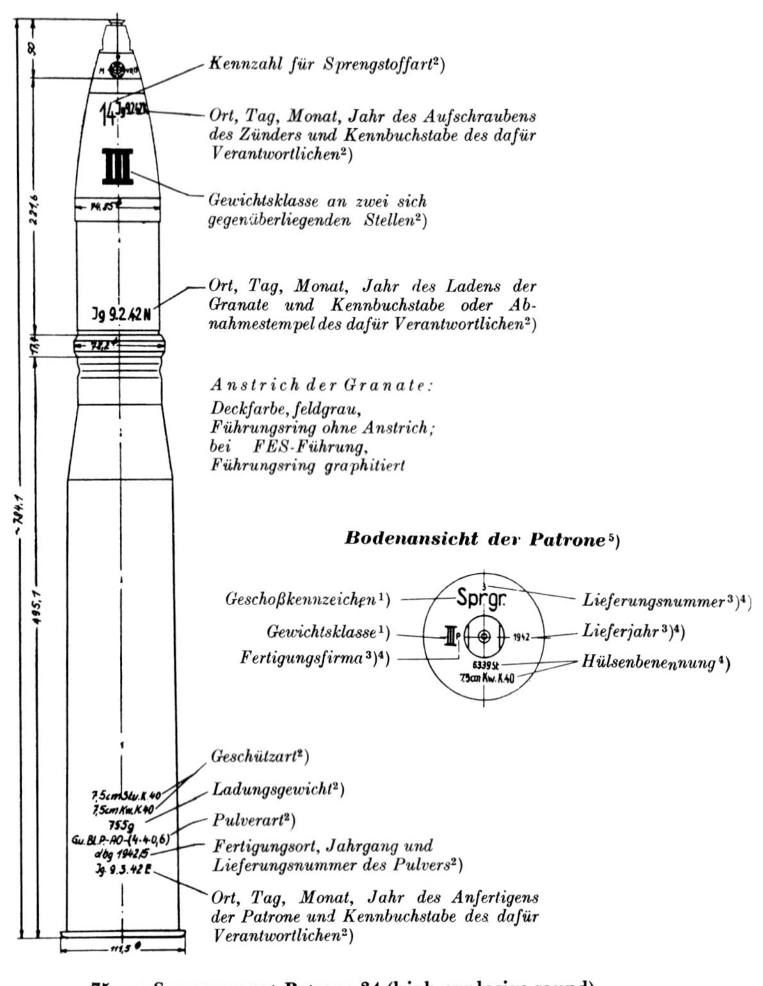

75mm Sprenggranat-Patrone 34 (high-explosive round).

¹) weiß aufgetragen
²) schwarz aufgetragen
³) der Patronenhülse
⁴) eingeprägt
⁵) Patronen, die Geschosse mit FES-Führung haben, tragen auf dem zylindrischen Teil des Geschoßmantels an zwei sich gegenüberliegenden Stellen und auf dem Boden der Patronenhülse die zusätzliche Bezeichnung „FES" in weiß

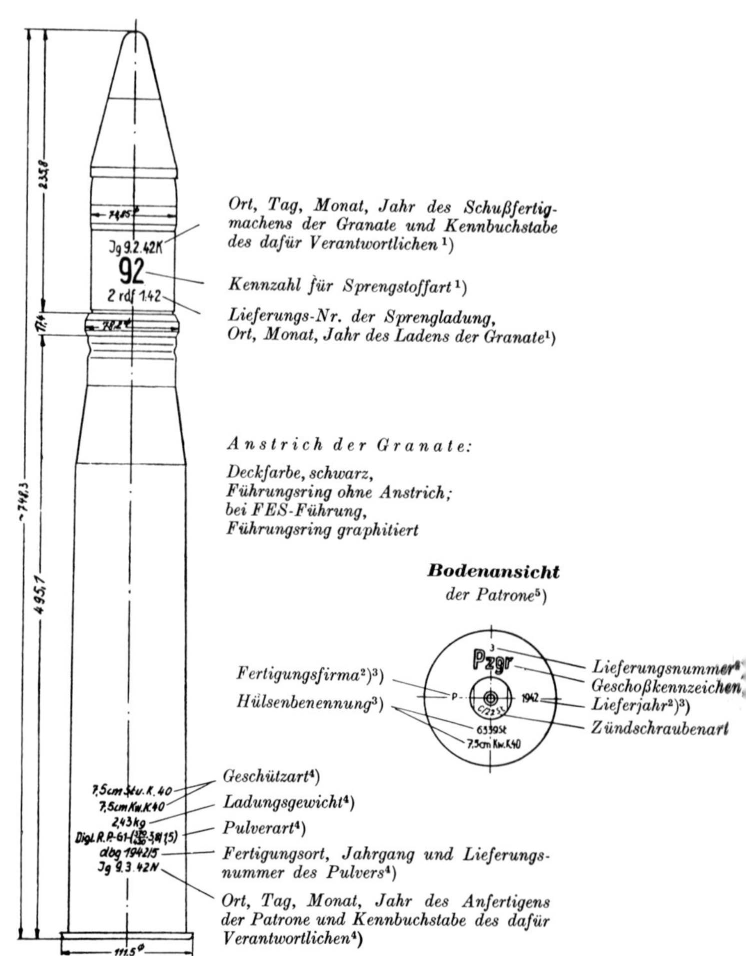

75mm Panzergranat-Patrone 39 (armor-piercing round).

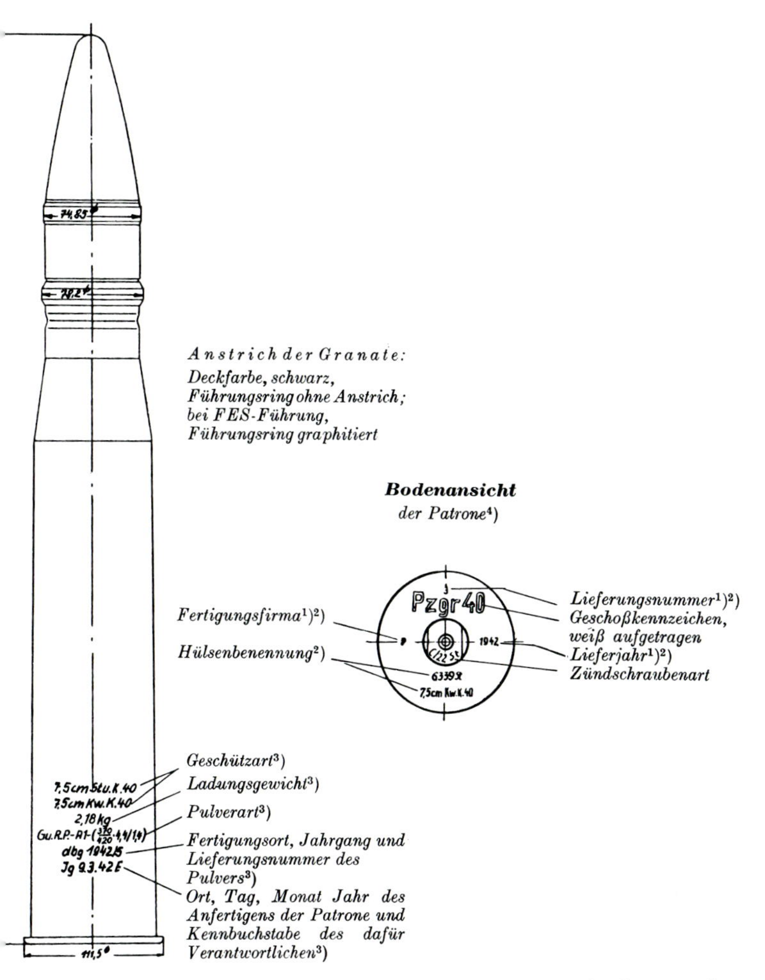

75mm Panzergranat-Patrone 40 (tungsten core armor-piercing round).

In a conference with Hitler on the 10th and 12th of August 1942, it was determined that all Sturmgeschütz undergoing major overhaul were to be refitted with the longer gun.

A comparative of the four guns designed for the Sturmgeschütz reveals:

	Gun length mm	Shell Casing length mm	Shell Casing width mm	Complete Round length mm	Armor Piercing Projectile weight (kg)
7,5cm Kan.L/24	1766.5	243	81.7	507	6.8
7.5cm StuK lg.L/40	3031	514	89	794	6.8
7.5cm StuK 40 L/43	3281	495	102	748	6.8
7.5cm StuK 40 L/48	3615	495	102	748	6.8

Penetration ability (mm) against armor plate, inclined at 30 degrees from vertical were predicted to be:

gun type	shell type (m/s)	muzzle velocity	At the following ranges: 100m	500m	1000m	1500m	2000m
7.5cm Kan. L/24	K.Gr.Patr.rot.Pz.	385	41	39	35	33	30
7.5cm StuK lg.L/40	K.Gr.Patr.rot.Pz.	685	65	64	62	59	55
7.5cm StuK 40 L/43	Pzgr.Patr.39	740	98	91	82	72	63
7.5cm StuK 40 L/48	Pzgr.Patr.39	790	106	96	85	74	64

Description of the 7.5cm Sturmkanone 40

The following data compare the 7.5cm Sturmkanone 40 L/48 (in use beginning June 1942) with its predecessor the 7.5cm Sturmkanone 40 L/43 (of which 120 were produced). The differences between the two weapons:

7.5cm StuK 40 L/43:

1. Several do not have the hydraulic safety switch.
2. On the upper gun carriage: roller support with roller bearings without a track ring.
3. Gunner's seat is not adjustable and has a fixed back support.
4. Electric firing circuit.
5. Rifling increases from 6 to 9 degrees.
6. No gun carriage lock on the roller grip.
7. All have a traverse lock on the traverse gear.
8. The gun sight is mounted lower to the left on all 120 examples.

7.5cm StuK 40 L/48:

1. All examples have hydraulic safety switches.
2. Partial roller support with roller bearings and track ring bearing; newer examples have track grips.
3. Gunner's seat is adjustable with a folding back support.
4. Electronic firing circuit.
5. After the 400th gun, constant rifling of 7 degrees; up to then the same as the L/43.
6. Some have no travel lock on the support; on the newer guns there are travel locks on both sides.
7. Some of the guns have a traverse lock on the traverse gear. Newer guns have a travel lock with bayonet lock.8. Beginning with the first L/48 gun, the sight is located higher and further to the right.

The 7.5cm Sturmkanone 40 was designed as a semi-automatic gun with electric firing. It fired armor piercing and high-explosive rounds (fixed singe piece cartridges). The Selbstfahrlafetten-Zielfernrohr 1a gun sight was used for direct sighting and the Rundblickfernrohr 32 or 36 was used for indirect targeting.

The gun barrel consisted of one-piece with an easily removable breech. It was mounted in a gun cradle. The rifled portion of the barrel had 32 lands and grooves, which increased clockwise from 6 to 9 degrees. The newer guns had a constant twist to the rifling of 7 degrees.

The muzzle brake was threaded onto the front end of the gun and assisted in braking the recoil. It had two chambers, each separated at the front with a baffle plate. The recoil was partly reduced by these baffles. The gases released during firing struck these baffles and was diverted out the sides, thereby buffering the recoil of the barrel.

The 7.5cm StuK 40 was not to be fired without a muzzle brake.

The breech was a semi-automatic falling block breech with electric firing. The breech opened automatically, shortly before the completion of its recoil run, and ejected the empty shell casing. A trigger was located on the hand traverse wheel. Armor protected the barrel cradle, recoil cylinder and the recuperator from hits from the front. It was bolted to the barrel cradle. A counterbalance was installed to equalize the weight forward of the gun trunnions.

The elevating mechanism was designed as a geared crescent. The individual parts of the geared sector and the pinion were mounted in their respective casings and bolted together. The elevation range extended from -6 to +17 degrees.

The traverse mechanism was also a geared device. The gun could be traversed from 10 degrees left through to 10 degrees right of center.

Targeting was slaved, meaning that the gun sights moved concurrently with the gun. The Sfl.ZF1a was used for direct firing at visible targets. The Rundblickfernrohr periscope and the Geländewinkelmesser (gunner's quadrant) was used to engage indirect targets when the gun was in a covered position.

The 7.5cm StuK 40, which was developed by Rheinmetall-Borsig, was chiefly produced by the firms Wittenauer Maschinenfabrik GmbH (Wimag), Berlin-Wittenau and Skoda in Pilsen, Czechoslovakia. The price per gun was 13,500 RM.

Sturmgeschütz, Ausf.F

On 4 February 1943, Daimler-Benz recorded that the Sturmgeschütz, Ausf.F came into being merely by upgrading the armament of the Ausf.E. This name change reflected the replacement of the 7,5cm Kanone L/24 with the 7,5cm Sturmkanone 40 L/43. From March to September of 1942, a total of 366 units were produced (chassis numbers 91035 to 91400).

The Ausf.F was simply created by modifying the Ausf.E. Changes included altering the design of the superstructure front and installing ammunition bins designed for the longer rounds. The superstructure frontal armor on both sides of the gun was further cut out to enable the new gun to traverse through an arc of 20 degrees. The superstructure roof remained unchanged when compared with the Ausf.E, with the sole exception of the rear sector. This had been redesigned to mount a fume extraction fan to aid in ventilating the fighting compartment.

The 22 ton combat weight of the preceding Ausf.E had increased to 23.2 tons for the Ausf.F.

The Sturmgeschütz equipped with the longer weapon (L/43 as well as L/48) were designated as "Sd.Kfz.142/1." The price for the entire system was about 82,500RM.

The Sturmgeschütz Abteilung "Grossdeutschland" was the first unit of the German Army to be equipped with 22 of the new Sturmgeschütz with the 7.5cm Sturmkanone 40 L/43 guns.

Further changes which took place throughout the course of the Ausf.F production run include:

The Nebelkerzenabwurfvorrichtung (smoke candle launcher system) and its armored shield, beginning in May 1942, were no longer to be mounted on the rear of the Sturmgeschütz.

In a conference with Hitler on 6 and 7 May 1942, an investigation was demanded to determine as far as possible, if the frontal armor of the Sturmgeschütz could be increased to 80mm. Hitler further demanded an accelerated investigation to determine if, as an interim measure, the frontal armor could be reinforced to a total of 80mm by welding on additional armor plates.

He was of the opinion that should the additional weight of approximately 450kg negatively effect the speed of the vehicle, that this would not be very detrimental. The Sturmgeschütz were intended to go into action in platoon strength, separate from the armor units. Therefore, in view of their intended role for infantry support, greater speed would not be necessary.

Sturmgeschütz, Ausf.F, experimental, with the 75mm Sturmkanone 40 L/43 and Sfl.-Zielfernrohr (gun sight) mounted.

Sturmgeschütz, Ausf.F, experimental, with Rundblick-Fernrohr (indirect fire gun sight) mounted. View of left and right sides.

On June 4 1942, Hitler decided that the strengthening of the frontal armor to 80mm was to be accomplished at an accelerated pace. The previously set goal for the middle of July would be too late. Hitler agreed with the intended short-term increase of the monthly Sturmgeschütz production to 100 units per month. He expected that the resulting decrease in Panzer III production would be recovered later. On 23 June 1942, Hitler declared that increasing the frontal armor thickness on the Sturmgeschütz was to be expedited.

Sturmgeschütz production at Alkett in April/May 1942 (in the background are 14 Panzerkampfwagen III).

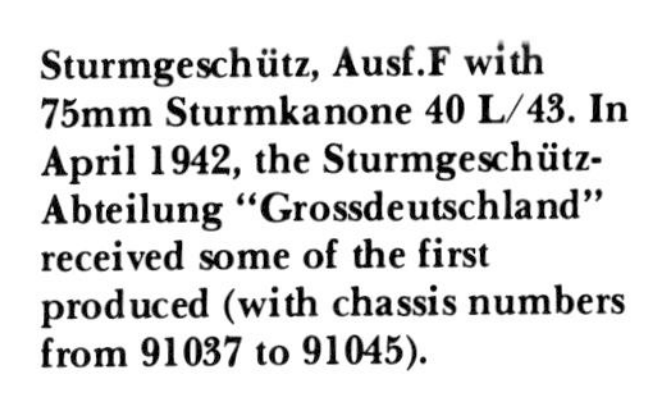

Sturmgeschütz, Ausf.F with 75mm Sturmkanone 40 L/43. In April 1942, the Sturmgeschütz-Abteilung "Grossdeutschland" received some of the first produced (with chassis numbers from 91037 to 91045).

Sturmgeschütz, Ausf.F with Sturmgeschütz-Abteilung 201. The smoke candle rack with armored guard is still mounted on the rear. (BA)

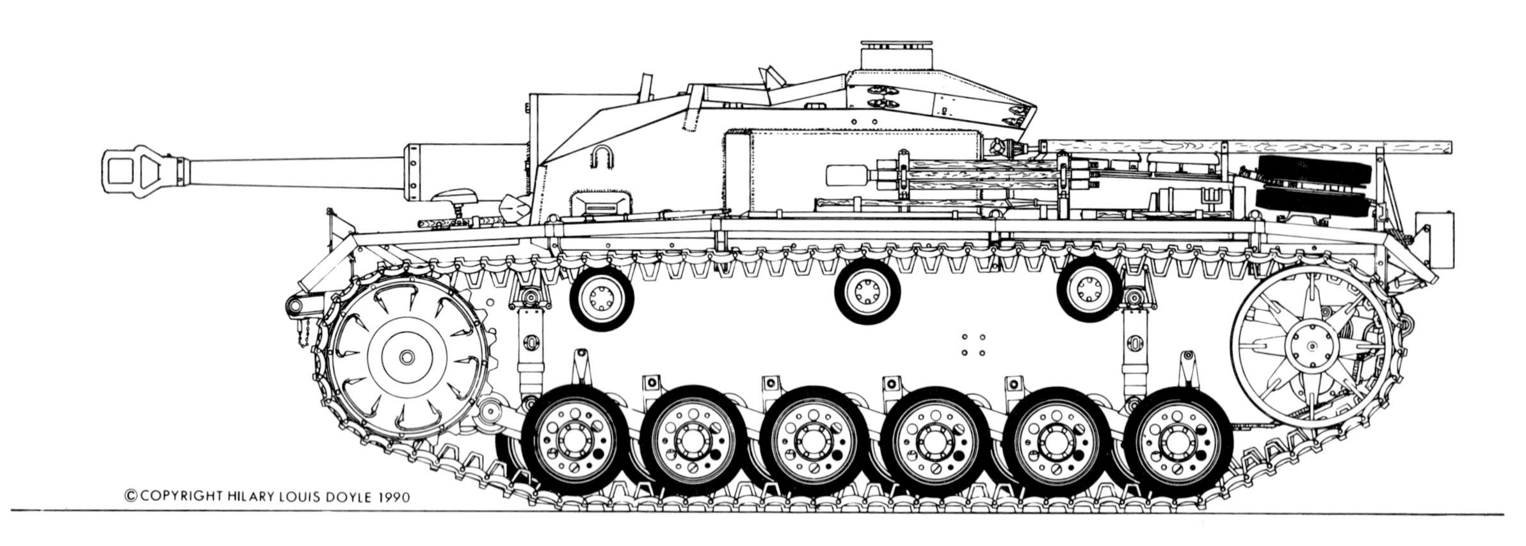

Sturmgeschütz, Ausf.F.

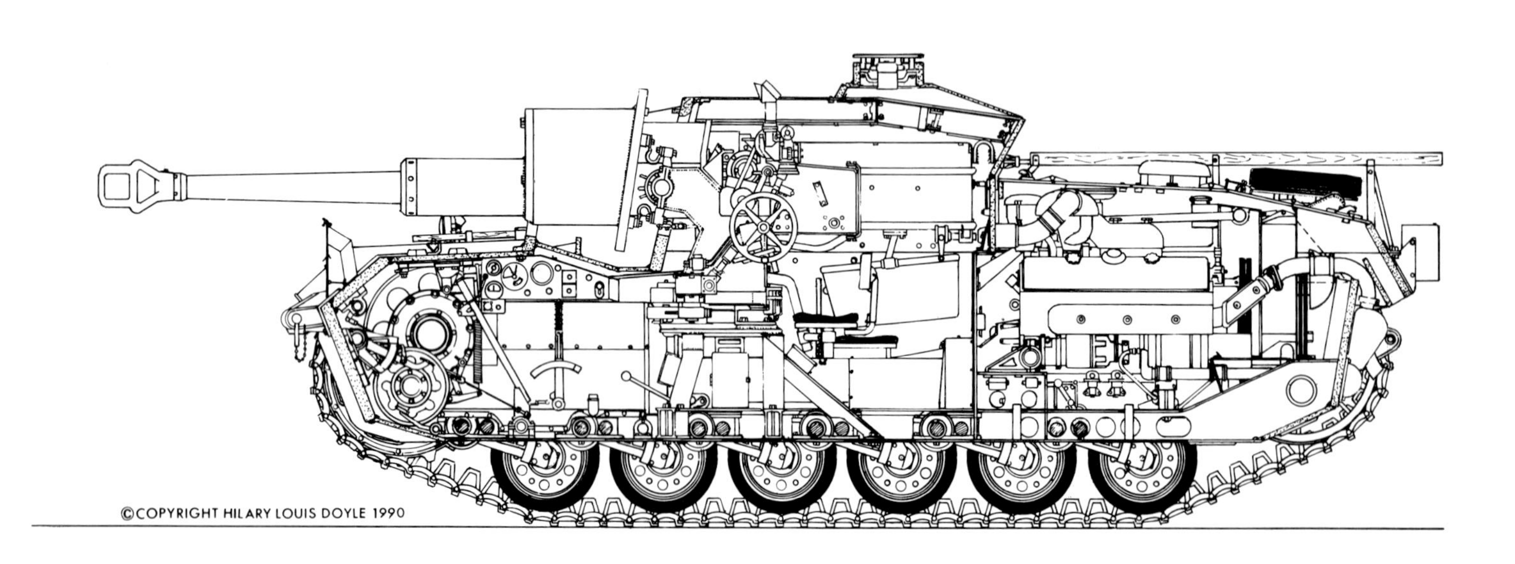

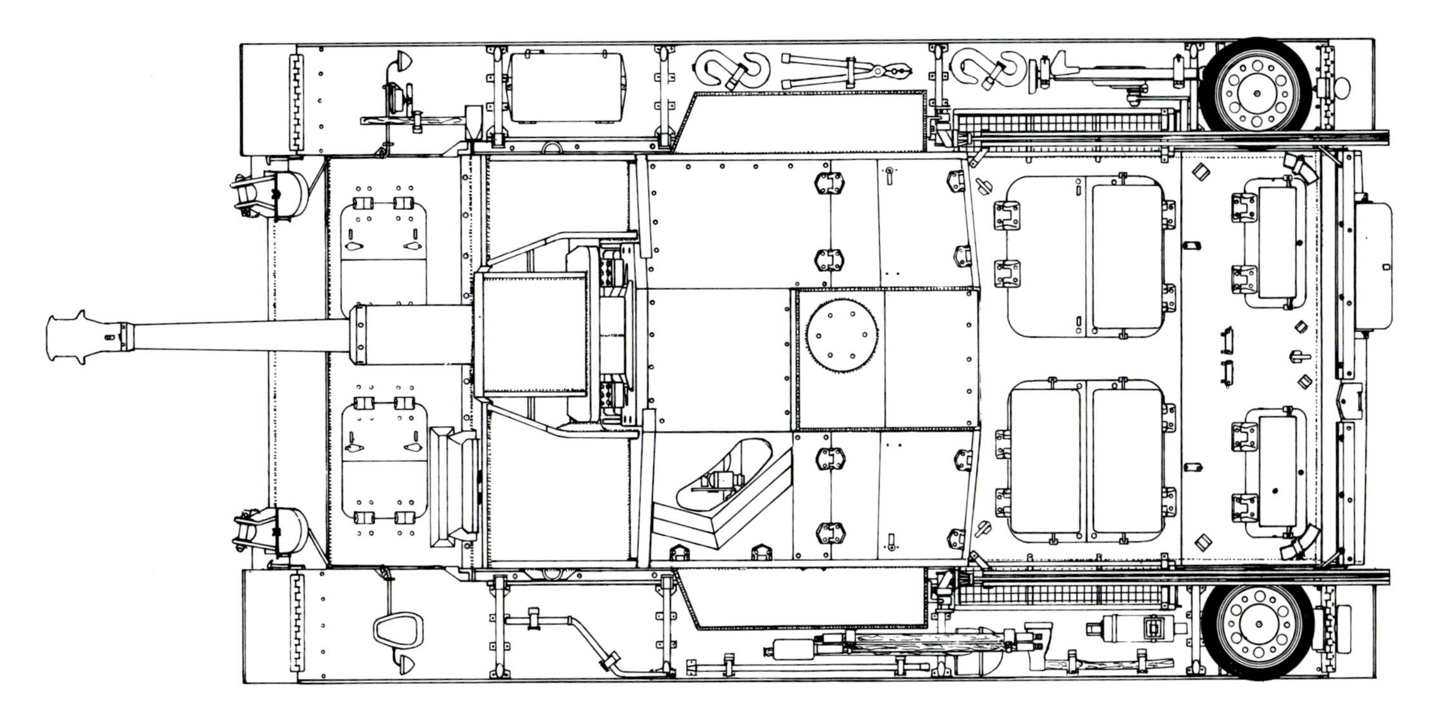

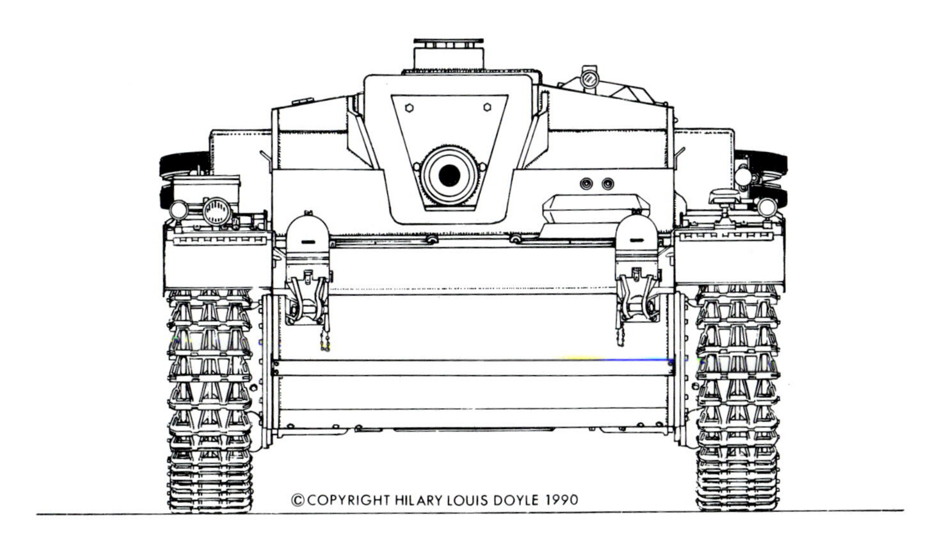
©COPYRIGHT HILARY LOUIS DOYLE 1990

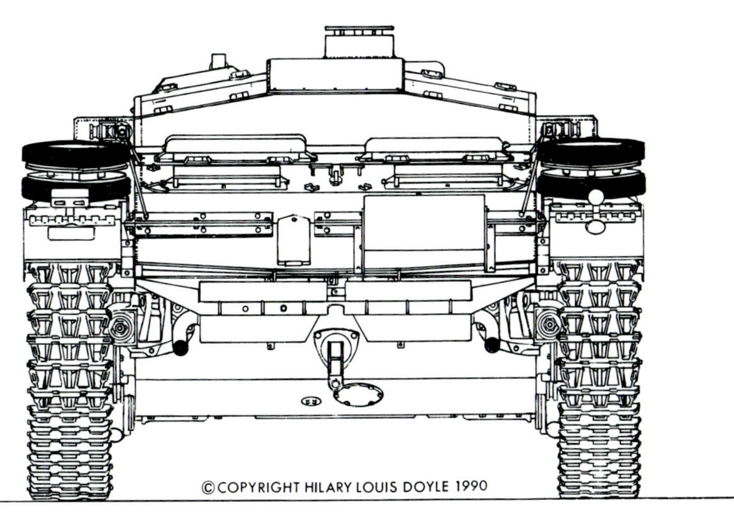
©COPYRIGHT HILARY LOUIS DOYLE 1990

Sturmgeschütz, Ausf.F, with winter tracks and the longer 75mm Sturmkanone 40 L/48. (StuG-Abt.197)

Sturmgeschütz, Ausf.F, chassis number 91284 assembled in July/August 1942. Note welded, supplemental armor plates.

Sturmgeschütz, Ausf.F with the longer 75mm Sturmkanone 40 L/48. Chassis number 91306 completed in August 1942. Note additional protection above the driver's position. (BA)

The June 28/29 conference confirmed, what had already been previously reported, that the strengthening of the frontal armor was to take place on all new production Sturmgeschütz.

Target Order Nr.8 dated 20 June 1942 specified: Sturmgeschütz with the 7.5cm StuK 40 are to be delivered with 30mm supplemental frontal armor. Deliveries of Sturmgeschütz with reinforced frontal armor will begin with 11 in June and continue from July with all further Sturmgeschütz produced. A report dated August 6th 1942, confirms the fulfillment of this order:60 Sturmgeschütz were produced with 80mm frontal armor and zero with 50mm armor in July of 1942.

Beginning with the last 11 Sturmgeschütz produced in June 1942, additional 30mm thick armor plates were welded to both forward 50mm base plates for the superstructure as well as to the upper and lower nose plates of the hull. At the same time, the two headlights (with armored covers that had been previously mounted on the front upper hull) were replaced by mounting a "Notek" headlight in the center of the upper hull (relocated from its previous position on the left fender).

Beginning in August of 1942, the slanted roof plate above the driver, as well as the same roof plate on the opposite side, appeared in a redesigned configuration. The angle of the plates was increased in order to achieve an unbroken transition from the front plate to the superstructure roof. This reduced the likelihood of a shot penetrating the 50mm plate at the upper superstructure forewall.

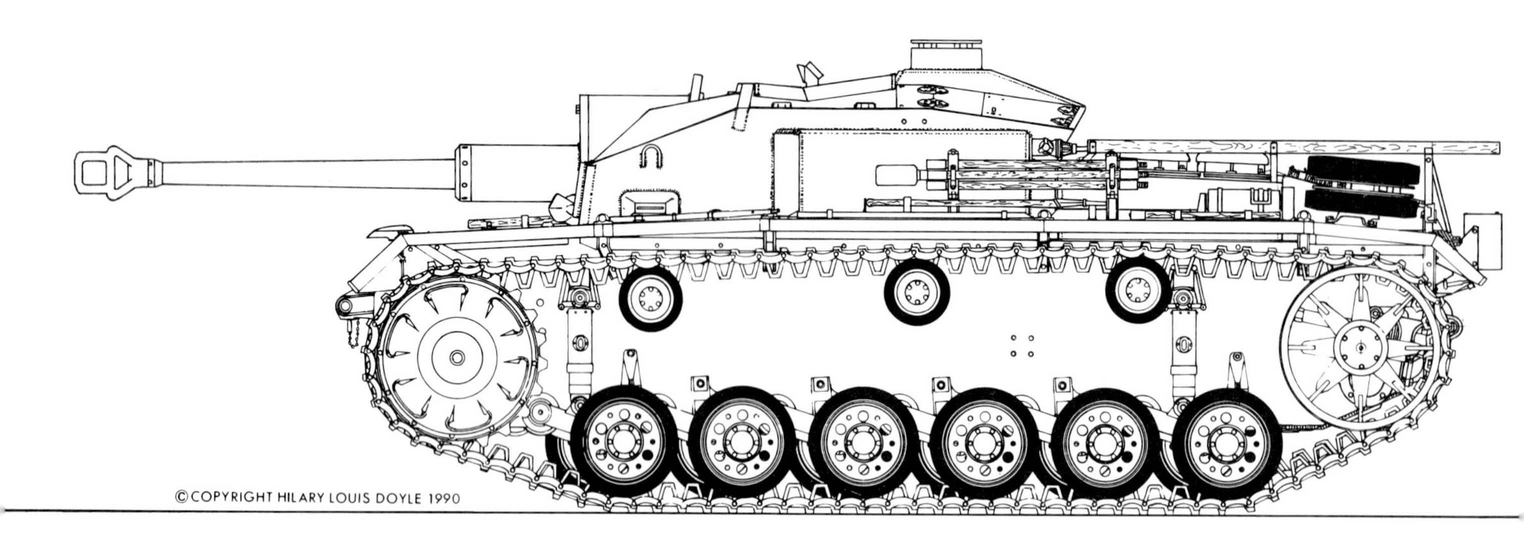

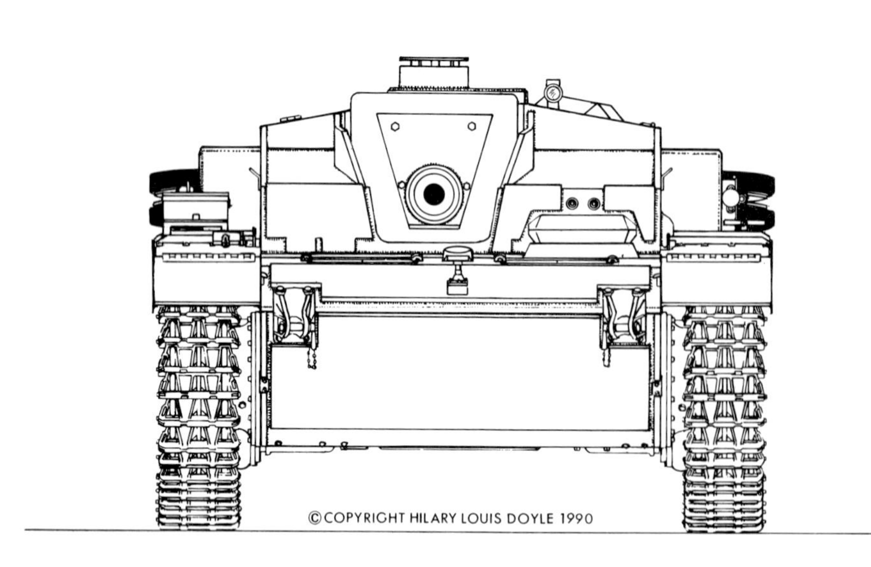

Sturmgeschütz, Ausf.F with the 75mm Sturmkanone 40 L/48 and welded supplemental armor, both introduced as improvements in June 1942.

bins were altered to increase stowage to 54 main gun rounds. As was often the case, units would entirely remove the factory-installed ammunition stowage bins and racks and by using a jury-rigged stowage system, decidedly increased the amount of ammunition that they carried.

Several Sturmgeschütz units, with earlier production Sturmgeschütz, tried to aid themselves by poring a concrete deflector into this critical position above the driver's head and on same location on the far side of the gun.

During the production of the Ausf.F, the ammunition

Sturmgeschütz, Ausf.F/8

In September 1942, the Sturmgeschütz, Ausf.F/8 followed the Sturmgeschütz 40, Ausf.F off the assembly line. It remained in production until December 1942. A total of 250 vehicles were produced with chassis numbers from 91401 to 91650.

The Ausf.F/8 chassis quite closely corresponded to the 8./Z.W. chassis of the Panzerkampfwagen III, Ausf.J through N. The superstructure remained basically unchanged from its predecessor on the Ausf.F. The size of the opening in the hatch to accommodate the gun sight was reduced and a wire mesh cage was secured over this opening.

On the glacis plate both of the two piece, hinged maintenance access hatches were replaced, by two one-piece hatches taken over from 8./Z.W. hull design for the Panzerkampfwagen III.

Up to and including the Ausf. F, two towing brackets bolted to the full front and two on the hull rear, were used to fasten cables or tow bars. Beginning with the Ausf.F/8, (again adopted from the 8./Z.W. hull) towing brackets were created by extending the hull sides and drilling holes into the extensions at both the front and rear.

Beginning with the Ausf.F/8, the radio antennae mounts, which up until now could be pivoted to lay down toward the rear, were replaced with fixed mounts.

The deck over the engine compartment received larger openings, protected by armored cowlings, designed to improve engine cooling. This same modification was also introduced on the Panzerkampfwagen III.

The fender extensions were shortened and no longer hinged at the front. In November of 1942, both single-piece maintenance hatches in the glacis plate, again reverted to the earlier two-piece, side-hinged hatch design.

Starting in October 1942, welding of the supplemental 30mm armor plates to the hull nose and superstructure front ceased. In keeping with the plan for increasing production rates, these additional plates were drilled and fastened with bolts.

An experimental machine gun shield was introduced and installed on the roof of several Sturmgeschütz in December 1942. This offered frontal protection for the loader while he was half exposed, firing the machine gun. The later design for the machine gun shield was backfitted to Sturmgeschütz Ausf.F/8 starting in early 1943. A fastener, on the front half of the loader's hatch, was used to secure the machine gun shield in its upright position.

Sturmgeschütz, Ausf.F/8. Side and top views of the hull. The basis was the hull from the Panzerkampfwagen III (8./Z.W.).

Photos above: Sturmgeschütz, Ausf.F/8. Interior views during assembly.

Sturmgeschütz, Ausf.F/8. Hull on a boring machine. This is chassis number 91500, delivered in October 1942.

Above: Sturmgeschütz, Ausf.F/8. Because of material bottlenecks, several of the guns were fitted with single-chamber muzzle brakes.

Center: A Sturmgeschütz, Ausf.F/8 with StuG-Battr.90 in Tunisia.

Left: A Sturmgeschütz, Ausf.F/8 produced in November 1942 and back-fitted with Schürzen side skirts and a machine gun shield. (BA)

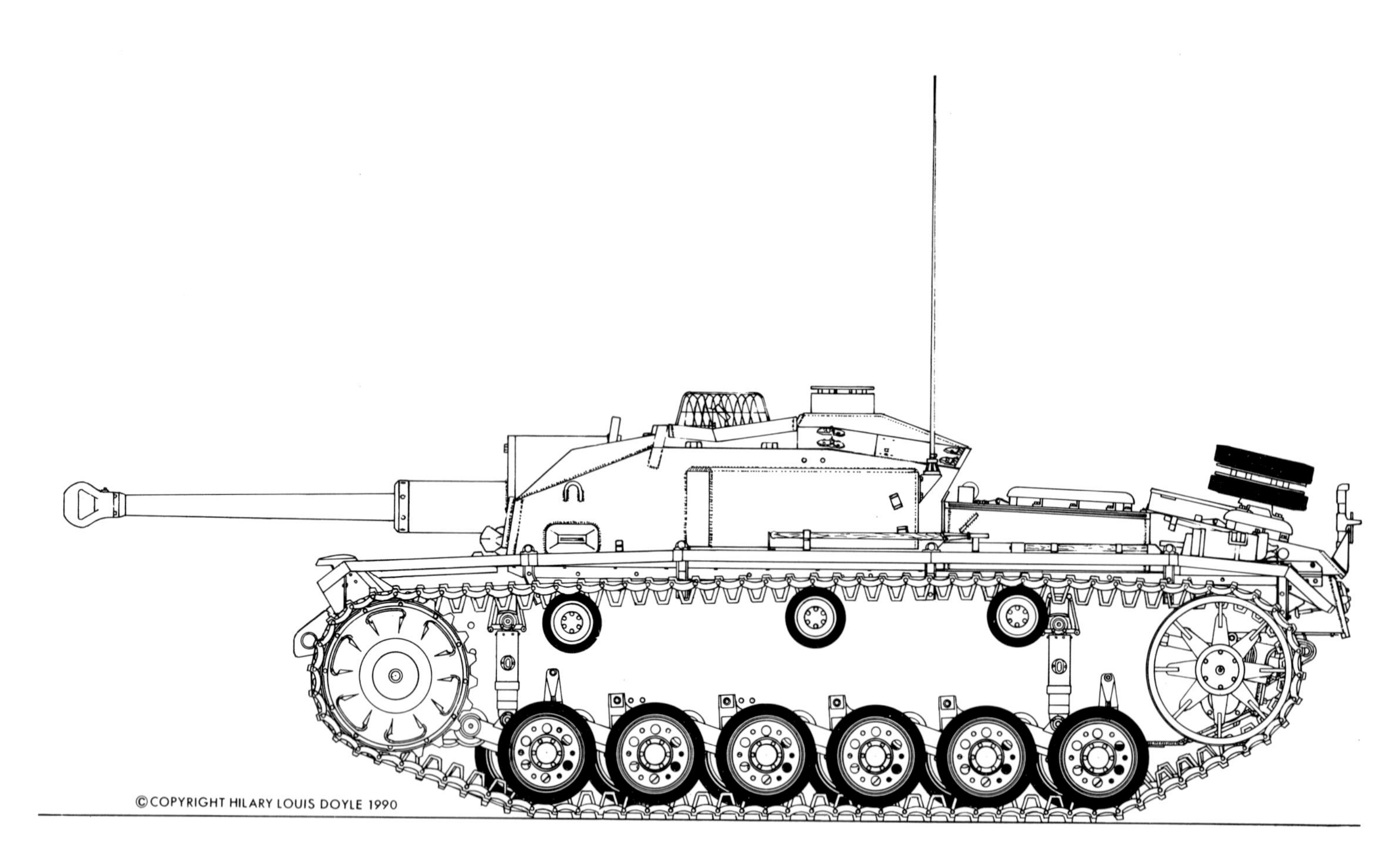
©COPYRIGHT HILARY LOUIS DOYLE 1990

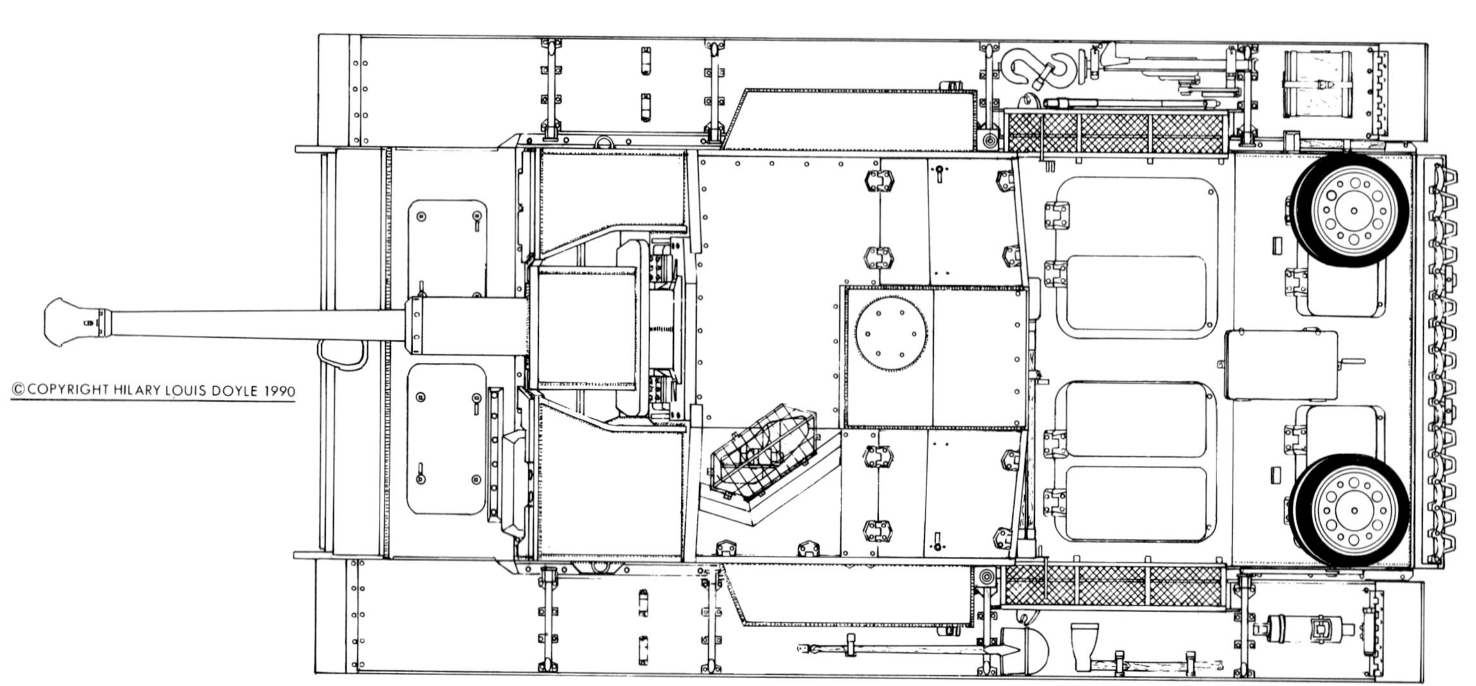
©COPYRIGHT HILARY LOUIS DOYLE 1990

Sturmgeschütz, Ausf.F/8, with the single-chamber muzzle break (an exception).

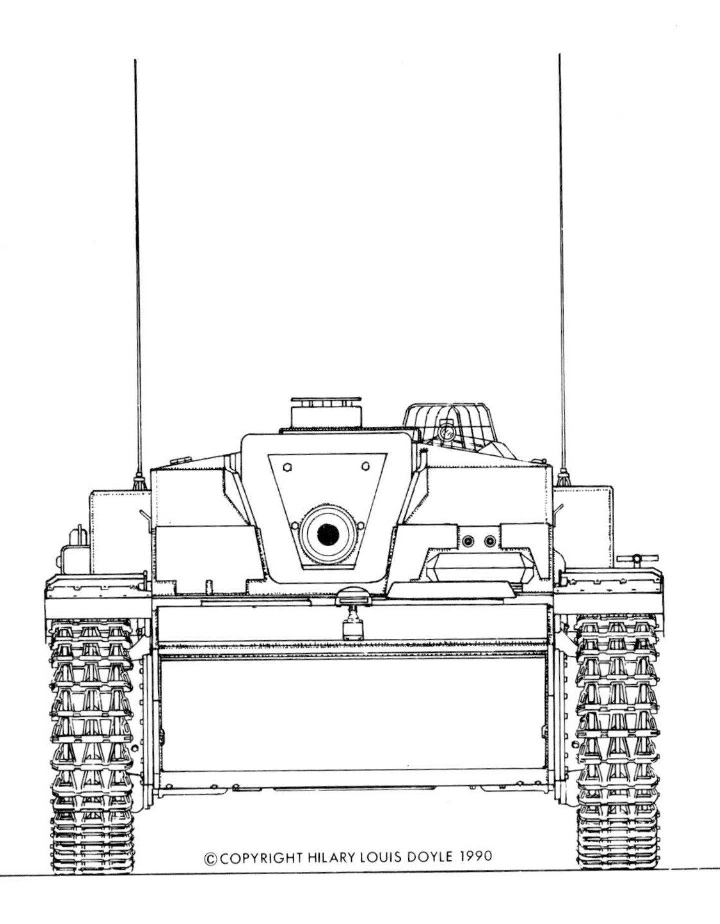

A Sturmgeschütz, Ausf.F/8, being supplied with ammunition. The supplemental armor plates have been bolted on.

Sturmgeschütz produced in November 1942 were issued to the StuG-Lehr.Battr.901. An experimental design for the machine gun shield has been mounted on the superstructure roof.

Sturmgeschütz, Ausf.F/8. The antennae were in fixed mounts.

Sturmgeschütz, Ausf.F/8 with bolted on 30mm supplemental armor plating and shot deflector in front of the driver's visor.

Connections for the Pz.-Kühlwasserheizgerät 42, a coolant heating system.

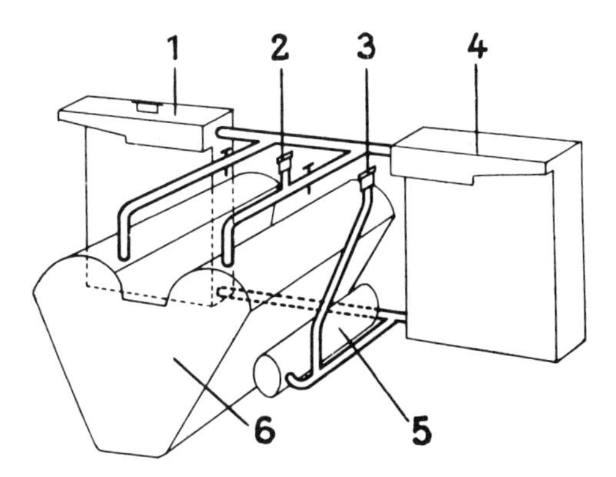

Anschlüsse für Pz-Kühlwasserheizgerät 42
Schema

1 Kühler
2 Anschluß für Kaltwasserrücklauf vom Motor
3 Anschluß für Warmwasserzulauf zum Motor
4 Kühler
5 Ölkühler
6 Zylinderblock

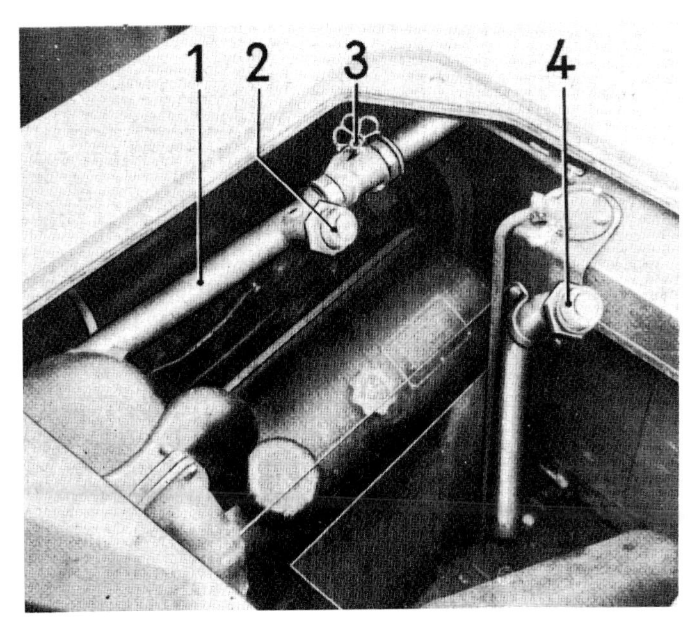

View of the connections for the coolant heating system.

Beginning in May 1943, Sturmgeschütz Ausf.F/8 were backfitted with "Schürzen" steel plates as side-skirts. These last two improvements were accomplished as backfits after the production run of the Sturmgeschütz Ausf.F/8 had been completed in December 1942.

General Army communication Nr.95 dated 21 October 1942 addressed the installation of a warm water transfer system for Sturmgeschütz's: All Sturmgeschütz vehicles are to have a connector for a warm-water transfer system installed. This connector will allow the coolant fluids to be heated, either through connections with coolant transfer hoses to another Sturmgeschütz with its engine running or by using the Panzer-Kühlwasserheizgerät 42 (engine coolant heater). The connections were to be made in accordance with plans Nr. BSKB 262. O.K.H. (Ch H Rüst u. BdE), 7.9.1942-5332/42-AHA Ag K/In 6 (Ing).

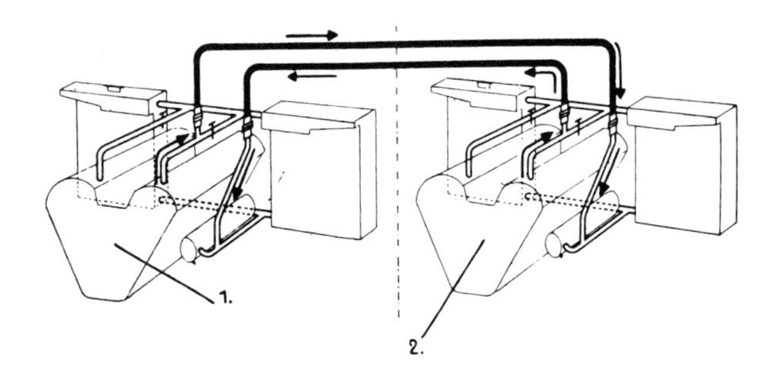

Coolant transfer.

Kühlwasserübertragung
von Pz Kpfw zu Pz Kpfw
Schema

1 Anzuwärmender Motor
2 Warmer Motor

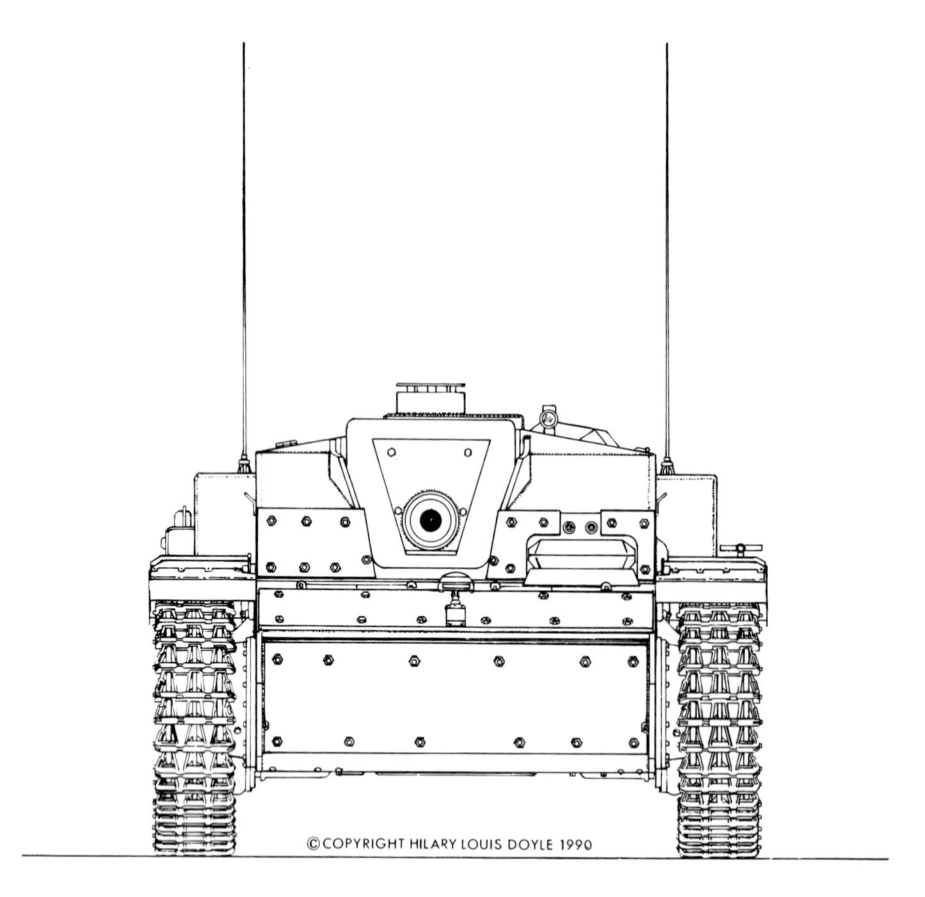

Sturmgeschütz, Ausf.F/8 (supplemental armor bolted, beginning in October 1942).

During a conference with Hitler on October 2 1942, it was reported that a considerable number of the Sturmgeschütz already deployed would be equipped with winter tracks. In accordance with plans, 75% of the vehicles were to be fitted before January 1st 1942. Hitler emphasized that all vehicles in use with Heeres Gruppe Nord and Mitte (Army Groups North and Center) were to be retro-fitted.

Hitler again emphasized the hinderance to target observation during firing caused by smoke forced out the sides of muzzle brakes. He fully realized that smoke is far more hindering in nature than a strong muzzle flash. Based on the assurance given him, that smoke could be reduced in favor of the muzzle flash, he expected immediate implementation of the appropriate measures.

Sturmgeschütz, Ausf.G

Production of the final model to be produced on the Z.W. chassis, the Ausf.G, began in December 1942 and ended in April 1945. Sturmgeschütz produced by ALKETT had chassis numbers from 91651 to 94250 and continued in a second series starting from 105001. MIAG started production in February 1943 with chassis numbers is a series beginning with 95001.

While maintaining an almost unaltered chassis design as the previous Ausf.F/8, the Ausf.G was outfitted with a wider superstructure featuring a commander's cupola. These changes increased the vehicle's height to 2160mm. Instead of the earlier box-shaped side panniers, the entire superstructure was now wider and the side armor plates were inclined. The rear wall of the fighting compartment was now vertical and the rear section of the superstructure roof was raised.

Sturmgeschütz, Ausf.G. Completed in December 1942. Original superstructure. Gun sight unprotected. Fume exhaustion ventilator mounted on the superstructure roof.

The fume extraction fan remained in its previous horizontal position in a centered rear section on the superstructure the roof. Only now the armored guard was somewhat shorter. The fan with electric motor was bolted to the superstructure roof from the inside, so that only the disc-shaped protective armor protruded above the surface. Despite the fact the rear portion of superstructure roof had been raised, the function of the loader was still encumbered by a lack of space. Therefore, beginning in January 1943, the extraction fan was removed from the roof and relocated to the middle of the vertical rear wall of the superstructure.

The roof above the gunner and that on the opposite side were expanded directly outward and the sides inclined at an angle. The ballistic protection for the gun sight was improved. The armor thickness of the slanted roof above the driver's head and on the far side of the superstructure were increased to 30 mm. The view port located to the left of the driver was initially retained from previous models. The fenders were now fastened to brackets along the superstructure sides. Since the bottom of the superstructure side pannier was higher than the top of the hull, the fenders were stepped upward to accommodate fastening to the superstructure.

Up until October of 1943, the commander's cupola could be rotated through 360 degrees on a ball-bearing race. As a result of the shortage of ball bearings, after September 1943 the cupola was fastened in a fixed position to the cylindrical base. Beginning in August 1944 (and continuing through the end of the War), a ball bearing race was again installed.

Modifications Introduced During Production

Unlike the earlier models of the Sturmgeschütz in which smaller numbers were produced in a relatively short period of time, the Ausf.G was produced in large numbers through April 1945. During this time, many modifications were introduced into the production run. These were intended to improve automotive performance, armor protection, mobility, or to simplify production methods. Many improvements were the direct result of troop experiences provided through after-action accounts.

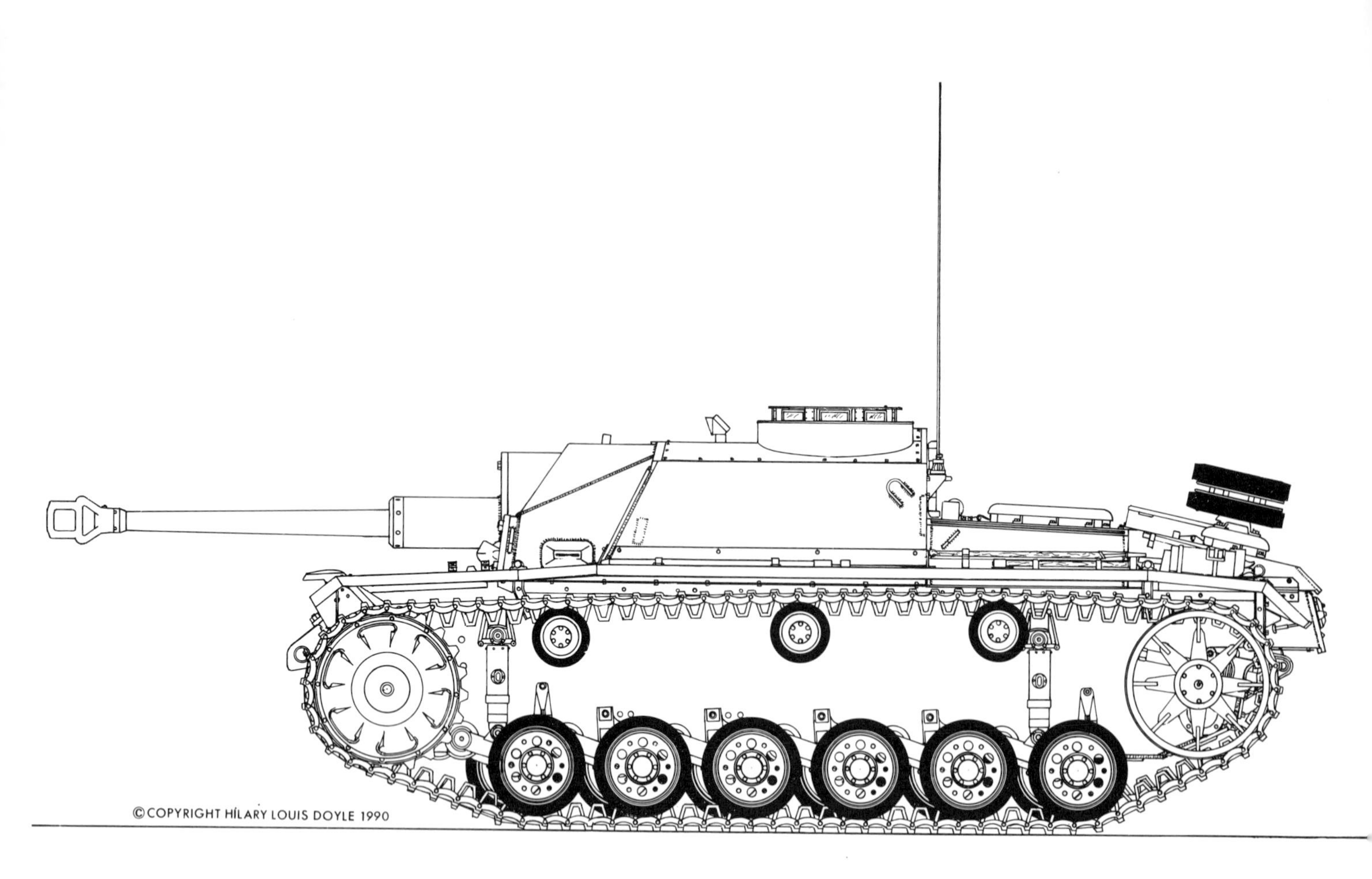
©COPYRIGHT HILARY LOUIS DOYLE 1990

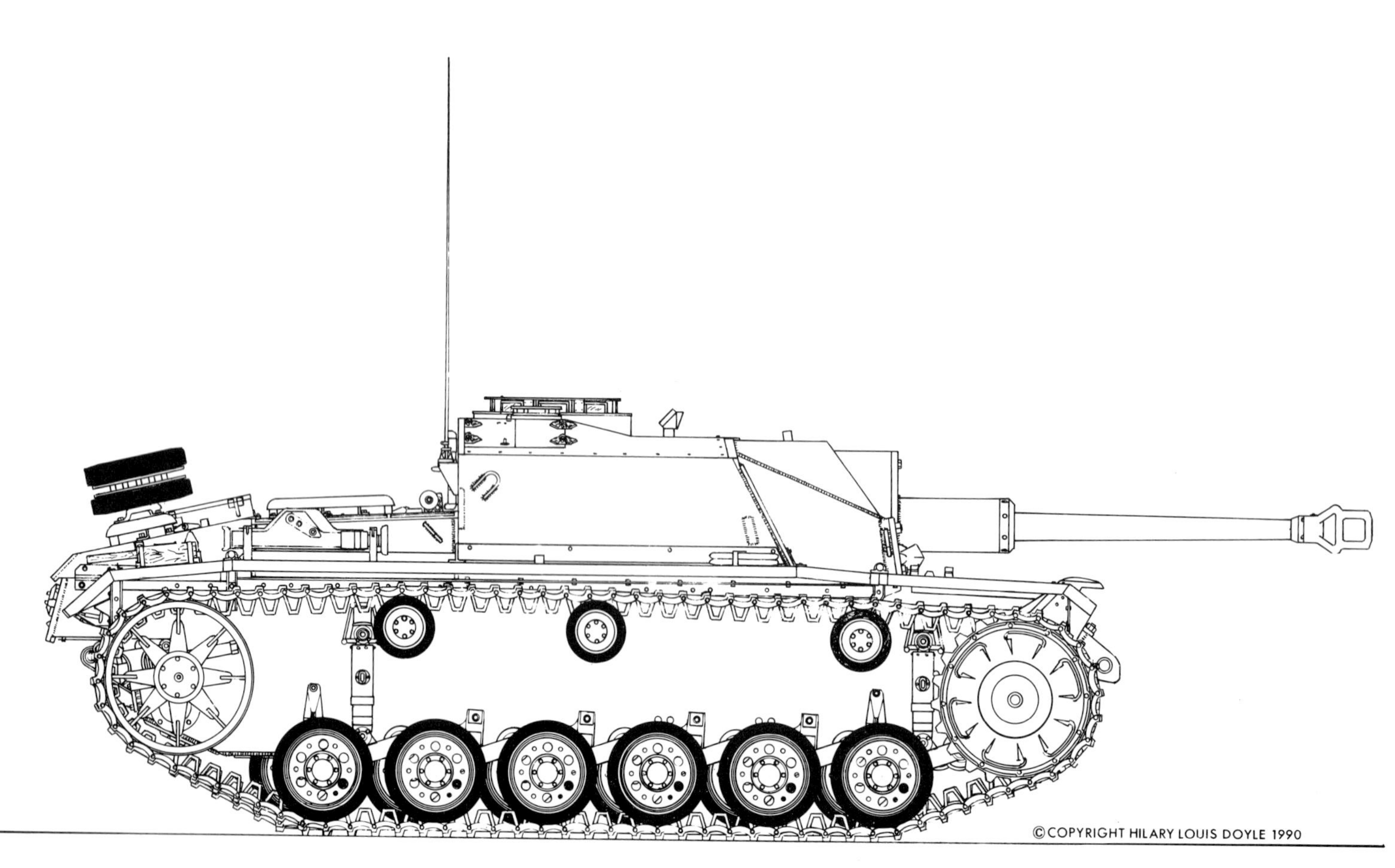
©COPYRIGHT HILARY LOUIS DOYLE 1990

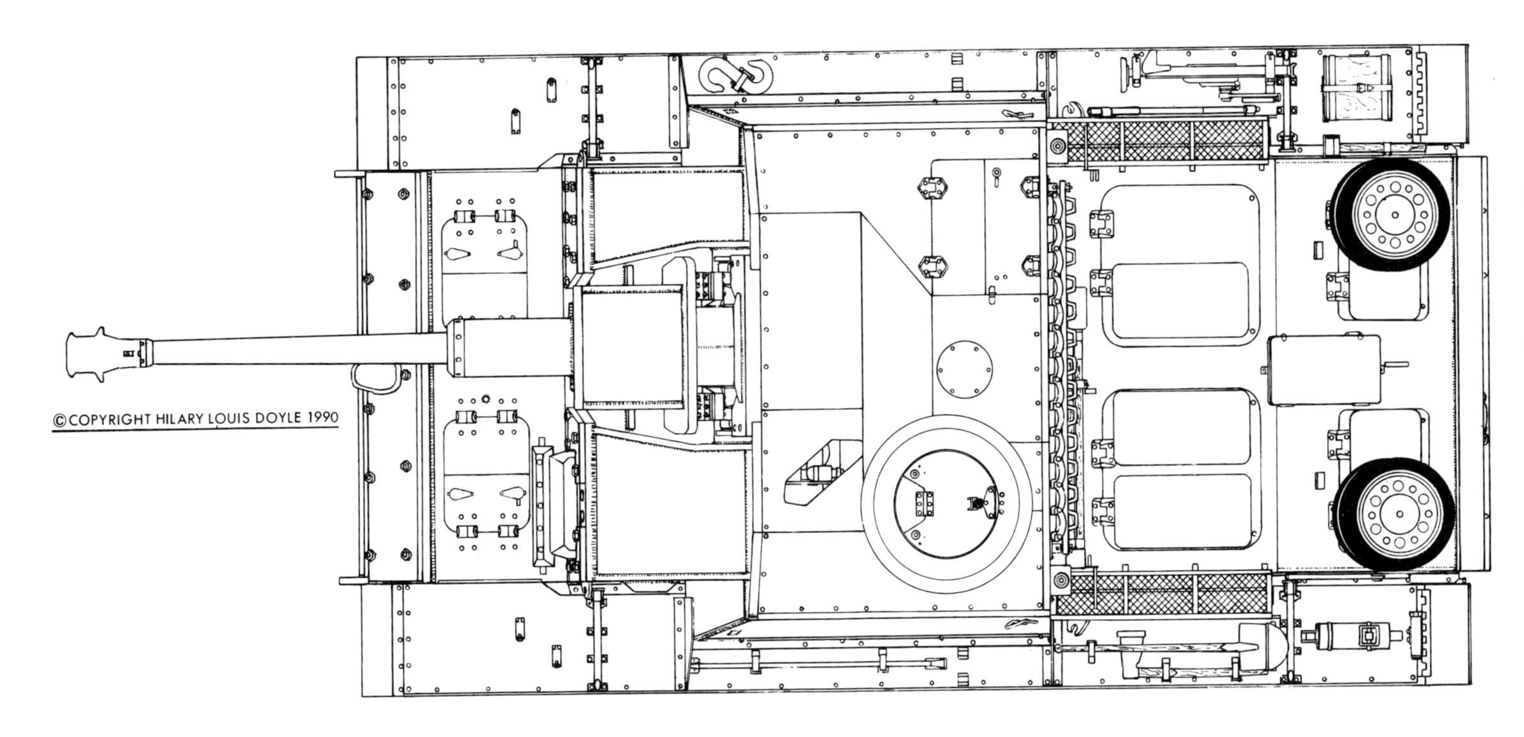

Sturmgeschütz, Ausf.G. Produced in December 1942. Original superstructure with unprotected gun sight and exhaust fan ventilator on the superstructure roof.

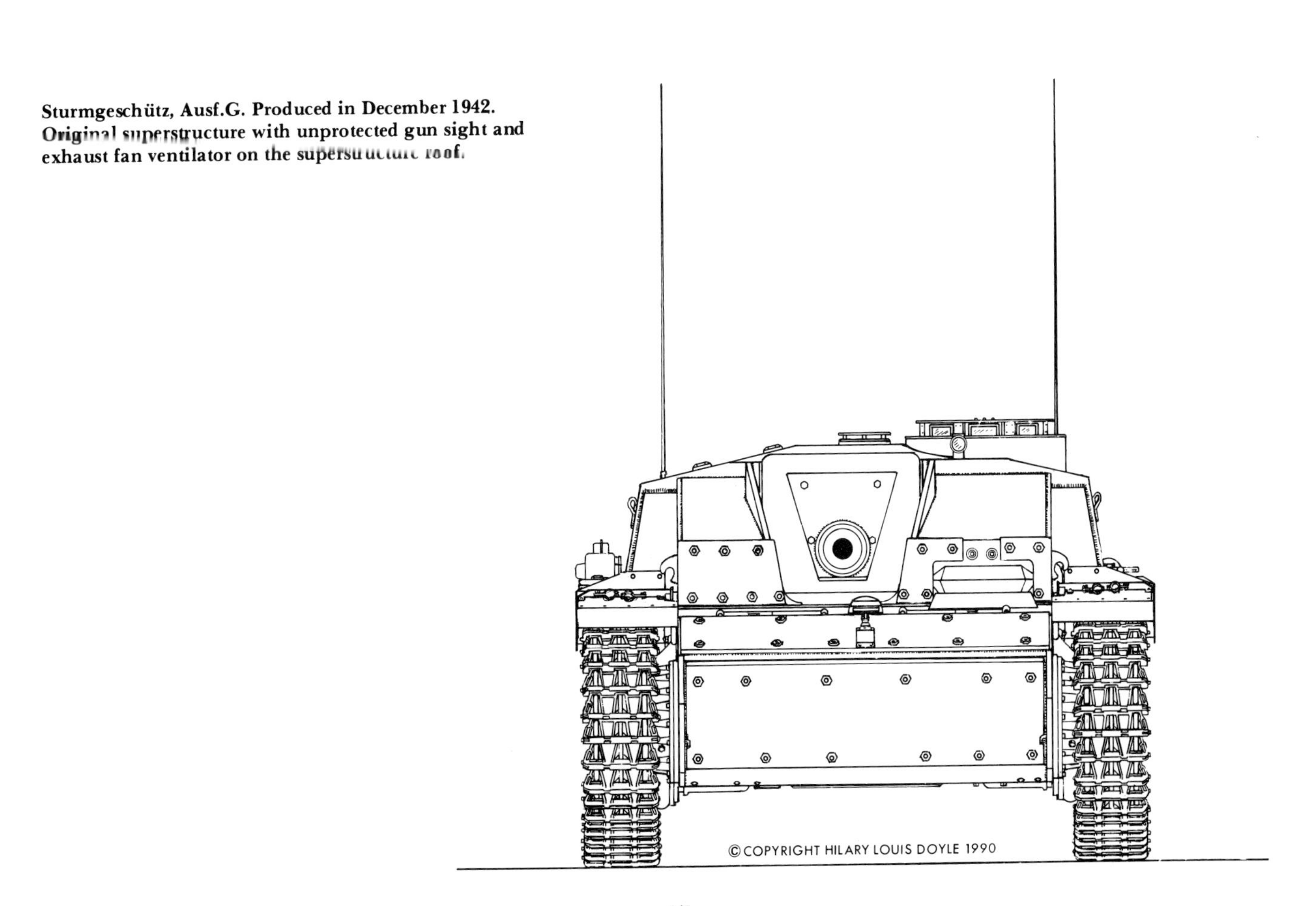

Sturmgeschütz, Ausf.G. Note the view port to the driver's left.

Sturmgeschütz, Ausf.G. The driver's view port has been replaced by a pistol port.

Sturmgeschütz, Ausf.G, chassis number 91675 completed in December 1942. StuG-Abt."Grossdeutschland."

A Sturmgeschütz, Ausf.G with winter tracks of the SS-StuG-Abt. 2 "Das Reich" (completed in December 1942.). (BA)

A typical example, for design changes originating from the troops, occurred in a meeting on April 8th 1944 with the Inspektor der Artillerie during which Hauptmann Metzger presented the following:
These following suggestions are aimed at improving the current Sturmgeschütz design. In part, these recommendations have already existed for months in the form of proposals for a new design as the "Sturmgeschütz (neue Art)." NOTE:This Sturmgeschütz (neue Art) was the original designation for the Jagdpanzer IV and the Jagdpanzer 38 (later popularly known as the "Hetzer").

a) Increase the stowed ammunition from the current 54 up to 70-75 rounds. The model at Alkett had burned. The new model currently in development should be accelerated.
b) Improvement and expansion of the heating system. Testing was promised.
c) The railing mounted on the rear deck (intended for the accompanying Grenadiers) was not present on all delivered vehicles. Quality control was promised.
d) Adding a handle, and redesigning the superstructure hatch so that it will lie flat, was approved for further investigation. Whether or not it was suitable for introduction on the Sturmgeschütz (neue Art), based on changes to periscopes, had to be checked.
e) Anti-dust belt on the superstructure. Delivery was approved.
f) A rest for the gas pedal (permitting the pedal to be held in place during warm-ups). This would be given consideration. (Subsequently designed and introduced beginning October 1944)
g) The mounting for the steering lever by the steering gear was too narrow. Checked into.
h) A Machine gun mount in the gun mantle. A representative of "In 6" referred to the differences between the cast and welded models of armor as well as to those between the rigid- or spring-mounted machine guns. Rigid mounts had led to absolutely no complaints from the Feldbrigade and therefore could not be the prime factor. This modification was to be expedited. At the very least, the corresponding hole in the gun mantle could be drilled for all future Sturmgeschütz. On the objection that the machine gun was a bottle-neck, Oberstleutnant Schaede made reference to major increases in production.
i) A closure for the rear deck hatches. This simple device was necessary in order to prevent screeching, etc., as has been previously experienced. (Introduced beginning August 1944).

j) Strengthening of the roof armor. Firing tests with Russian mortars had not yet been conducted. Strengthening meant a considerable increase in weight and steel was in short supply. The use of a wire net was suggested and testing was approved.
k) Cylindrical armor guard for the periscope cupola. Hauptmann Metzger's proposed a sheet metal ring filled with concrete. The accepted solution was welded steel bars reinforcing concrete.
l) Close-combat device. The Sturmgeschützschule had procured 100 projectiles with time-delays from Kiel. Consideration was given to the fact that a normal resupply of this type ammunition could not be assured.
Counter-proposal: There was currently a flare pistol in development that could be used to fire grenades. Further tests and selection of equipments was to be considered in investigating this possibility.

There were a host of reasons why all of the above requested modifications did not find their way from the drafting table into the production line. Even in cases of the most insignificant changes, often more than six mounts elapsed from the time a proposal originated until the modification was designed, approved, and finally installed.

A Sturmgeschütz with its many individual parts was not the product of a single factory. The Sturmgeschütz were completed at assembly plants, which were dependant upon deliveries from 10 manufacturers in the case of major components, and more than 50 manufacturers of significant parts. Even when changes were approved, ordered, scheduled, and delivered, it was simply not possible to have them installed into all the Sturmgeschütz at the same time. One could not afford to simply scrap or change the old parts as long as inventories of outdated parts were still available. Other problems arose when, due to circumstances beyond their control, all the manufacturers could not adhere to the agreed deadlines for changes. The ability to meet deadlines was often controlled by the availability of raw materials, the relocation of machining plants, and interruptions from air raids. Therefore, in the period between 1943 and 1945, several months would lapse between the time a new modification was installed and when it was present on every Sturmgeschütz which left the factory.

The numerous and constant changes, which came into being during the production of the Ausf.G, are presented in the following sections, arranged in accordance with the categories of improvements in protection, armament and mobility.

Improvements in Protection

Pistol Ports

The view port located to the left of the driver had already been deleted in December 1942, but was replaced by a pistol port with a plug closure. This pistol port was positioned farther forward than view port it had replaced. The relocation was necessary to provide additional space in order to increase the incline of the supplemental armor plates at the front of both superstructure side panniers. This modification was introduced beginning in January 1943.

Machine Gun Shield

Starting in late December 1942, a protective shield for a machine gun was installed on the roof, directly in front of the loader. Mounted on a hinged base, the shield could be laid flat when not in use. The shield only offered protection toward the front. On the back of the shield were two mounting brackets for the machine gun, one for engaging ground targets and the other for aerial targets.

The machine gun shield could be retro-fitted to all previously delivered Sturmgeschütz, either by the units themselves, or during maintenance overhauls.

Gun Sight Guard

Beginning with the Ausf.C, the gunner's periscope protruded through an opening in the superstructure roof. Beginning in January 1943, protection was improved by covering this comparatively large opening with a hinged, sliding, armored guard. An arm connected this guard to the gun mount so that the guard moved with the gun and the sight. The gunner and gun sight were better protected and no interference occurred when the gun traversed or elevated.

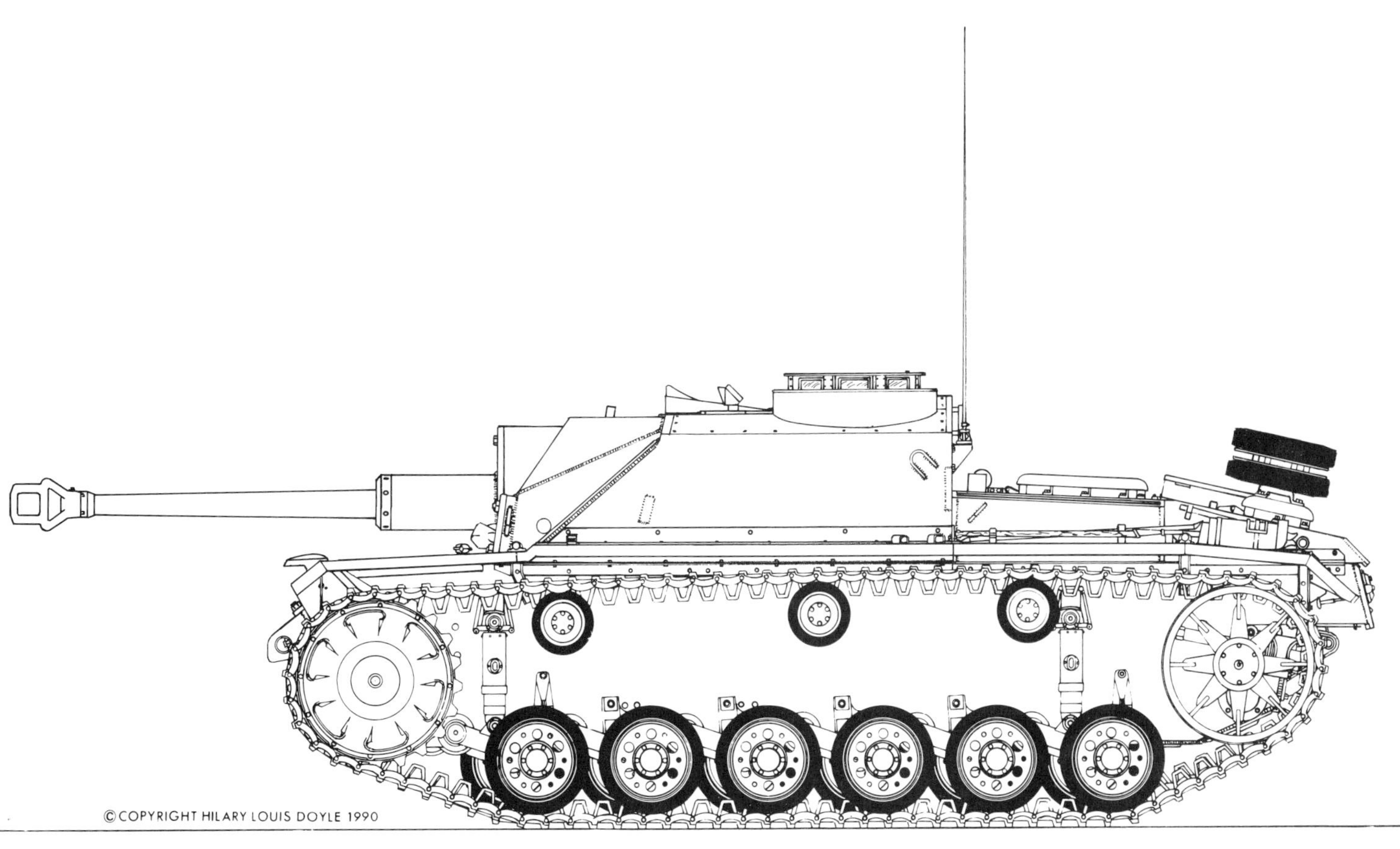

Sturmgeschütz, Ausf.G. Increased slope to the front of the superstructure pannier introduced as an improvement in December 1942.

Sturmgeschütz, Ausf.G with experimental superstructure. Increased protection, machine gun shield for the loader, a steel deflector in front of the commander's cupola and a single-chamber muzzle brake (1943).

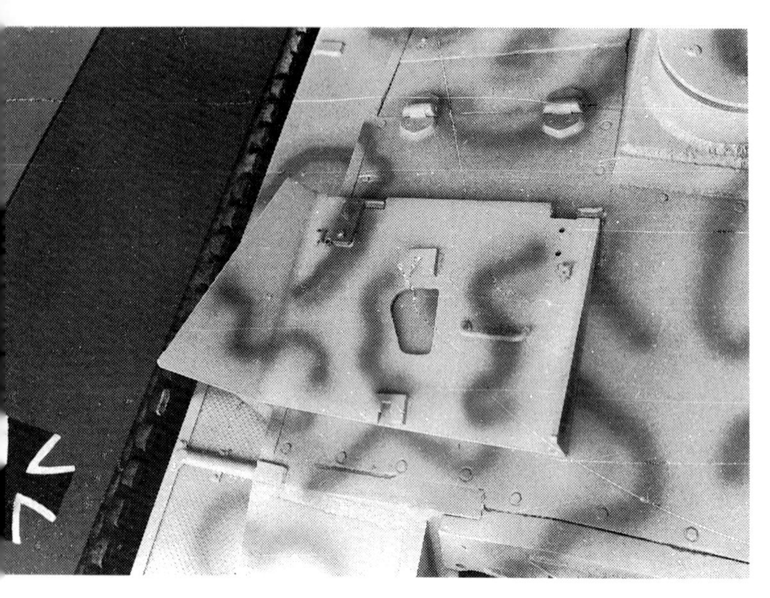

The loader's machine gun shield on the superstructure roof (Backfitted to this Ausf.F/8 with chassis number 91507 produced in October 1942).

Machine gun shield with weapon mounted (improved shield design introduced in Autumn 1943).

Driver's Visor and Front Plates

Basic to all Sturmgeschütz vehicles was the "Fahrersehklappe 50" driver's visor. Its glass vision block protected the driver from small arms fire. For additional protection, the visor was closed, and the driver used the "Kraftfahrfernrohre" (KFF, driver's periscopes). The periscope heads were positioned opposite two holes bored through the driver's front plate. Starting in February and ending in March of 1943, the KFF was dropped from production and no longer issued.

Instead of designing a new "Fahrersehklappe 80" that could be mounted to a 80 mm thick basic armor plate, or replacing the driver's visor with a periscope, the driver's front plate was retained at the basic thickness of 50 mm. This was supplemented by 30 mm plate bolted to its face. Originally this supplemental armor consisted of two pieces, one bolted on each side of the driver's visor, with an opening in the center for the two holes needed for the "KFF" driver's periscopes. However, when the "KFF" was dropped, the space was covered by a small plate welded between the two side supplemental plates. A final solution, introduced in April 1943 and continuing to the end in April 1945, was fabricated from a single piece of 30mm thick armor plate, cut to fit around the "Fahrersehklappe 50" and bolted to the driver's front plate.

Schürzen Side Skirts

The previously mentioned Schüzen side-skirts became a topic of discussion during the Führer's conference on 6 and 7 February 1943. Hitler was quite in agreement with mounting skirts on the Panzer III, IV and Sturmgeschütz to provide protection against Russian anti-tank rifles.

Test firings on Schürzen protective skirts (wire and steel plates) were reported on February 20, 1943. Firing tests utilizing the russian 14.5mm anti-tank rifle at a distance of 100m (90 degrees) showed no tears or penetration of the 30mm side armor, when protected either by plates or wire mesh. When testing was conducted with the 75mm high explosive shell (Charge 2) from a field gun, there was no damage to the sides of the hull armor when protected by the wire or plates. Wire mesh and plates had indeed been penetrated and even torn away, but, they still remained usable.

The decision to utilize the plates as opposed to the wire mesh (although both had proven effective and the mesh

Sturmgeschütz, Ausf.G, chassis number 95219 with Schürzen side skirts. Completed at MIAG in May 1943.

was lighter) was based on the fact that the wire mesh required the design of a new mount, which would have required additional time to be developed.

Additionally, the procurement of wire mesh for the side skirts was difficult. The skirts were not tested against shaped charges, nor were they intended as protection against this type of shaped charge (HEAT) shells.

On March 6 1943, Hitler indicated that he was satisfied with the favorable results of the firing tests against the Schürzen side skirts. In addition to outfitting all newly produced Sturmgeschütz, Panzer IV and Panthers with side skirts, all armored vehicles of these types currently deployed and those undergoing maintenance, were to be backfitted with them. The schedule for fitting Schürzen was to be expedited.

The manufacturing firms beganof Schürzen side skirts for Sturmgeschütz for the purpose of retro-fitting had already been sent to the Eastern Front. In early June 1943 the first front-line units retrofitted their Sturmgeschütz. With this modification, the Sturmgeschütz were ready to begin the Kursk offensive.

Numerous complaints by the field units attested to the fact that the Schürzen side skirts, although successfully providing protection against anti-tank rifles, mountings were entirely inadequate, resulting in the frequent loss of the Schürzen.

Beginning in March 1944, a new mount was introduced. The new mounting brackets were located inside the plates, and secured on triangular or rectangular tabs welded to the rails mounted on each side of the Sturmgeschütz.

Hull Armor

Beginning with the Ausf.F, 30mm supplemental armor plates were welded to the 50mm thick, basic frontal armor. In order to reduce the manufacturing time needed to attach this supplemental armor, it was bolted to the basic armor and no longer welded. This change occurred during the production of the Ausf.F/8 and continued during the production of the Ausf.G. In February 1943, the armor manufacturers were ordered to introduce frontal armor consisting of single 80 mm thick plates. The first Sturmgeschütz with basic frontal armor, 80mm thick plates, left the assembly plants in May 1943.

A backlog of already completed hulls existed, 50 mm

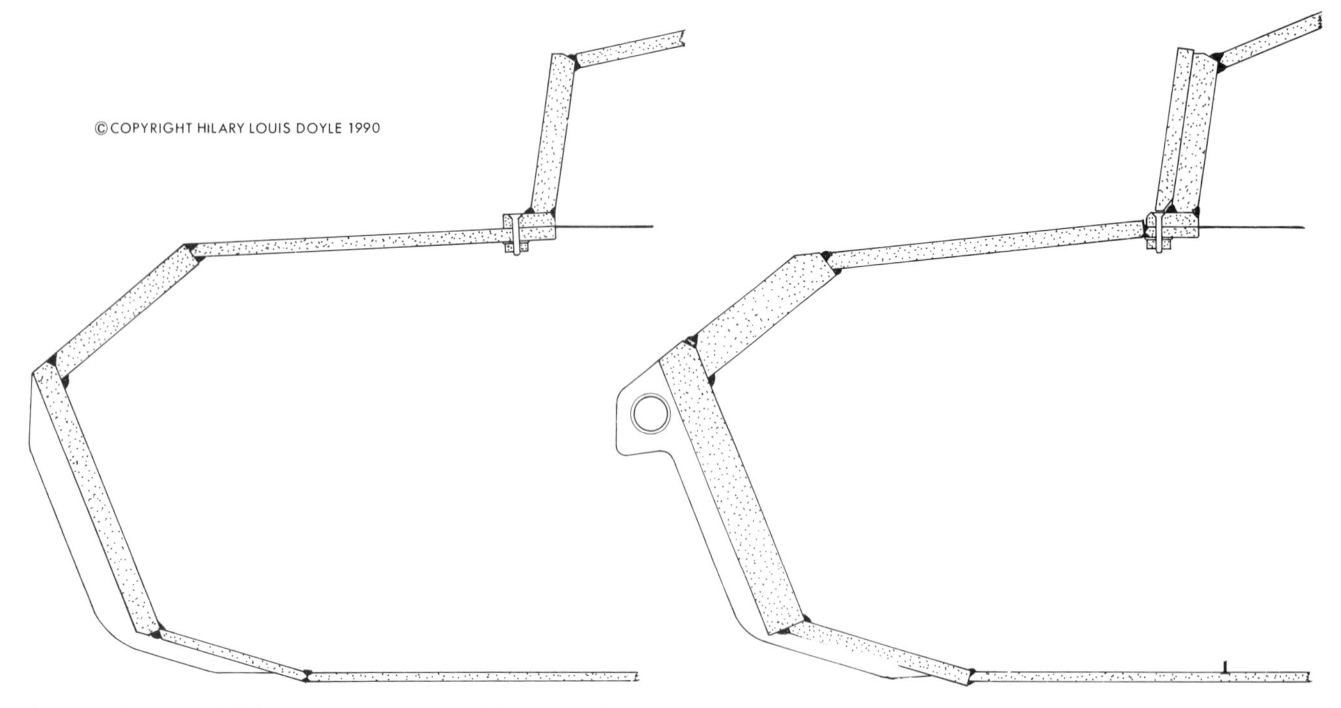

A comparison of the thickness and arrangement of the armor plates, between an Ausf.F with 50mm frontal armor and an Ausf.G with 80mm front hull plates.

armor plates that were already rolled and cut to size were still available, and the three armor suppliers changed to 80 mm thick plates at different times. This resulted in MIAG still delivering newly completed Sturmgeschütz with 50mm hulls and 30mm supplemental plates bolted to the front until October 1943.

In an attempt to quickly increase production rates, beginning in January 1943, two additional manufacturers started production of Sturmgeschütz armored components. Because many parts had already been produced for the manufacture of the Panzerkampfwagen III (8./Z.W.) hulls, there were hulls on Sturmgeschütz completed between January and March 1943 which had one-piece glacis hatch covers (with and without rear hinges) until the supply ran out. Then, even the new firms produced the normal two-piece, hinged sides, glacis hatch covers.

In order to achieve higher production rates, not only was production increased at ALKETT, and MIAG called upon to switch production from the Panzerkampfwagen

At the Brandenburger Eisenwerke GmbH, Werk Kirchmöser in May 1944. The thickness of the basic frontal hull armor had been increased from 50mm to 80mm beginning in April 1944.

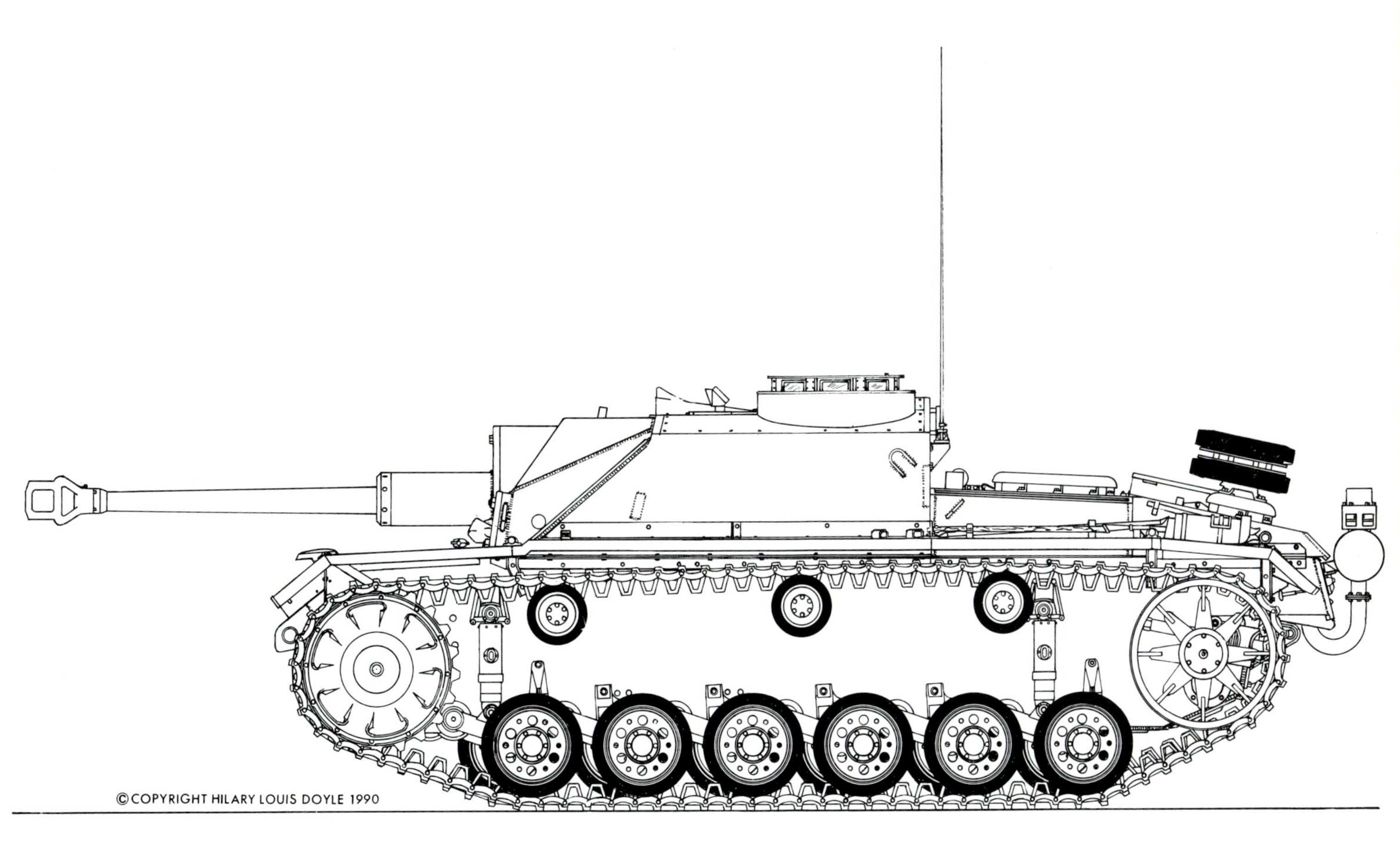

Sturmgeschütz, Ausf.G superstructure on a Panzerkampfwagen III Ausf.M chassis.

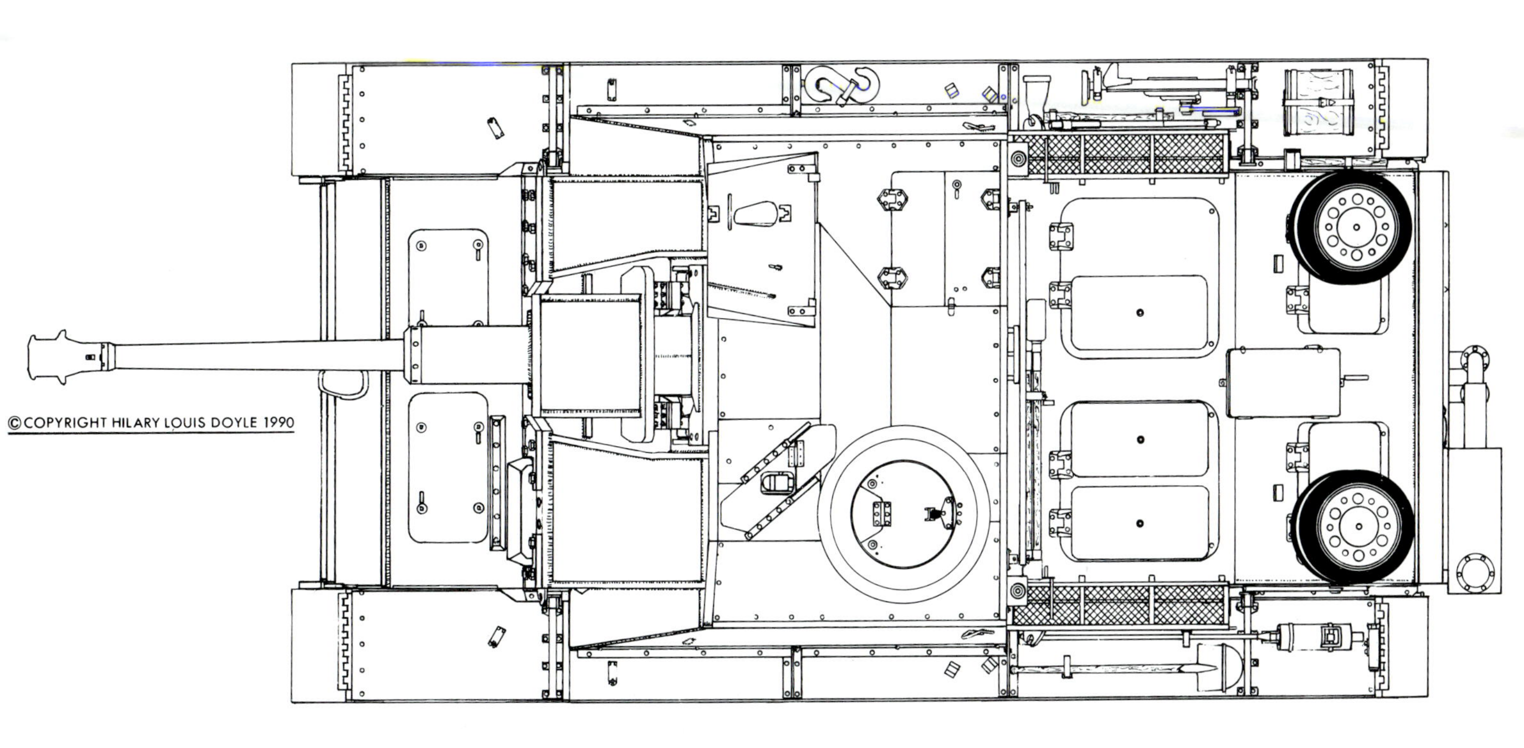

Sturmgeschütz, Ausf.G, with a Panzerkampfwagen III chassis, welded supplemental armor plating on hull front, with 30mm supplemental armor bolted to the front of the superstructure. Chassis completed by M.A.N.

Sturmgeschütz, Ausf.G with a Panzerkampfwagen III, chassis number 76171. Chassis completed by M.A.N. in March 1943. (BA)

This Sturmgeschütz, Ausf.G, with a chassis number in the range from 91769 to 91865, was completed in January 1943. It features the stepped fenders, still hinged at the rear. The rail on the rear deck was added after it was issued to the unit.

Sturmgeschütz, Ausf.G, with a Panzerkampfwagen III chassis from M.A.N.. The driver's KFF (periscopes) was no longer installed. Modified fenders have been mounted.

Sturmgeschütz, Ausf.G, with a Panzerkampfwagen III Ausf.M chassis with a deep-fording exhaust system mounted on the rear.

III to the Sturmgeschütz, a third firm, the Maschinenfabrik Augsburg-Nürnberg (M.A.N.) was called upon to produce 142 chassis for the Sturmgeschütz. M.A.N. produced their chassis on the Panzer III assembly line and delivered them to either ALKETT or MIAG for final production. These M.A.N. chassis are notable in that they were built according to the specifications for the "8./Z.W." Panzerkampfwagen III, Ausf.M and therefore differ from the normal chassis for a Sturmgeschütz. The M.A.N. chassis still had the shot deflector for protection of the 8./Z.W. driver's visor protection, 8./Z.W. glacis hatches and welded 30mm supplemental armor instead bolted standard, and 8./Z.W. fenders, hinged at the front and rear. Still mounted on the rear was the deep-fording exhaust muffler for the Panzerkampfwagen III, Ausf.M.

Beginning in March 1943, some of the Panzer III 8./Z.W. chassis, returned for major overhaul and rebuilt, were used for Sturmgeschütz chassis. They retained the characteristics of the original Panzer III chassis, but received the Sturmgeschütz superstructures with the features current for the time. 30mm armor plates were bolted to the basic 50 mm thick hull frontal armor of these overhauled, converted Panzer III chassis.

Zimmerit Protective Coating

As related in the following report, measures were introduced to prevent the enemy from effectively using magnetic charges against armored vehicles.

A teletype message dated June 30, 1943:

Subj: Protective Measures Against Magnetic Charges
Ref: Demonstration at Kummersdorf on 5 and 8 June 1943
Field testing of the effectiveness of protective coatings against magnetic charges and projectiles will be conducted using a protective coating invented by the ZIMMER firm. Field testing has been ordered by the 7.Panzer Division in Heeres Gruppe Mitte and the 4.Panzer Division in Heeres Gruppe Süd.

Chef H Rüst and BdE are asked to send the appropriate

materials and installation personnel to the 7. and 4. Panzer Divisions.

Gen. St. d H/Ord. Abt. (III b) Nr 35590/43 geh.

The Panzer officer on staff with the Chef, General Stab des Heeres summarized in the following report dated 2 July 1943:

Subject: Protective Substances for Countering Magnetic Charges

I) Trial by the Heereswaffenamt in January of 1943 have determined the following to be effective countermeasures against magnetic charges:

1.) Coating the armor to a thickness of 3 to 5 mm with cement milk or cement mixed with fine gravel.

2.) Covering with bitumen or tar which has fine gravel spread on top (not particularly suited to hot climates).

3.) Thickly applying oil products (bitumen, tar, oils, etc.) keeping in mind the same limitations as in number 2.

4.) A very thickly applied coat of paint (2-3mm) can suffice. This method may not be effective.

5.) In the winter, when it is very cold, water can be poured over the vehicle causing ice to thickly form. This makes it impossible for magnetic charges to stick.

The troops were informed of these methods in a message dated 9 February 1943 from the *General der Schnellen Truppen*.

The only truly workable method as listed under number 1 above, that concerning tar, was rejected by the troops due to the danger of fire.

II. Further testing has resulted in a protective substance which was shown during a demonstration on 5 and 6 June 1943 in Kummersdorf.

Protective Coating for Fully-Tracked Armored Vehicles

In order to prevent magnetic charges being attached, a "Zimmerit" protective coating is to be applied to the armored vehicles on the Pz.Kpfw.III and Pz.Kpfw.IV chassis, to the Pz.Kpfw.Panther and Pz.Kpfw.Tiger, to the Sturmgeschütz, to the Pz.Jäg 38 for the 7.5cm Pak 40/3 (Sf) and Pz.Jäg.Hornisse as well as on the G.W.II, G.W.38 and G.W.III/IV.

The following are general application guidelines:

a) Zimmerit is to be applied to all sloped and vertical armor surfaces on the chassis and superstructure, even under the Schürzen side skirts.

b) Portions to remain without protective coating include: Schürzen side skirts on the hull and turret, turret and superstructure roofs and the commander's cupola, bottom of the hull, horizontal fenders, removable equipment and spare parts, mufflers, lights and blackout headlights, etc., as well as portions of the armor which are covered by permanent or removable parts (such as the running gear and externally mounted tools and equipment).

The Zimmerit protective coating is to be applied as follows:

1. Thoroughly clean the areas to be covered. If applied directly onto the rust-preventing undercoating, the undercoating must be completely dry.
2. Thinly apply a first coat of Zimmerit with a putty knife.
3. Apply the spackling, creating small squares, leaving surrounding borders approximately 5 mm wide (Sketch 1).
4. Allow the mixture to dry for four hours.
5. Heat with a blow torch to harden the surface. Stroke the surface lightly with the blow torch flame, not holding it too close to the surface, so that the coating is not burned (about 5 mm distant).
6. Using the putty knife, apply a second, approximately 4mm thick layer in squares, as in sketch number 2.
7. Use the putty knife to create closely spaced ridges, by pressing the putty knife down at short intervals to form a waved surface pattern.
8. Let the protective coating dry overnight.
9. The next day, harden the surface once again with the blow torch, as in number 5.

Approximate amounts of Zimmerit needed for each vehicle:

Sturmgeschütz approx. 70kg

Panzerkampfwagen IV approx. 100kg

Panzerkampfwagen Panther approx. 160kg

Panzerkampfwagen Tiger approx 200kg

Zimmerit is not to be thinned before working, unless it has mistakenly been allowed to thicken. Only when thickening has occurred should the accompanying thinning agent be utilized. Camouflage paint can be sprayed directly onto the surface of dried Zimmerit without any further preparations necessary. The containers should be stored closed. A mixture which has become hard cannot be used, because the application will crumble. Containers which have been broken open or cracked are to be covered with wet cloth or sacks.

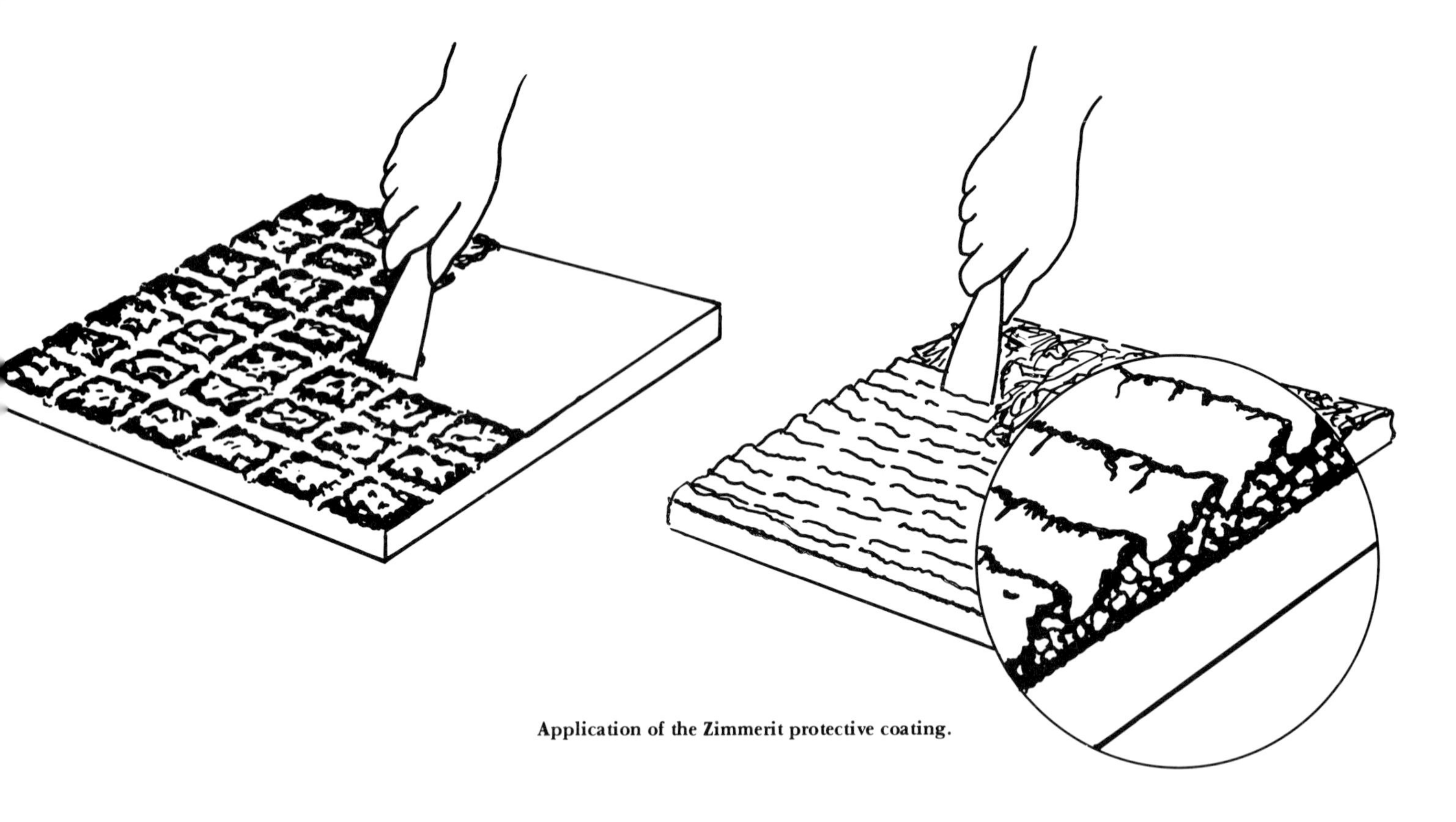

Application of the Zimmerit protective coating.

Zimmerit is to be obtained from the Heeres-Panzer Zeugamt Königsborn-Magdeburg, as well as at assigned armor spare parts depots.
O.K.H. (Ch H Rüst u. BdE), 23.12.1943 76 g Nr. 16975/43 In 6 (Z/Ing).

MIAG started applying Zimmerit protective coating toward the end of September 1943. Contrary to instructions, the coating was applied in a pattern of small squares with dimple impressions made in each square. For unknown reasons, perhaps due to time constraints, ALKETT delayed the application of the Zimmerit until the end of November/beginning of December of 1943. ALKETT also deviated from the instructions by creating "waffle" patterns in the Zimmerit.

An order to the assembly firms dated September 9 1944 stated: Effective immediately, the Zimmerit protective coating against magnetic charges was no longer to be applied to newly manufactured armored vehicles.

Sturmgeschütz with Zimmerit (MIAG pattern). Additional protection through track sections around the commander's cupola.

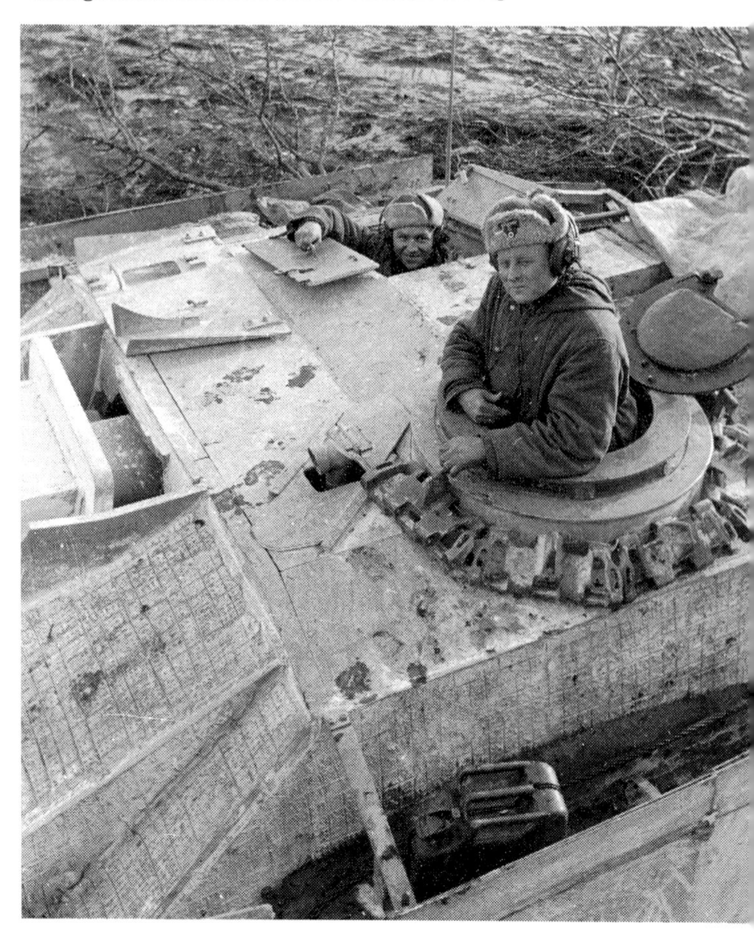

Zimmerit protective coating, pattern applied at MIAG. (BA)

Zimmerit protective coating, Alkett waffle pattern.

Zimmerit protective coating, Alkett waffle pattern.

Sturmgeschütz, Ausf.G. Steel deflector in front of the commander's cupola introduced in October 1943 and the Saukopf gun mantle introduced in November 1943, with and without Schürzen side skirts.

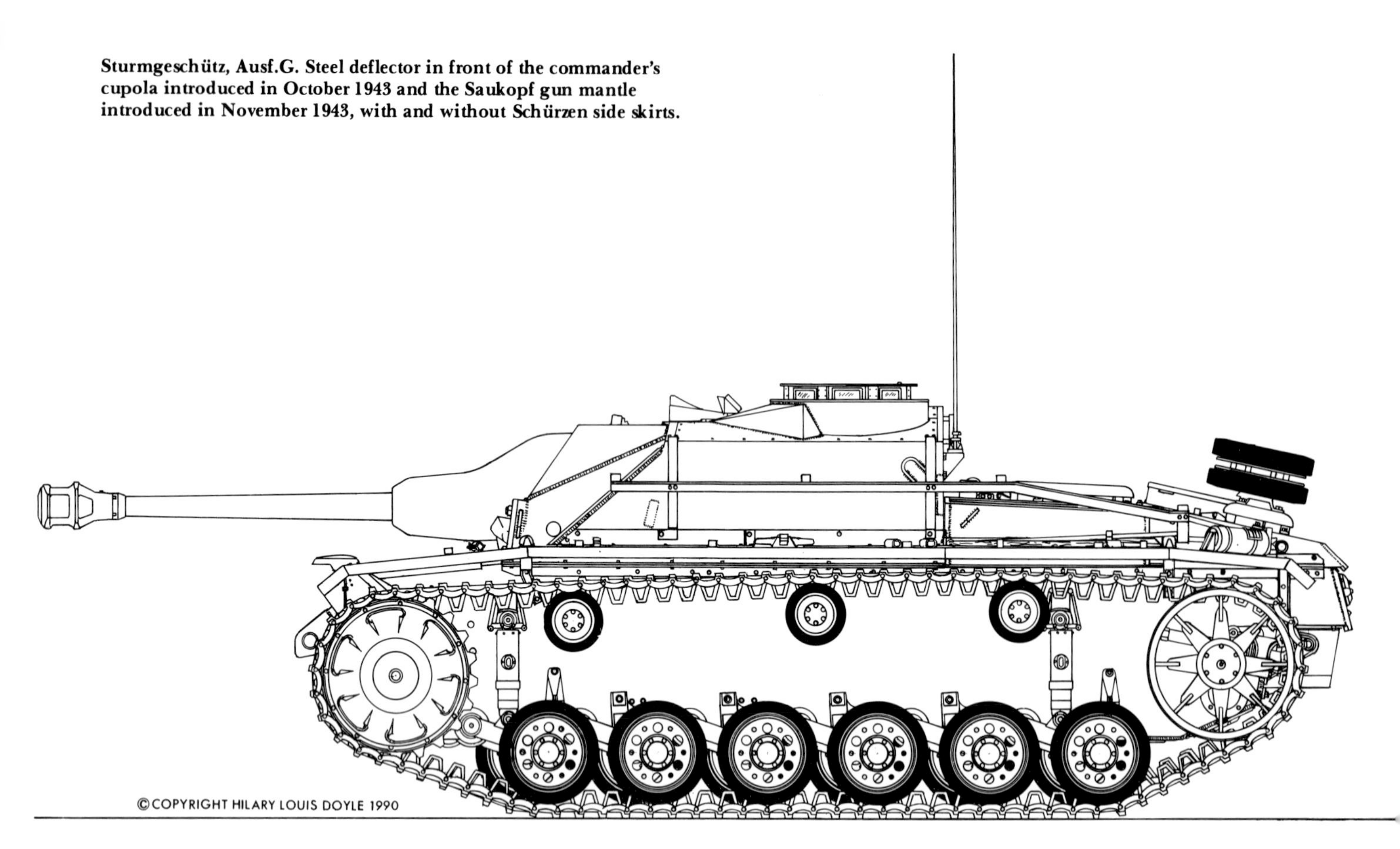

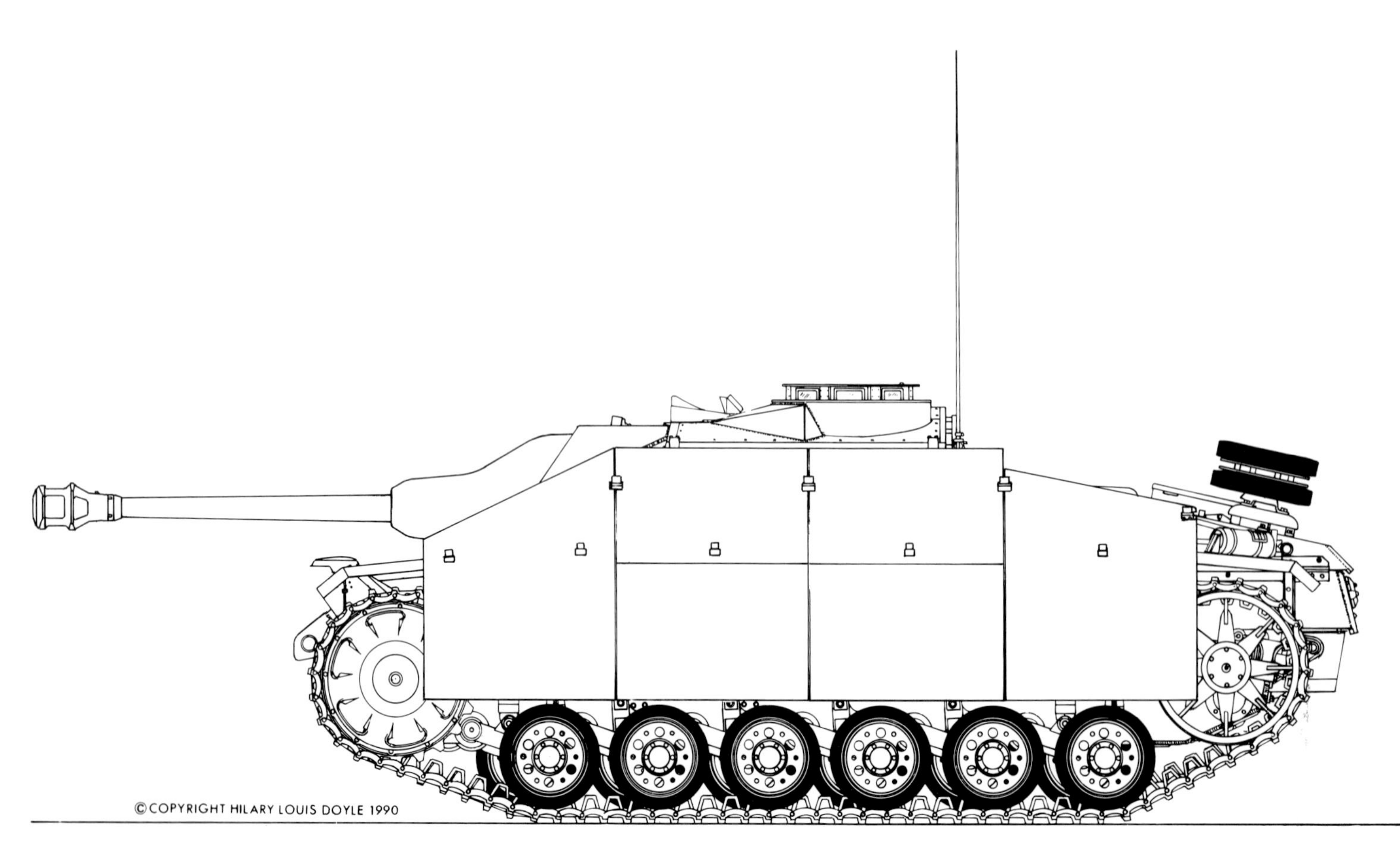

Shot Deflector for the Commander's Cupola

In order to better protect the commander's cupola against hits, a steel deflector was welded to the superstructure roof in front of the cupola. This modification was introduced by ALKETT beginning October 1943 and was present on all new Sturmgeschütz after February 1944. Previous production Sturmgeschütz which did not have this modification were ordered to install a shot deflector for the cupola during major repairs or overhauls. This field modification consisted of steel rods welded to the basic armor, surrounded by a molding of concrete.

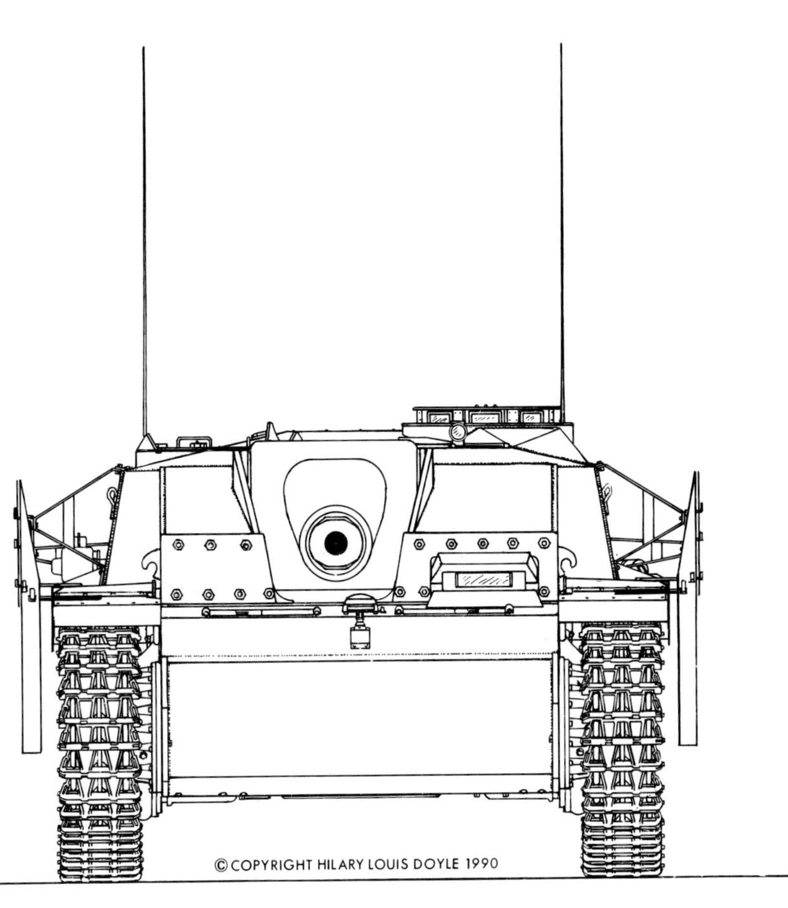

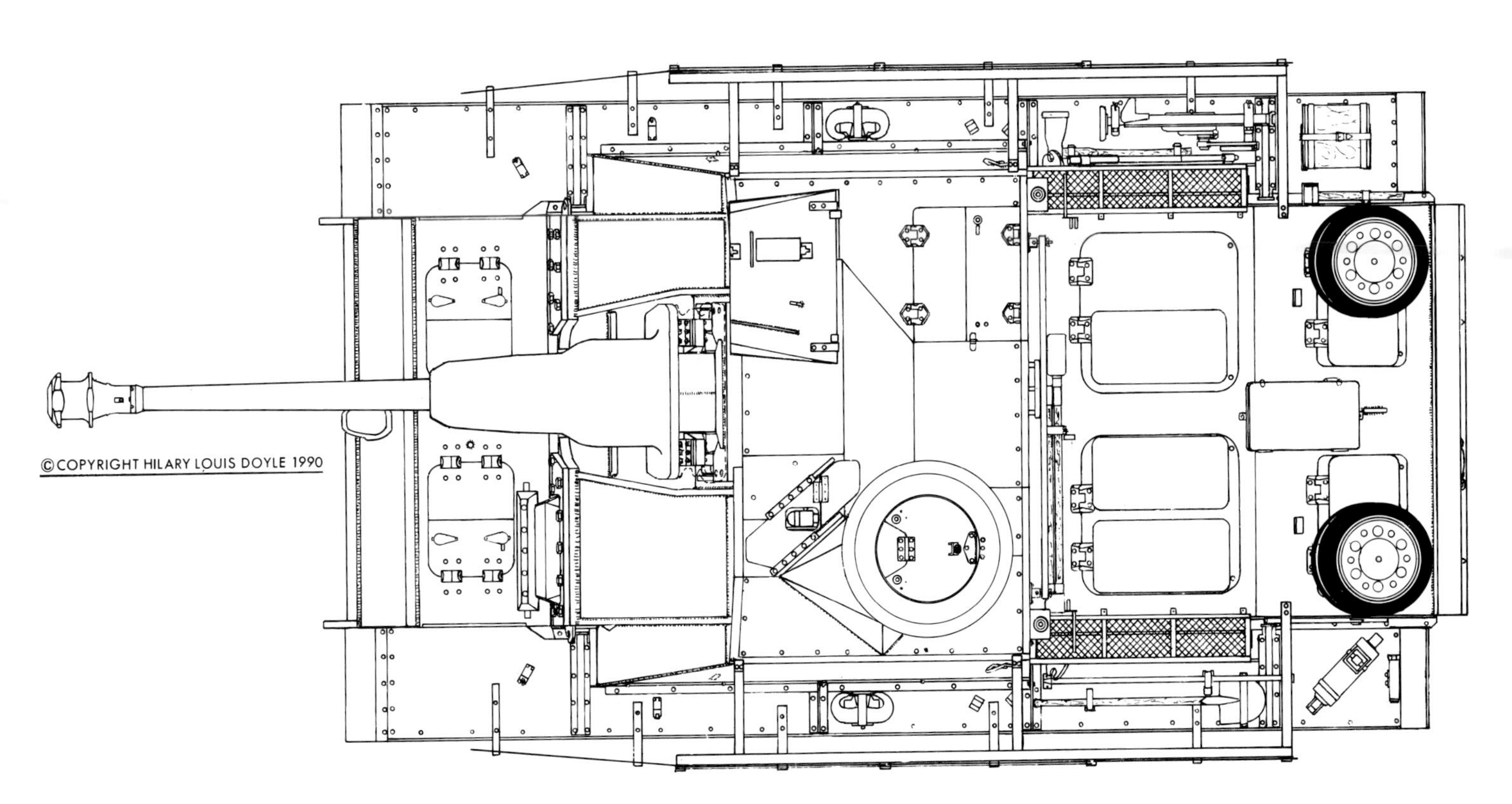

Steel deflector in front of the commander's cupola.

This view from above clearly shows the shape of the steel deflector in front of the commander's cupola.

Gun Mantle

After the basic frontal armor of all the Sturmgeschütz had been increased to 80mm thickness, the one remaining weak point was the gun mantle. The thickness of the welded gun mantle armor varied between 45 and 50mm. Beginning November 1943, a cast gun mantle was introduced which was called the Saukopfblende (sow's head mantle) by the troops. This cast mantle was also introduced for the 10.5cm Sturmhaubitze (105mm howitzer). Due to the difficulty in producing large castings, an insufficient number were produced to outfit all of the new Sturmgeschütz. Therefore many Sturmgeschütz were still produced to the end of the War, still originally outfitted with welded gun mantles.

A Sturmgeschütz, Ausf.G, produced by Alkett in the Fall of 1943. Note the Saukopf mantle and the reinforced concrete above the driver's position and on the opposite superstructure side.

Sturmgeschütz, Ausf.G, chassis number 94050. Completed by Alkett in March 1944. Chevron cleats on the face of the track sections. (BA)

Superstructure Armor

In order to better protect the Sturmgeschütz, the basic armor plate was also increased to 80mm for the right, front superstructure (replacing the previous base 50mm with supplemental 30mm). The last Sturmgeschütz to be built with the 50mm base armor and supplemental 30mm armor plate were produced in June of 1944. Due to the fact that the "Fahrersehklappe 50" driver's visor was retained, the left front superstructure basic armor plate remained at 50mm, reinforced with a 30mm plate.

Improvements in Armaments Muzzle Brakes

Starting in May 1943, the two-chamber muzzle brake on the 7.5cm Sturmkanone 40 was replaced with an improved model with side flanges. These side flanges were introduced to aid in reducing dust and smoke vortices that created difficult target observation.

After May 1944, a further change was introduced on some of the muzzle brakes, when full-disc, forward flanges started appearing. (Further investigation has revealed that the full-disc were the same type as muzzle brakes with only side flanges. The difference being the amount that the forward disc-shaped casting had been milled away in the shop).

On 27 July 1944 it was determined that the muzzle brake must be retained on the 7.5cm Sturmkanone 40. They could be dropped from:

— the Sturmkanone 40 n.A.

Sturmgeschütz, Ausf.G (MIAG chassis number 95219 assembled in May 1943). Improved muzzle brake design with side flanges.

— the 10.5cm Sturmhaubitze (howitzer) and
— the 7.5cm Sturmkanone auf 38(t).
At this time, this proposed change had not been tested with guns mounted in vehicles. All of the newer guns were designed without muzzle brakes from the very start. Muzzle brakes were very costly to manufacture and required a relatively large amount of materials.

In order to inhibit misalignment of the gun and sights caused by vibrations while the vehicle was on the move, a barrel support was introduced beginning in July 1944. This external barrel support was installed on the upper hull front and with the gun elevated at 6 degrees, supplemented the internal travel lock.

An investigation was conducted into adopting a longer gun with better range and armor penetration capability. A report from May 1944 stated: Sturmkanone L/70 — installation into the current Sturmgeschütz designs is not possible. A test chassis for the new design, the "Panzer-fahrgestell III/IV" is available, but not yet tested, and the interior layout has not been determined.

A barrel support mounted on the hull front was introduced in July 1944. This Alkett Sturmgeschütz, chassis number 106579, with 80mm armor plate on the right superstructure front, was completed in September 1944. Zimmerit was no longer applied at the assembly plant.

A spring fastened with a chain to the barrel support, pulled the support to the side as soon as the gun was sufficiently elevated to clear the rest.

Smoke Grenade Launcher

From February to May 1943 each Sturmgeschütz was equipped with two smoke grenade launchers, fitted on both sides of the superstructure. Each launcher had three fanned-out launch tubes, which were operated from the

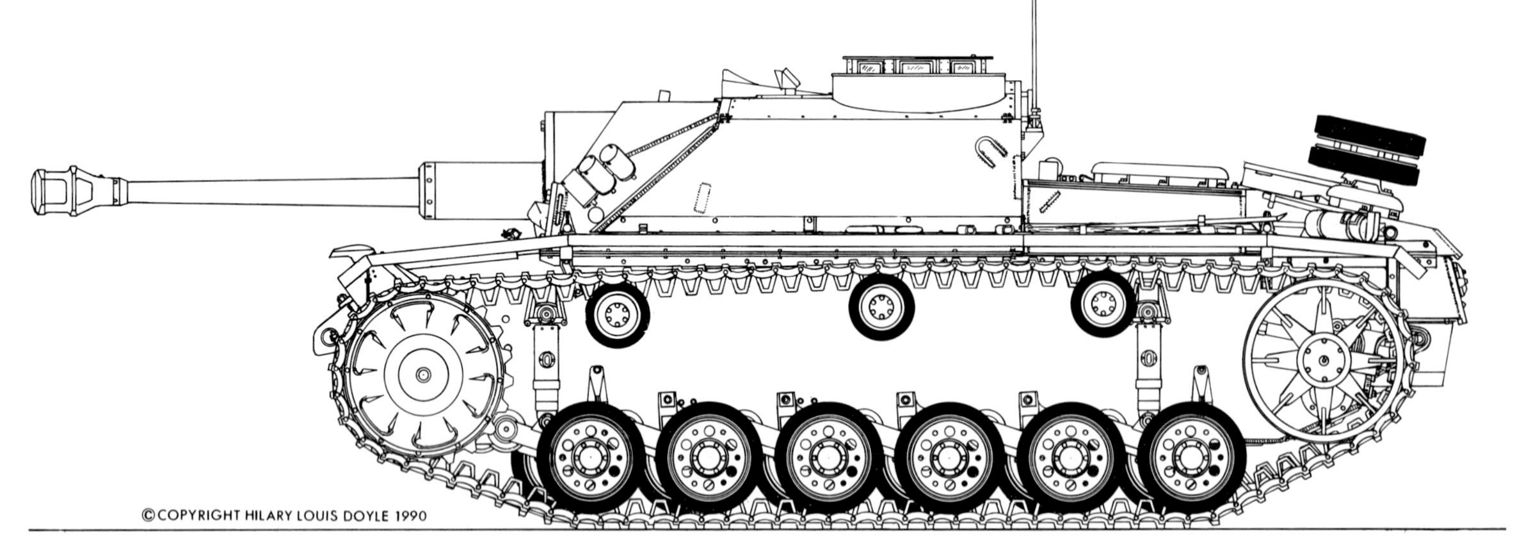

Sturmgeschütz, Ausf.G (as assembled in early 1943, with smoke grenade launchers).

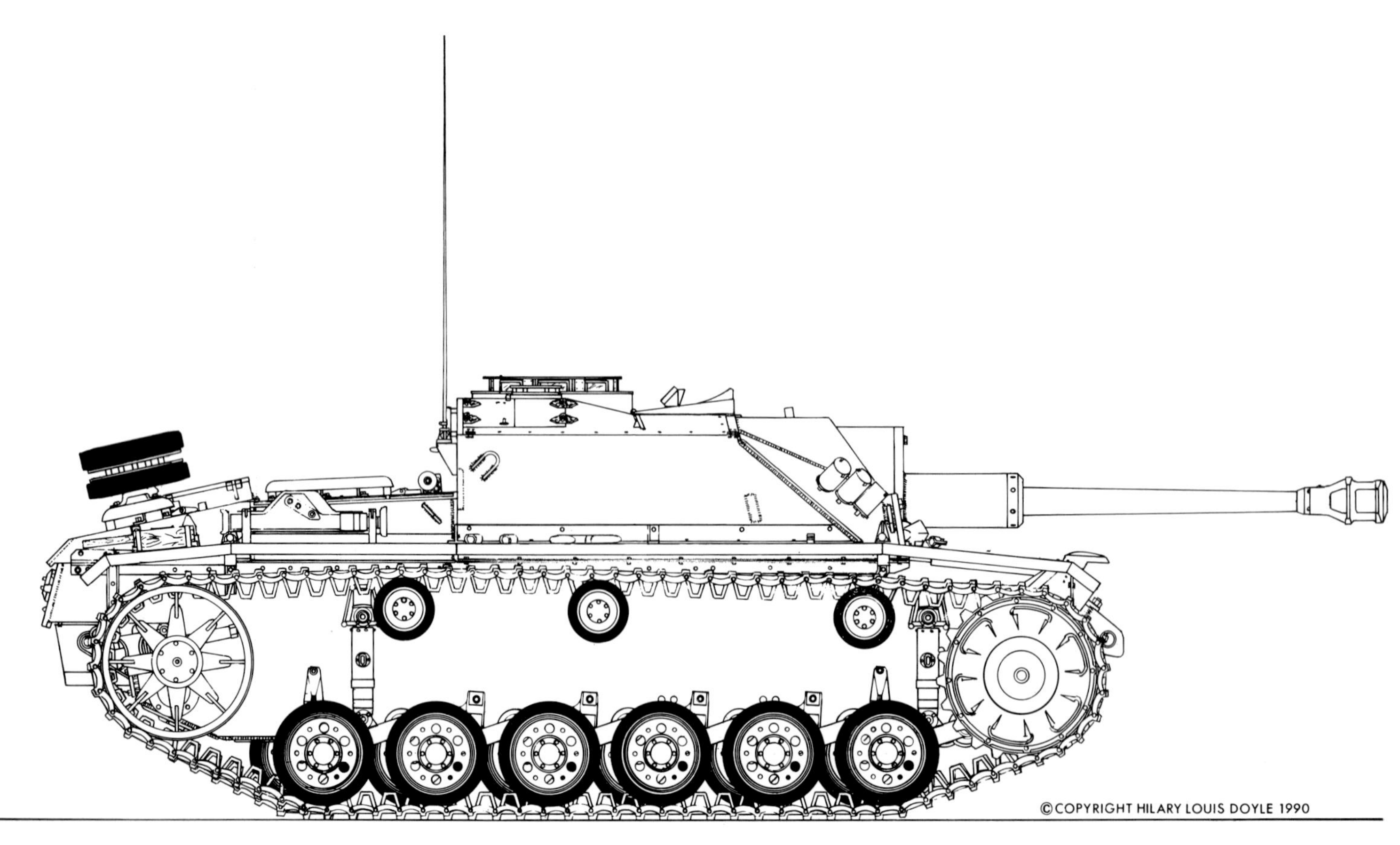

inside of the vehicle. The Heereswaffenamt ordered that their installation cease after field units reported that enemy small arms fire pierced the launch tubes, setting off the smoke grenades, which blinded the vehicle crew in the process.

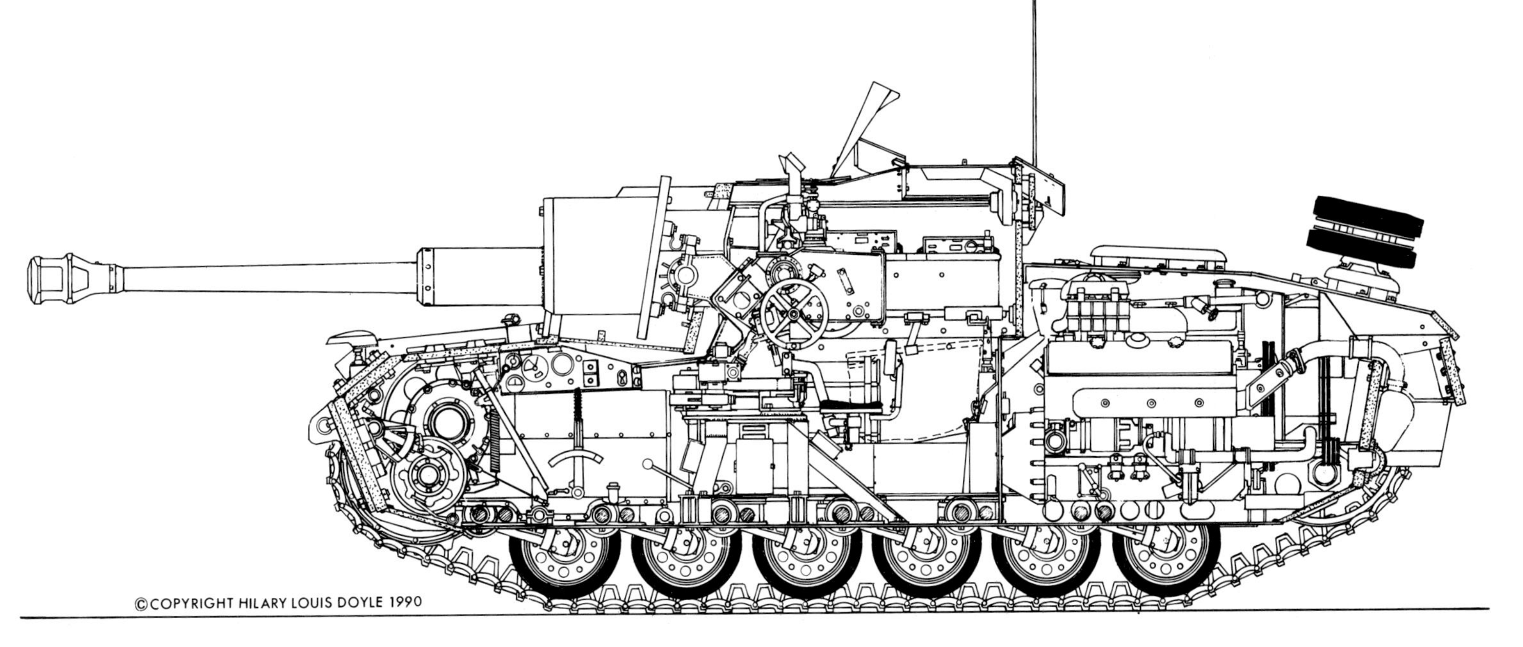

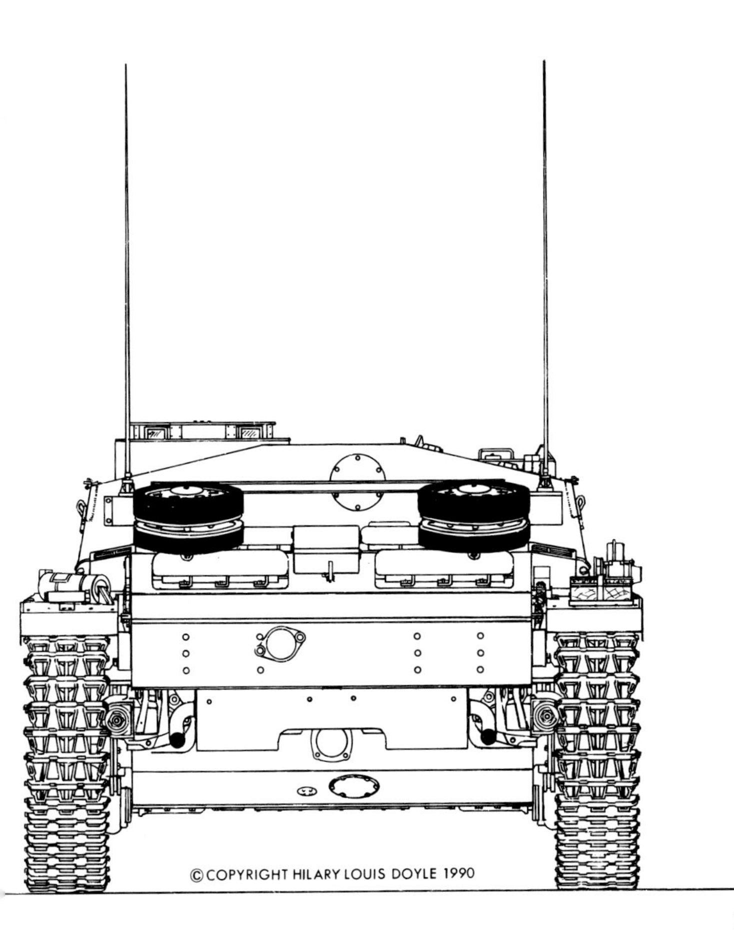

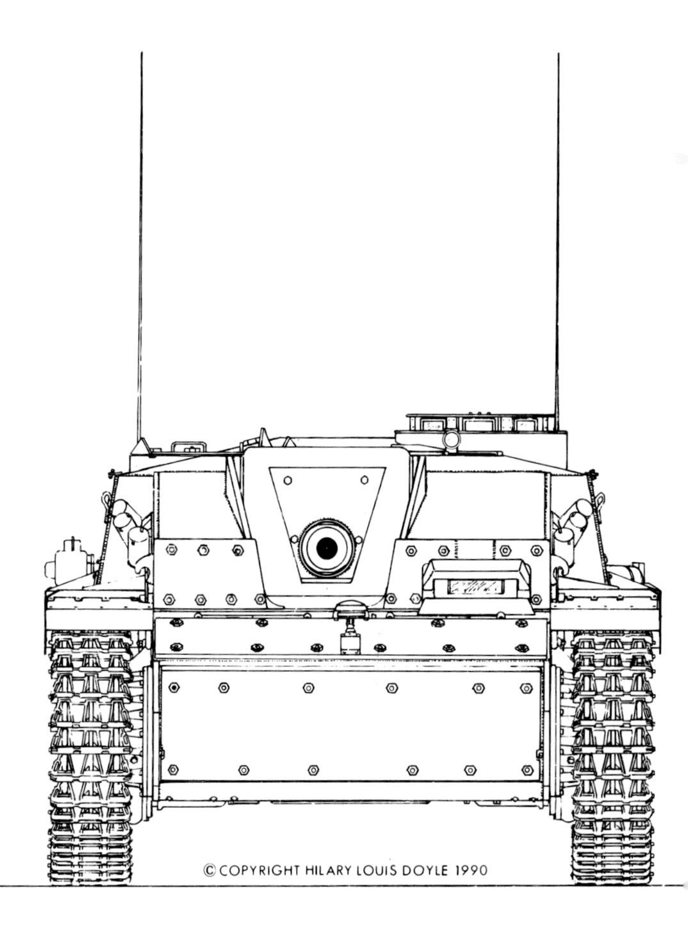

Smoke grenade launcher design for the Panzerkampfwagen III.

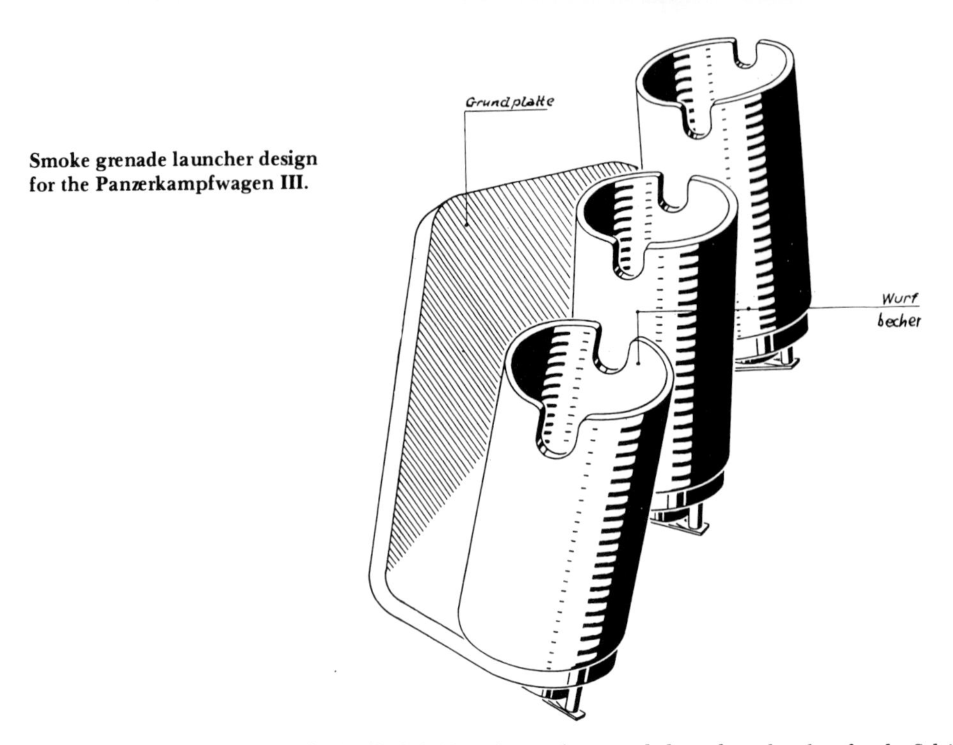

Sturmgeschütz, Ausf.G, chassis number 95078, completed by MIAG in April 1943. Note the smoke grenade launchers, brackets for the Schürzen side skirt mounts, loader's machine gun shield and concrete reinforcement over the driver's position.

Rotatable Machine Gun Mount

In the conference on 16 and 17 December 1943 Hitler gave his full support to the new, "Rundumfeuer" machine gun mount which could be operated from a protected position and fully rotated to fire in all directions. A large number were to be immediately procured and quickly distributed, installed in all Sturmgeschütz and Panzerjäger which did not possess a mounted machine gun.

On April 3rd, 1944 it was reported that there were 27 "Rundumfeuer MG" mounts (produced by Daimler-Benz) undergoing field testing at the Front. The first report spoke favorably about them.

The machine gun mount was designed to be mounted in the turret or superstructure roof and operated from a protected position inside the vehicle. A periscopic sight with 3x magnification and a field of view of 8 degrees was installed with the mounting. This "Rundumfeuer" mount interfered with the fore and aft opening loader's hatch. Therefore, the hatch was redesigned. The new hatch design had two halves which were hinged and opened toward the sides of the vehicle.

The Rundumfeuer MG (rotatable machine gun) mount with its protective shield (weapon dismounted).

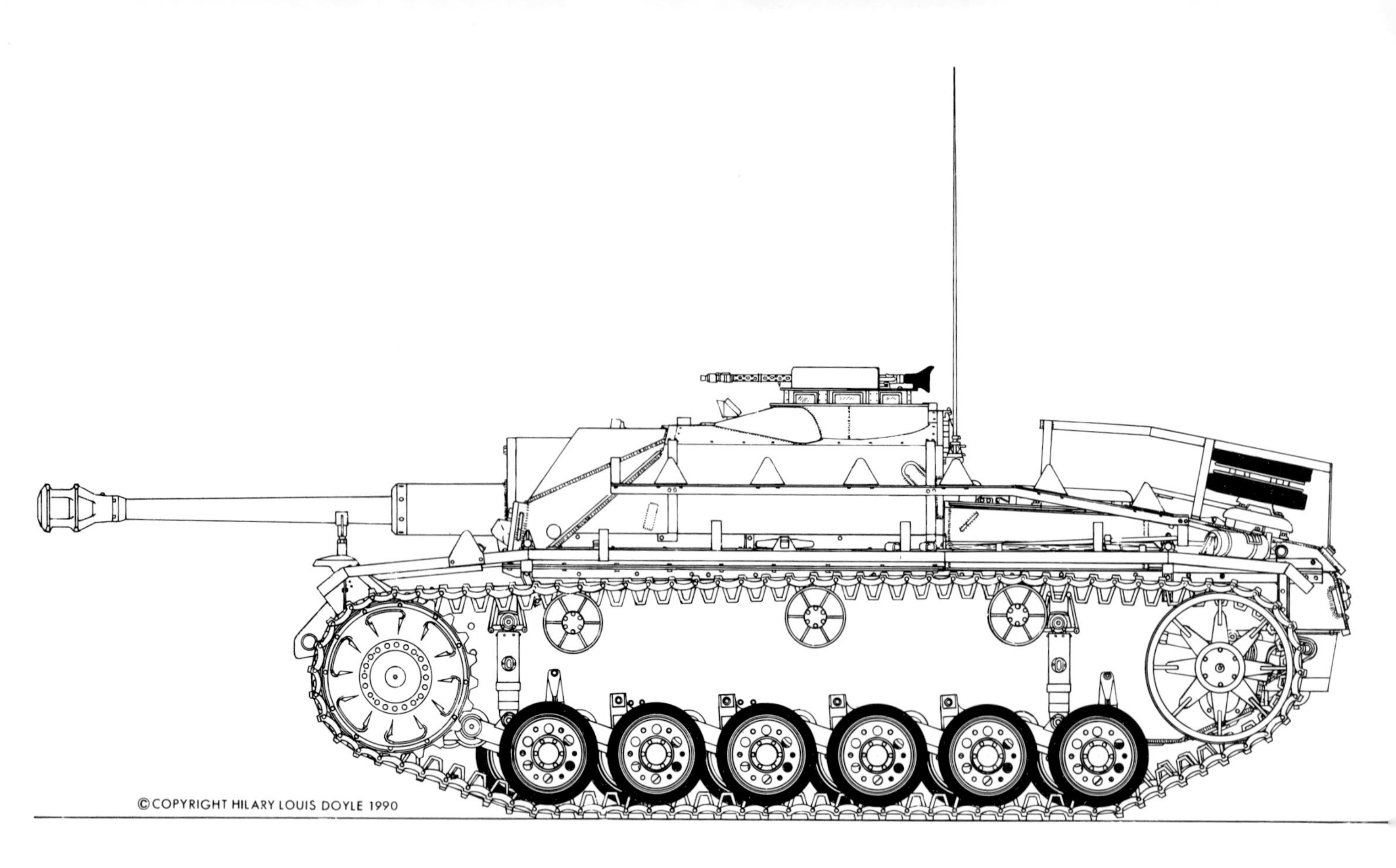
©COPYRIGHT HILARY LOUIS DOYLE 1990

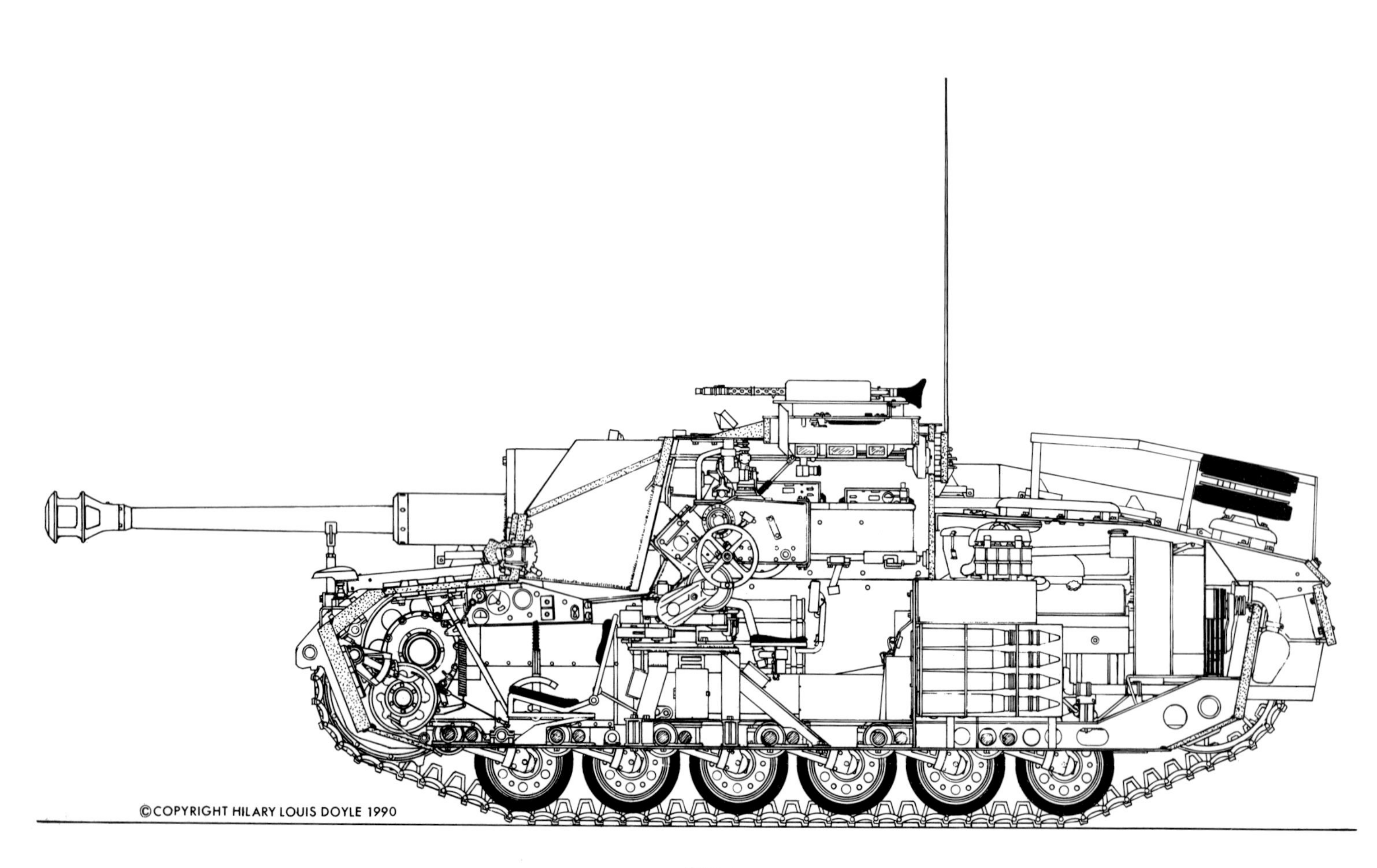
©COPYRIGHT HILARY LOUIS DOYLE 1990

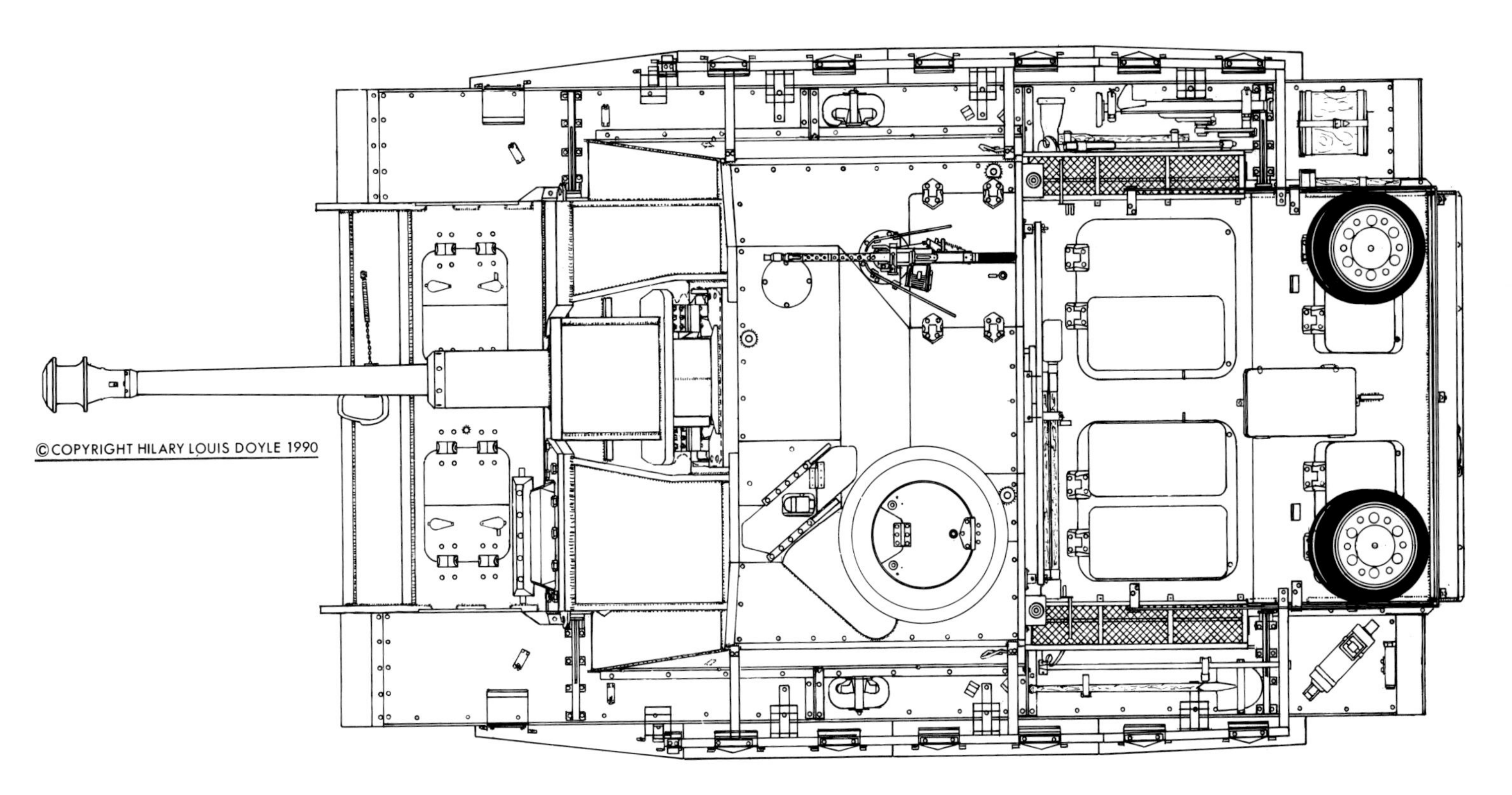

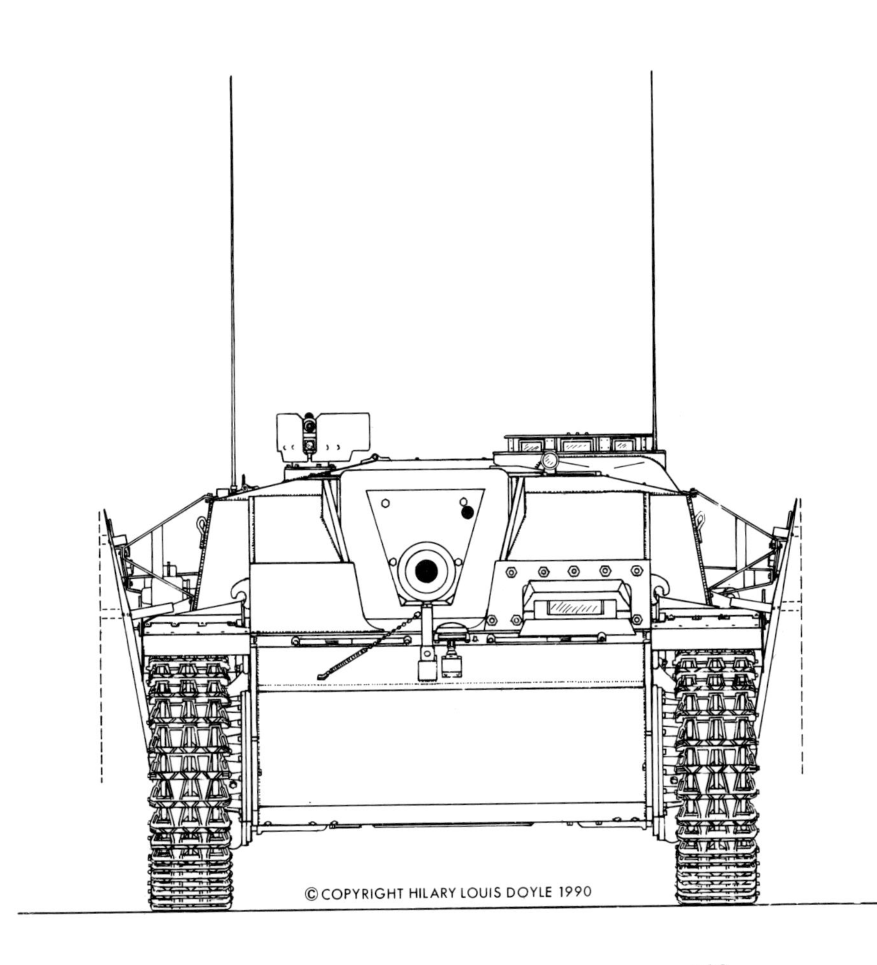

Sturmgeschütz, Ausf.G, with the Rundumfeuer M.G. mount introduced in April 1944, coaxial machine gun introduced in June 1944, and three Pilze mounts for the 2-ton jib crane.

The Nahverteidigungswaffe (close-combat defense weapon) on the superstructure roof can be seen near the lower edge of the photo.

Nahverteidigungswaffe (Close Quarters Defense Weapon)

Although this weapon had been installed on other types of Panzerkampfwagen and Jagdpanzer by the end of 1943, the Nahverteidigungswaffe (close quarters defense weapon) only appeared on the Sturmgeschütz beginning in May 1944. The limited production resulted in shortages, up to October 1944. Therefore, most of the new Sturmgeschütz left the factory without this weapon. The opening in the superstructure roof intended for this weapon was covered with an armored disc secured by four bolts.

Coaxial Machine Gun

Beginning in 1943, field units had requested a machine gun for the gunner. In March 1944, a modification was suggested to mount the machine gun so that it fired through a hole cut in the gun mantle.

An advantage was that this machine gun would be effective for engaging troops at close range, therefore saving 75mm main gun ammunition. Because it was easier to modify, beginning in June 1944, the coaxial machine gun was first installed in the welded gun mantle. But, it was not installed in the cast "Saukopf" mantle until after October 1944.

The Zielscheinwerfer 43/1 (targeting spotlight) was developed and produced for Sturmgeschütz and Sturmhaubitze. Its range was 600 meters. Electrical power, supplied by a 12 volt battery, fed a 100 watt bulb. Installed shutters were used to adjust the light intensity.

The installation for the coaxial machine gun, seen as a tube to the left next to the main gun.

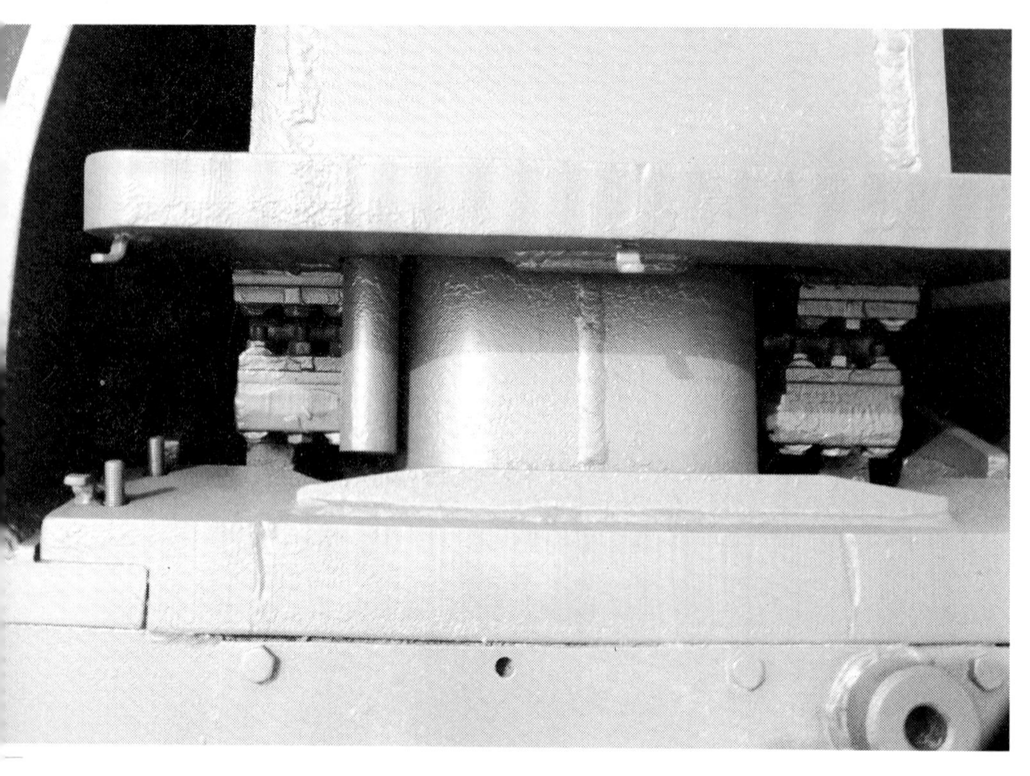

Mounting for the coaxial machine gun.

Welded gun mantle with opening for the coaxial machine gun.

Cast gun mantle with opening for the coaxial machine gun.

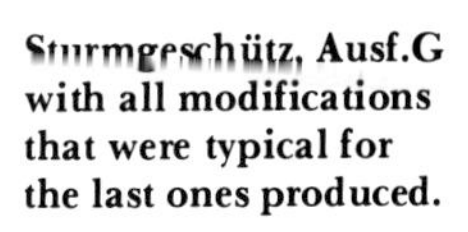

Sturmgeschütz, Ausf.G with all modifications that were typical for the last ones produced.

Improvements to Mobility

The lessons learned during the first winter in Russia in 1941/1942 were reflected in the form of improvements to winter equipment made available beginning in the fall of

In action, the Schürzen side skirts were very often lost. The troops modified them by rounding off the top, turned them upside down, and suspended each plate from the middle of it upper edge. The upper plates were mounted at an angle on the superstructure sides.

Also, this was possible: Sturmgeschütz, Ausf.C, with original superstructure. It was back-fitted with the long 75mm gun mounted with a Saukopf mantle.

1942. Sturmgeschütz Abteilungen were to be issued the following cold-weather gear:

Per Sturmgeschütz:

1 viscometer

1 warm water transfer device (Sturmgeschütz)

1 fighting compartment heater for the Sturmgeschütz

1 coolant fluid heater (Fuchs) for the Sturmgeschütz

1 set of Winterketten (winter tracks) for the Panzerkampfwagen III

1 warmer for the Pz.III and IV inertial starter

For each Abteilung:

1 drain viscometer

3 Panzer-coolant fluid heater model 42

2 hot air blowers, type H.B.50

1 motorized crankshaft starter

800 individual track sections for the left winter track

800 individual track sections for the right winter track

2000 Schneegreifer snow traction devices for the Panzer III and IV

2 hot air blowers (Zwerg)

4 snowplows for the Panzerkampfwagen III

1 pre-warmer (Tecalemit)

Winter experiments in St. Johann in 1943/44 with a snowplow and squad sled.

Sturmgeschütz (Panzerkampfwagen III, Ausf.M chassis) during the Winter of 1943/44 with Mittelstollen (cleats) bolted to the tracks for traction on ice and packed snow. (BA)

A Sturmgeschütz with Ostketten in the Summer of 1944. (BA)

Tracks

In addition to the Winterketten (winter tracks), which were to be returned to the Heereszeugämter in the spring, along with the other winter gear, the normal track permanently issued with the Sturmgeschütz was redesigned.

In order to increase traction on ice or packed snow, three pairs of raised chevrons were cast onto the face of each track Kgs 61/400/120 link (six chevrons per track link). Despite this, most of the Sturmgeschütz left the factory with the old, flat profile tracks.

In early 1944, a wider track was introduced, which proved to be effective during the muddy periods in Russia and during the winter:

Army technical order Nr.256 dated 1 May 1944, addressed Ostketten for armored vehicles.

For armored vehicles on the Panzerkampfwagen III and IV chassis, a wider track, called the Ostkette, is to be supplied to units on the Eastern Front. The Ostkette is intended to inhibit armored vehicles from becoming stuck in snow or soft terrain. They are solely intended for use on the Eastern Front. Prior to moving toward the west, to the Balkans or to Italy, they are to be exchanged for normal tracks. The following points are to be observed when using the Ostkette:

1. Any Schürzen side skirts on armored vehicles must be mounted vertically. The first and last Schürzen plates should be bowed slightly outward, so that the track pins cannot rub against them.
2. When driving along sunken lanes, in terrain with protruding tree stumps, and the like, caution is advised. Due to the track extending on the outside, the tracks can tilt, causing the road wheels to climb onto the center guides. This easily results in the vehicle throwing a track.
3. When loading onto rail cars the Schürzen side skirts and their mounts must always be removed. The loading width of vehicles with Ostkette are 3266 mm for Panzerkampfwagen III and variants, and 3206mm for Panzerkampfwagen IV and variants. This is quite close to the maximum allowable width of 3270mm for rail loads. Great care should be taken to ensure that armored vehicles with Ostkette are placed in the exact center of the rail car. Centering is to be checked by measuring the track overhang on both sides of the railcar.

OKH (Ch H Rüst u. BdE), 15 April 1944 Az 76g Nr. 3734/44 AHA/In 6 (Z/Ing.)

Running Gear Components

Running gear components were modified in order to simplify manufacturing processes, but were also caused by raw material shortages and shortages of key parts resulting from the bombing campaign.

A first measure taken was to replace the rubber-tired return rollers with all steel return rollers. This considerably raised the rolling resistance as well as the noise level. Both were accepted without objection.

Beginning in November 1943, the rubber tired return rollers were replaced with steel wheels. (MIAG chassis number 96105 completed December 1943).

Another version of the steel return rollers with 6 holes.

As long as there were still rubber-tired return rollers available, they were installed at the assembly plants (scantily through the end of 1944). The Kraftfahrversuchsstelle of the Heereswaffenamt Prüfwesen in Kummersdorf, under contract number 3/44 dated 16 August 1944, was requested by the Referat Pz.III SE to conduct tests on sliding bearings in road wheels and return rollers for the Panzerkampfwagen III chassis.

On the road wheels, the cylinder ring bearings (roller bearings) were replaced with encased bearings (ball bearings) or sliding bearings. The rolling-contact bearings in the return rollers were replaced with sliding bearings.

Sturmgeschütz, Ausf.G. Experimental vehicle equipped with rubber-saving road wheels.

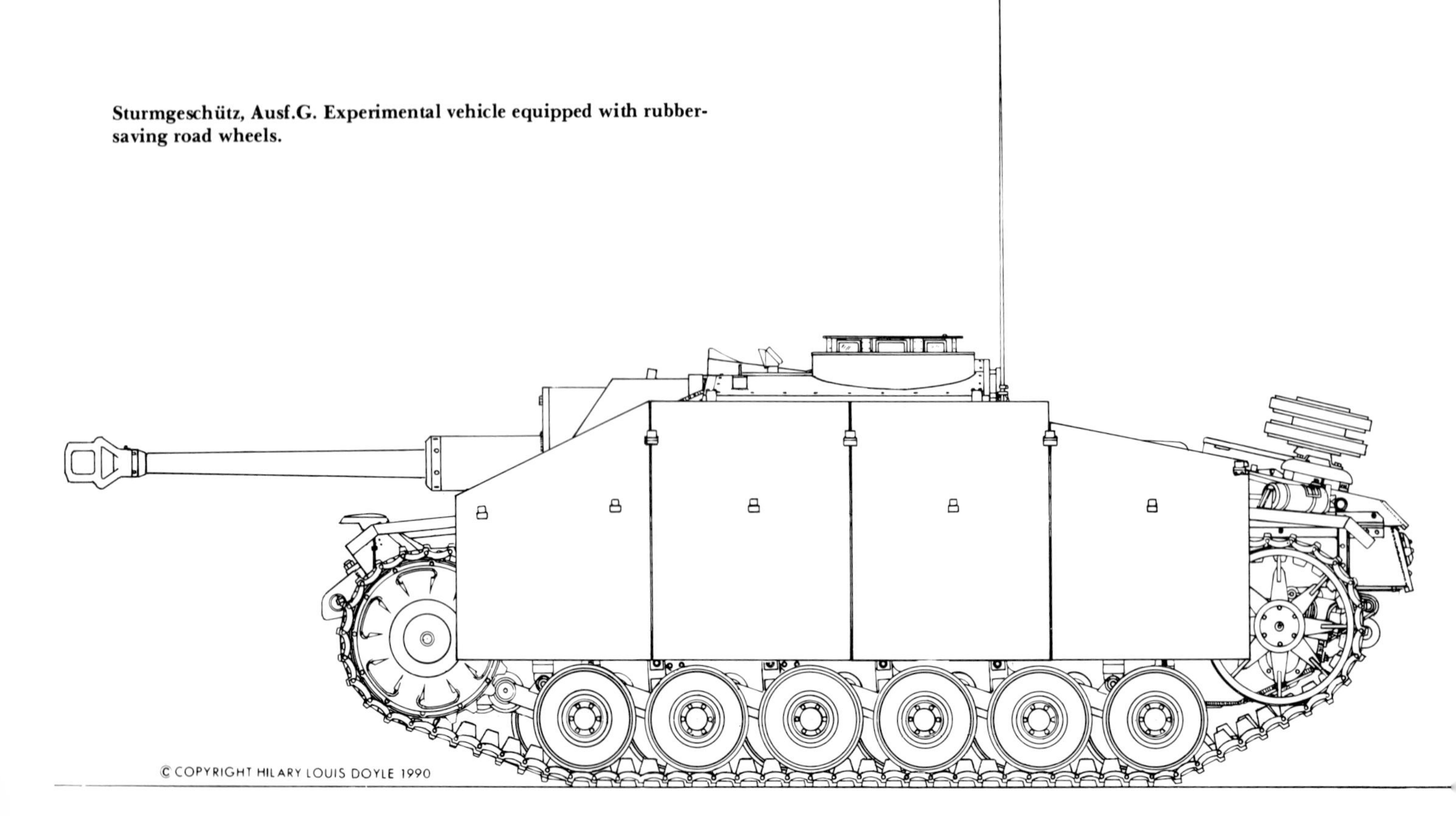

An experimental Sturmgeschütz in Kummersdorf with rubber-saving road wheels. (Alkett chassis number 92112 completed in March 1943) Also used for designing and locating the mounts for the 2-ton jib crane.

The scarcity of rubber and the high wear rate of road wheels instigated a new design for road wheels in 1942. The design was freely copied, but improved, from the roadwheels on the Russian KW I tank. These steel-tired road wheels saved rubber and increased the life of running gear components. The firm Deutsche Eisenwerke engaged in their design. Two steel discs compressed two rubber rings against the steel rim. At Berka near Eisenach, a Panzerkampfwagen III with steel-tired, rubber sprung road wheels underwent testing from 8 through 14 November 1942.

Trials in Kummerdorf revealed a decrease of rolling resistance by approximately 10%. There, a Sturmgeschütz underwent testing at the end of 1944. This modification never matured to the point that these steel-tired road wheels were mounted on a production run of Sturmgeschütz.

Another running gear problem encountered on the Panzerkampfwagen III chassis was the shearing off of track pins on the hull armor. Beginning December 1944, a track pin deflector was welded to the hull side inside the idler wheel. This modification was authorized for the troops to back-fit to previously issued Sturmgeschütz.

Rear view of the experimental Sturmgeschütz. The tow coupling is of the type introduced in the production series in December 1944.

Final Drives

The final drives were basically a weak point in the power train in fully-tracked vehicles. The over-taxing demands, made on them during turns and driving in difficult terrain, frequently exceeded the material strengths specified by design. Initially designed for a much lighter Panzerkampfwagen III, their use in the Sturmgeschütz was the cause for many breakdowns. The "Panzertruppen" news pamphlet reported in August 1944 that the final drives had been strengthened and referred to the following changes:
For the Sturmgeschütz and Pz.Kpfw.III chassis new final drives are immediately available. Comparison of the old and new designs reveals the following changes:
Old spurn gears 36 teeth, new spurn gears 35 teeth
Outside diameter of the old pinion 115.88mm
Outside diameter of the new pinion 118.00mm.

Maintenance Equipment

An order from June 6, 1944 stated that all the C-type towing hooks on all Panzer III and IV were to be replaced by S-type hooks. This notification was supplemented by an order dated 10 June 1944 stating that only one S-hook would be delivered to replace the two previous C-hooks for all Panzer III and IV chassis.

Army technical order Nr.422 dated 1 July 1944 addressed the 2-ton jib crane for Panzerkampfwagen, Sturmgeschütz and Panzerjäger: In the near future, the 2-ton jib crane for Panzerkampfwagen III, IV, Panther and Tiger, Sturmgeschütz (on the Panzer III chassis), and Panzerjäger 38 will be issued to the field units. Crane is to be mounted on "Pilze" mounts, welded onto the vehicle. In any case, where these mounts have not previously been installed on the named vehicles, they are to be welded in accordance with the referenced drawings. The "Pilze" mounts and drawings for their installation can be ordered from spare parts depots.

Pilze mounts on Sturmgeschütz: One of Pilze model I and two of Pilze model II are to be welded to the superstructure roof in accordance with drawing 745-14-C1. When welding model II, care must be taken to ensure that the upper surface is parallel to the roof.

The welding electrodes appropriate for armor are be used. The welding bead around the "Pilze" mount are not to close the drain channel on the bottom (to allow grease or rainwater to flow through). To prevent rust, fill the threaded hole with grease.
OKH (Ch H Rüst u. BdE), 22 June 1944, 76g-A Nr. 77444/44 AHA/In 6 (Z/Ing).

Beginning in October 1944, five Pilze mounts were welded to the superstructure roof of the Sturmgeschütz, one at each corner and the fifth in the middle. This arrangement allowed flexibility in the use of the 2-ton jib cranes. It was then possible mount the crane in various configurations that enhanced removal of components from other vehicles or to removal/installation of its own

Sturmgeschütz with the 2-ton jib crane.

The Pilze mounts for the 2-ton jib crane were welded to the superstructure roof. This MIAG-produced vehicle (chassis number 96722, completed in May 1944) was shipped to Finland in 1944.

rear deck and engine.

Broken down vehicles were predominantly towed with steel cables, attached to the towing brackets with C- or S-hooks.

Unfortunately, when using steel cables, it was very difficult to control the towed vehicle, especially downhill or when braking. In order to rectify this, a tow coupling with horizontal supports was welded to the center of the hull rear. This coupling was designed for tow bars to be attached, enhancing control of the towed vehicle.

Sturmgeschütz, Ausf.G, with features that were present on those produced toward the end (without the Schürzen mounted)

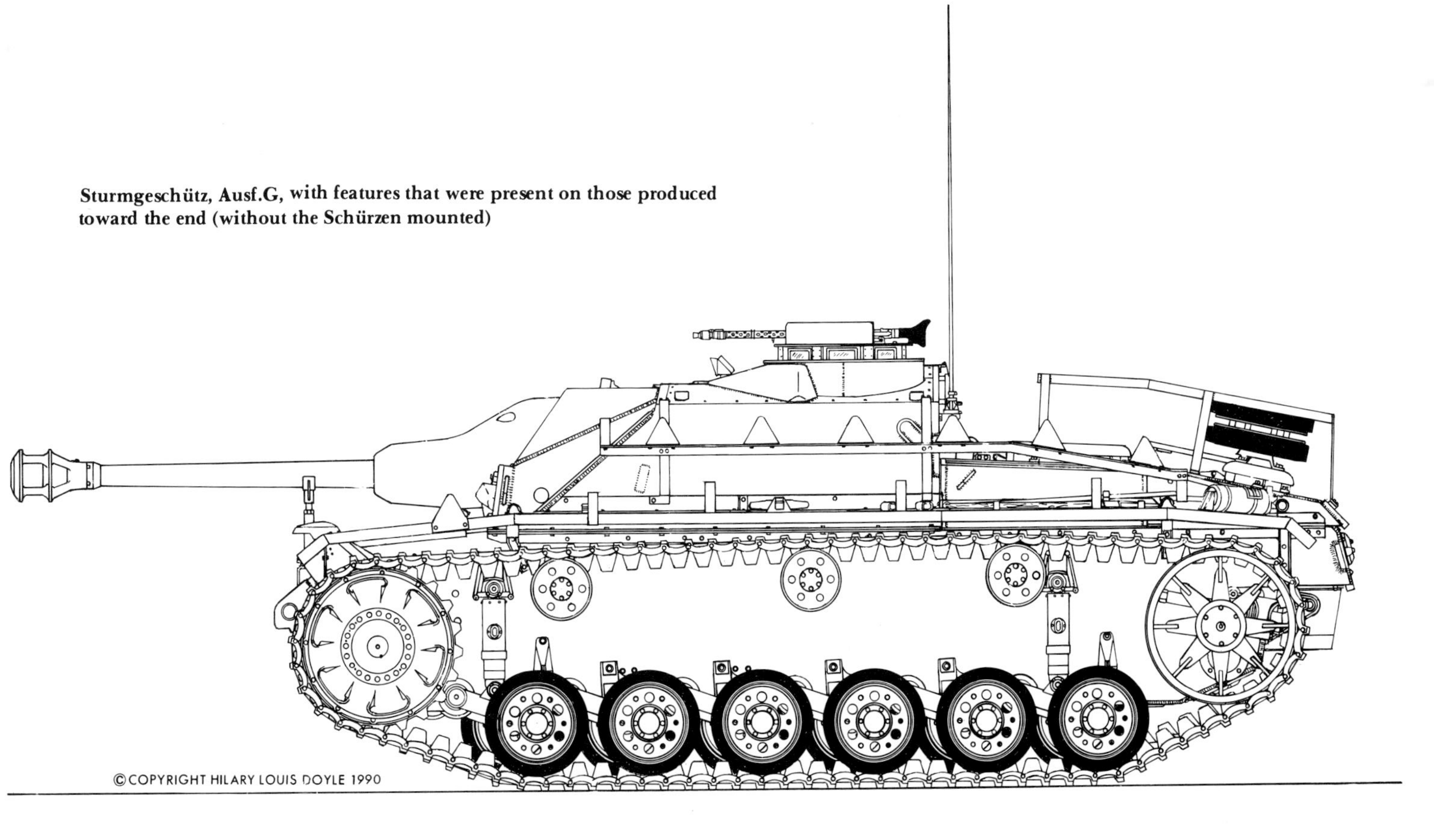

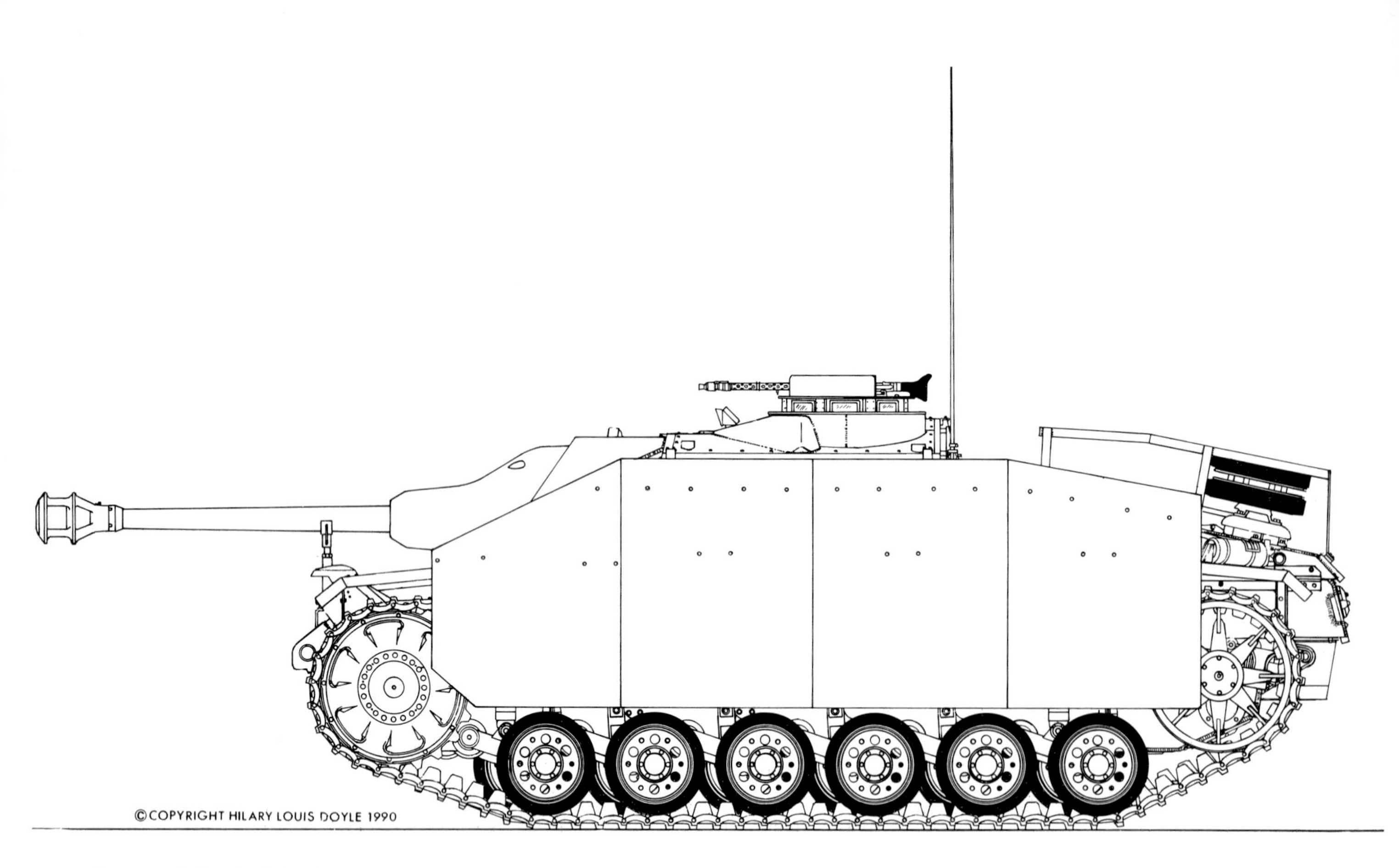

Sturmgeschütz, Ausf.G, with features that were present on those produced toward the end.

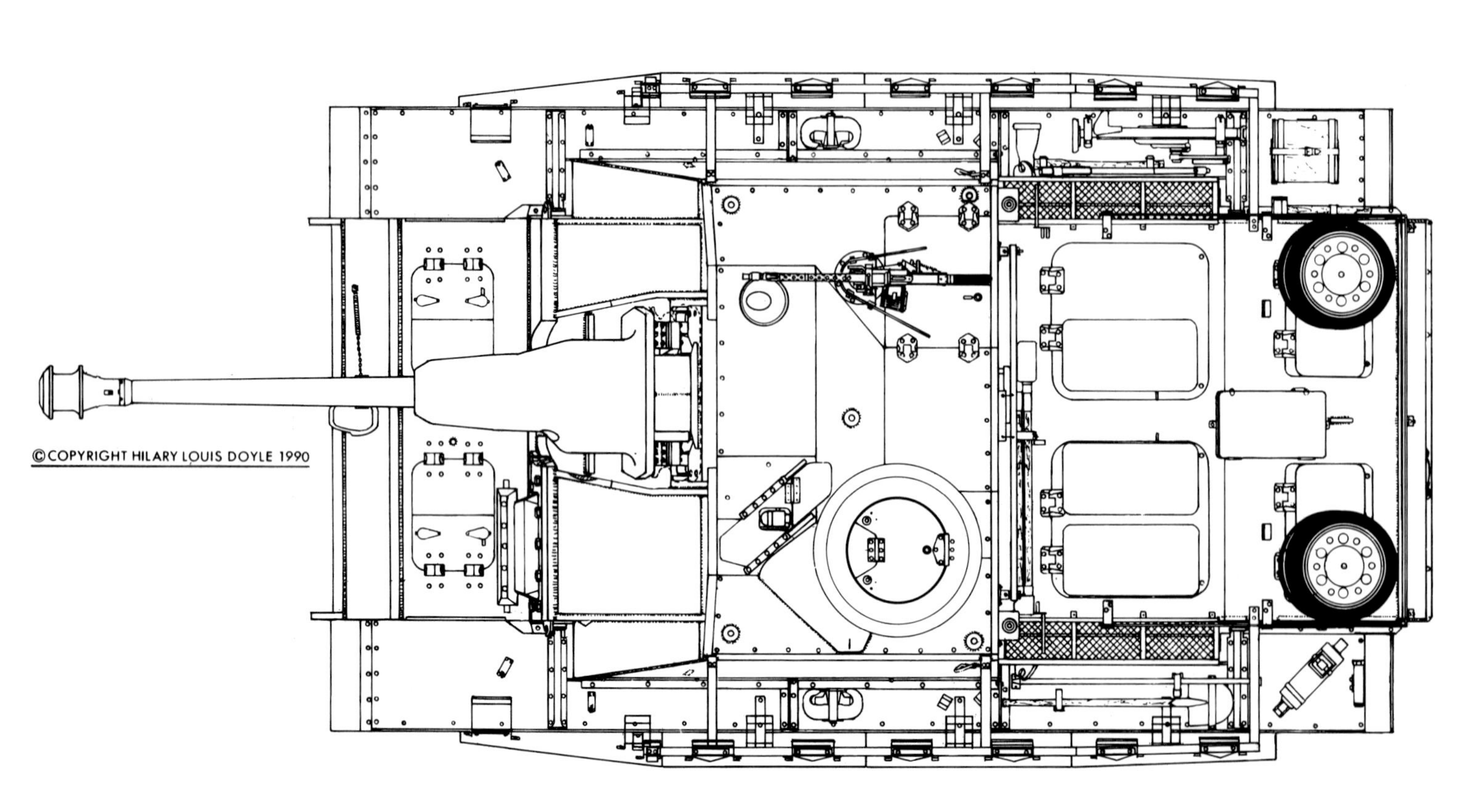

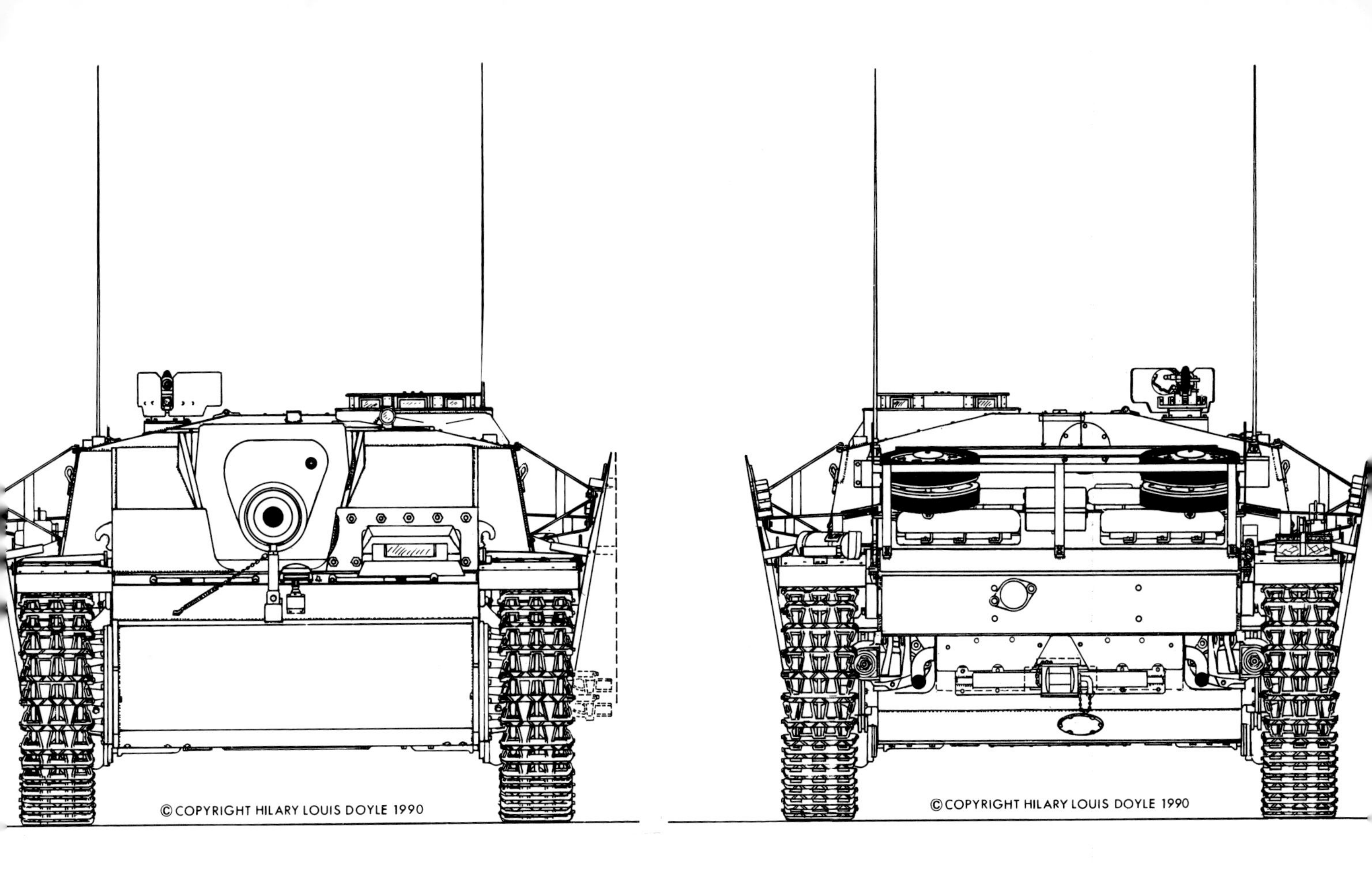
©COPYRIGHT HILARY LOUIS DOYLE 1990
©COPYRIGHT HILARY LOUIS DOYLE 1990

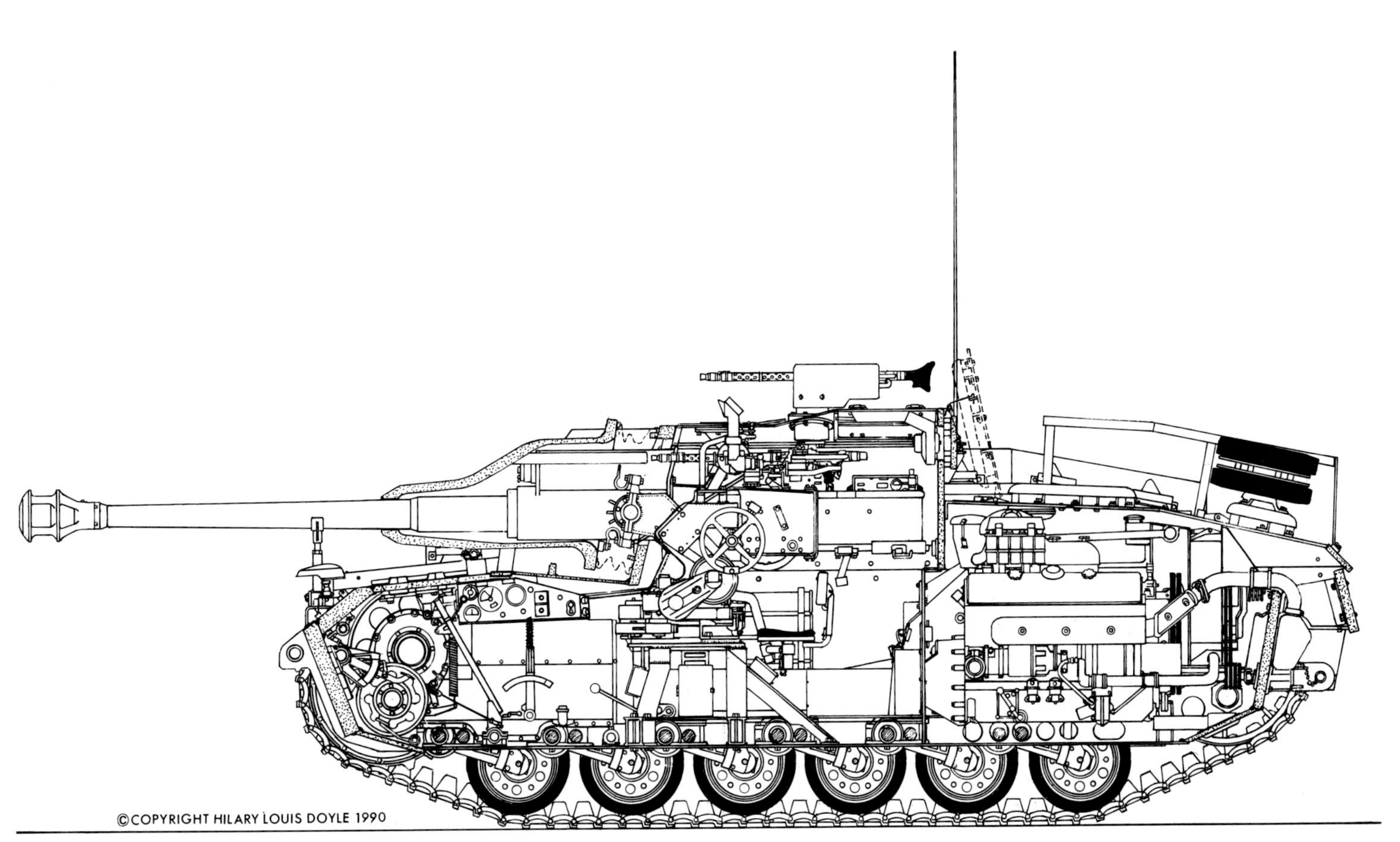
©COPYRIGHT HILARY LOUIS DOYLE 1990

Sturmhaubitze, Ausf.G

With the installation of the longer 7.5cm Sturmkanone 40, a tank destroyer was created in the Sturmgeschütz, which had initially been intended to engage soft targets. The primary mission of this tank destroyer was to protect the infantry from enemy armor. A howitzer mounted in a vehicle, as similar in construction to the Sturmgeschütz as possible, was deemed necessary for the effective engagement of soft targets, including those behind cover. The newly created Sturmhaubitze differed from the Sturmgeschütz solely in the installation of the 10.5cm Sturmhaubitze 42 L/28 and its with associated ammunition stowage.

This new weapon was first mentioned on December 2, 1941, in a report on the scheduled tank and weapons production. A delivery of 12 units of a trial series of light field howitzers for installation in Sturmgeschütz vehicles was planned with the following production numbers: Five in December 1941, five in January 1942 and two in February 1942.

A prototype was completed by March 1942. Five light field howitzers, modified for installation into the Sturmgeschütz, had been produced by the end of May 1942.

During the Führer's conference of October 2nd 1942, Hitler was shown a Sturmgeschütz with the light field howitzer installed and was quite impressed. He particularly appreciated its unusually low firing height — 1.55 meters. It was reported to him that a one-time test series of 12 vehicles were being produced. As reported, six had already been produced, three additional vehicles were expected on 10 October and the last three in about four weeks.

In the conference on 13 October, Hitler referred to the concept of a light field howitzer installed in the Sturmgeschütz as an ideal solution. He asked for an immediate report disclosing the schedule for producing 12 Sturmgeschütz per month, equipped with the 10.5cm le.F.H. (light field howitzer).

Nine Sturmhaubitze issued to the 3.Batterie/Sturmgeschütz Abteilung 185 first saw action on November 22nd, 1942. The chassis used for these nine test Sturmhaubitze were rebuilt and repaired earlier models of Sturmgeschütz (not the latest Ausf.F/8 chassis). On December 9, 1942 is was reported: Up to this time chassis taken from repair facilities have been used in the conversions for the 105mm and 150mm Sturmgeschütz. These do not meet the requirements for these weapons. Chassis from new production are required. Beginning in March of 1943, Sturmhaubitze were produced with the Sturmgeschütz, Ausf.G chassis and superstructure.

Once installed in the vehicle, the Rheinmetall-Borsig designed 10.5cm Sturmhaubitze 42 could be traversed through 20 degrees and was elevated through an arc from -6 degrees to +20 degrees. Its maximum range with "Charge 6" was 10,650 meters, with a special long range round, 12,325 meters. The howitzer length at 2940mm had the same internal design as the le.F.H.18. The tube itself had 32 rifling grooves with a right helical twist, increasing from 1/30 to 1/15.

Thirty-six rounds of ammunition (26 high explosive and 10 shaped charge (HEAT) shells) were carried on board.

During the Führer's conference of 1-3 December 1942,

Sturmhaubitze, experimental. Maximum barrel elevation (March 1942).

Sturmhaubitze, experimental. Fighting compartment roof raised in the rear with horizontal fume exhaust ventilator.

Sturmhaubitze, Versuchsserie (test series).

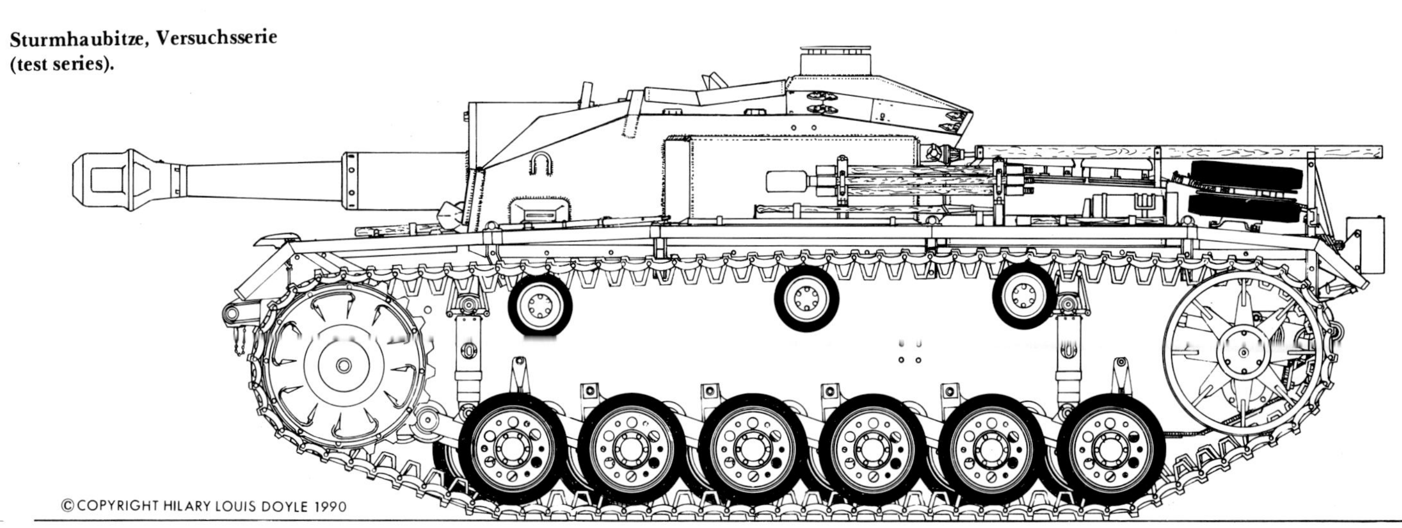

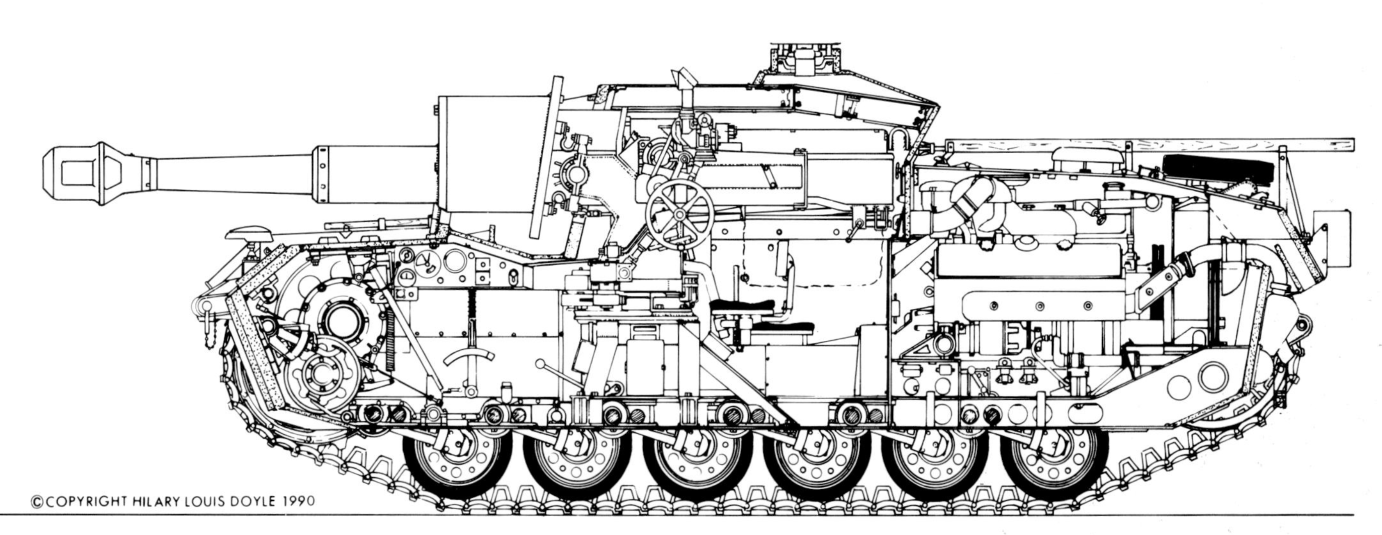

Sturmhaubitze, experimental, barrel at maximum depression.

Sturmhaubitze completed at Alkett in October 1942 converted from a overhauled Ausf.F with supplemental armor.

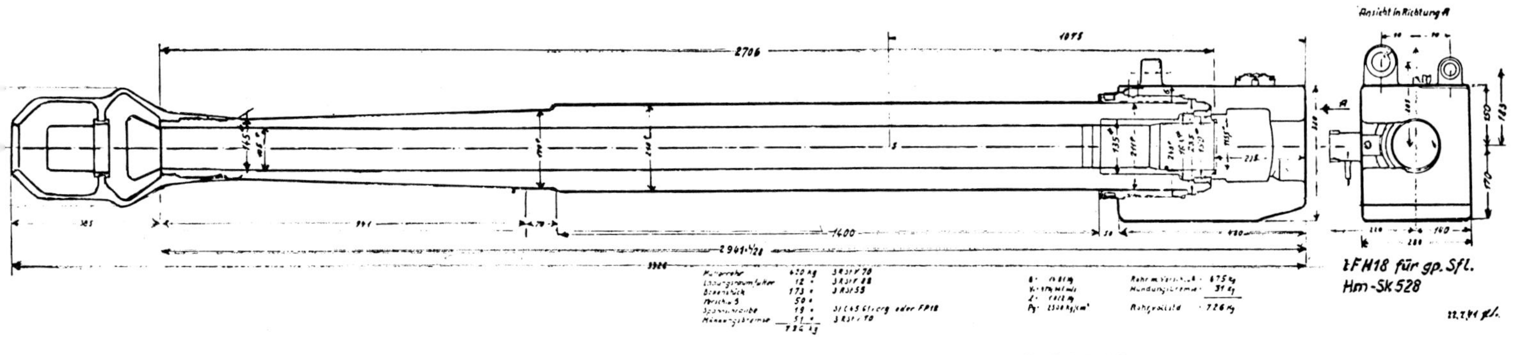

Cross-section of the 105mm barrel for the armored self-propelled artillery gun, based on the barrel from the l.F.H.18. (Cen)

Sturmhaubitze, view from above into the fighting compartment (an early version of the weapon with a long recoil).

Sturmhaubitze, Ausf.G.

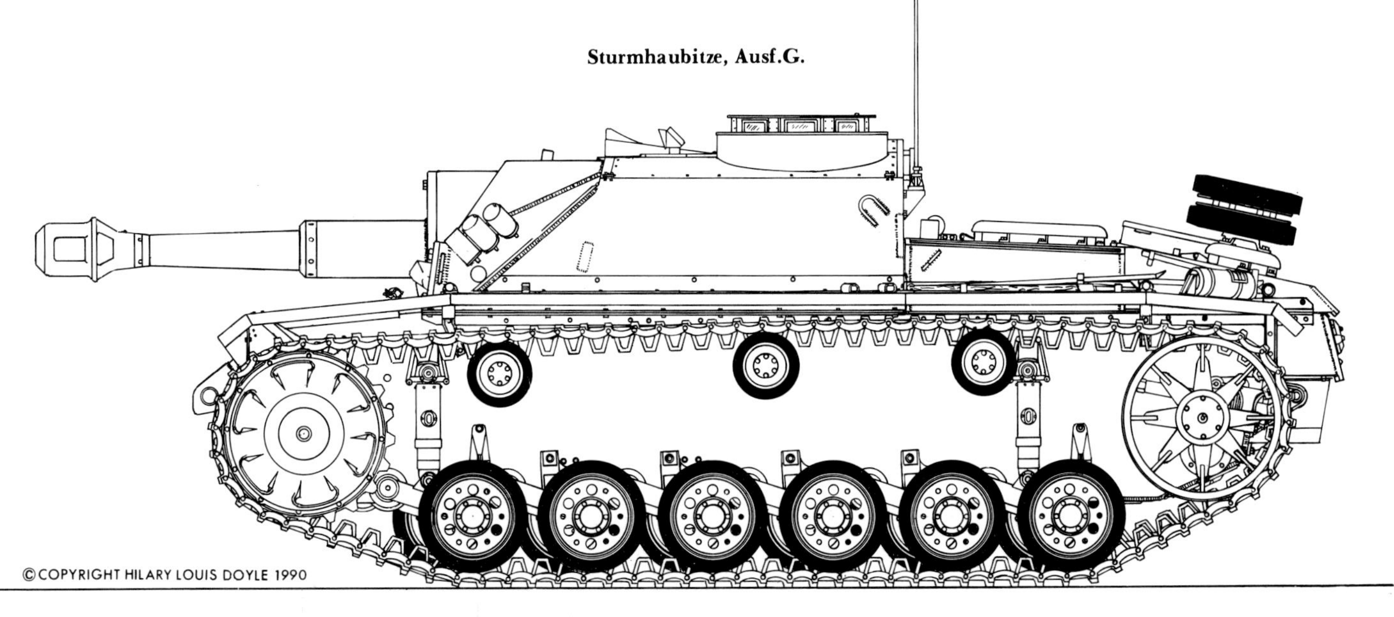

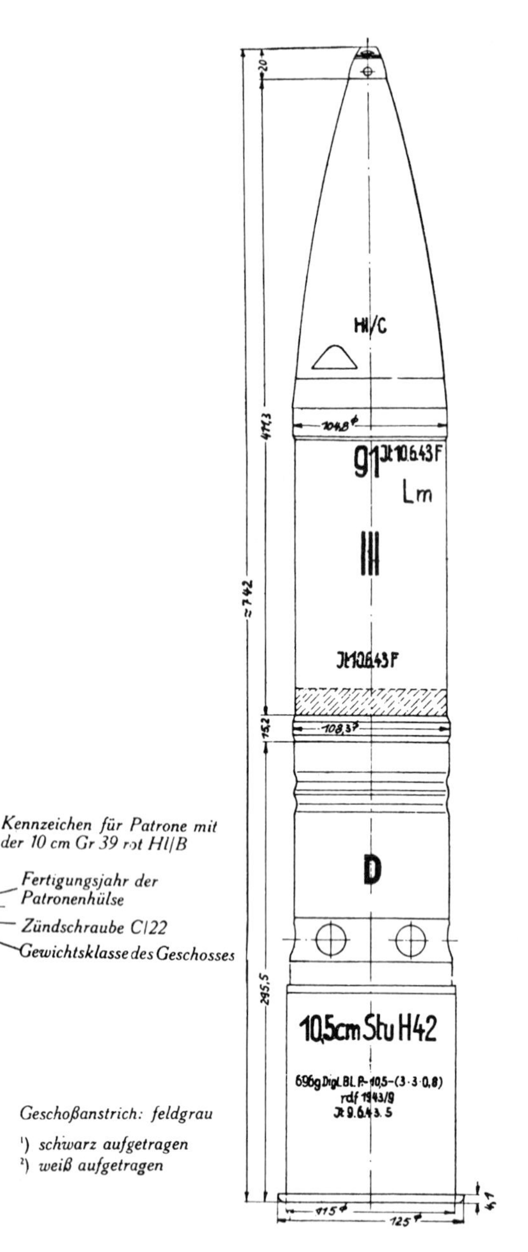

105mm Stu.H.42-HL/C (a shaped charge HEAT round).

production of the Sturmgeschütz with the le.F.H. field howitzer was ordered to increase from 12 to 24 per month. On 6 and 7 February 1943, Hitler noted that production was to start with 20 in March, by May the number of Sturmhaubitze should have increased to 30 per month. Hitler considered it important that at this point in time "Hohlladungs" (shaped charge HEAT) rounds were to be available for this weapon. In the middle of October 1943, based on experiences at the Front with the Sturmhaubitze, Speer proposed, to increase their scheduled production rate.

The official vehicle number for the Sturmhaubitze was Sd.Kfz.142/2.

The 10.5cm Sturmhaubitze 42 was originally equipped with a two-chamber muzzle brake, which allowed supplemental propellant charges to be fired. According to an Army technical paper (Heer. Tech. V.-Blatt 1944, Nr. 635) of September 1944, the muzzle brake was to be dropped from newly produced guns. If Sturmhaubitze had originally possessed muzzle brakes, that had been damaged, and no a replacement was not available; muzzle brakes from the le.F.H.18 or le.F.H.18/40 field howitzers could be used. In addition to the welded mantle, beginning in 1944, the so-called "Saukopf" mantle was introduced.

The Sturmhaubitze's were only assembled at the Alkett plants.

Changes to the chassis and superstructure in the series produced between March 1943 and April 1945 were identical to those modifications made on the Sturmgeschütz, Ausf.G. Only after the assembled chassis had been tested and accepted was it decided whether the chassis would used for a Sturmgeschütz or a Sturmhaubitze.

Sturmhaubitze, Ausf.G, produced in April 1943.

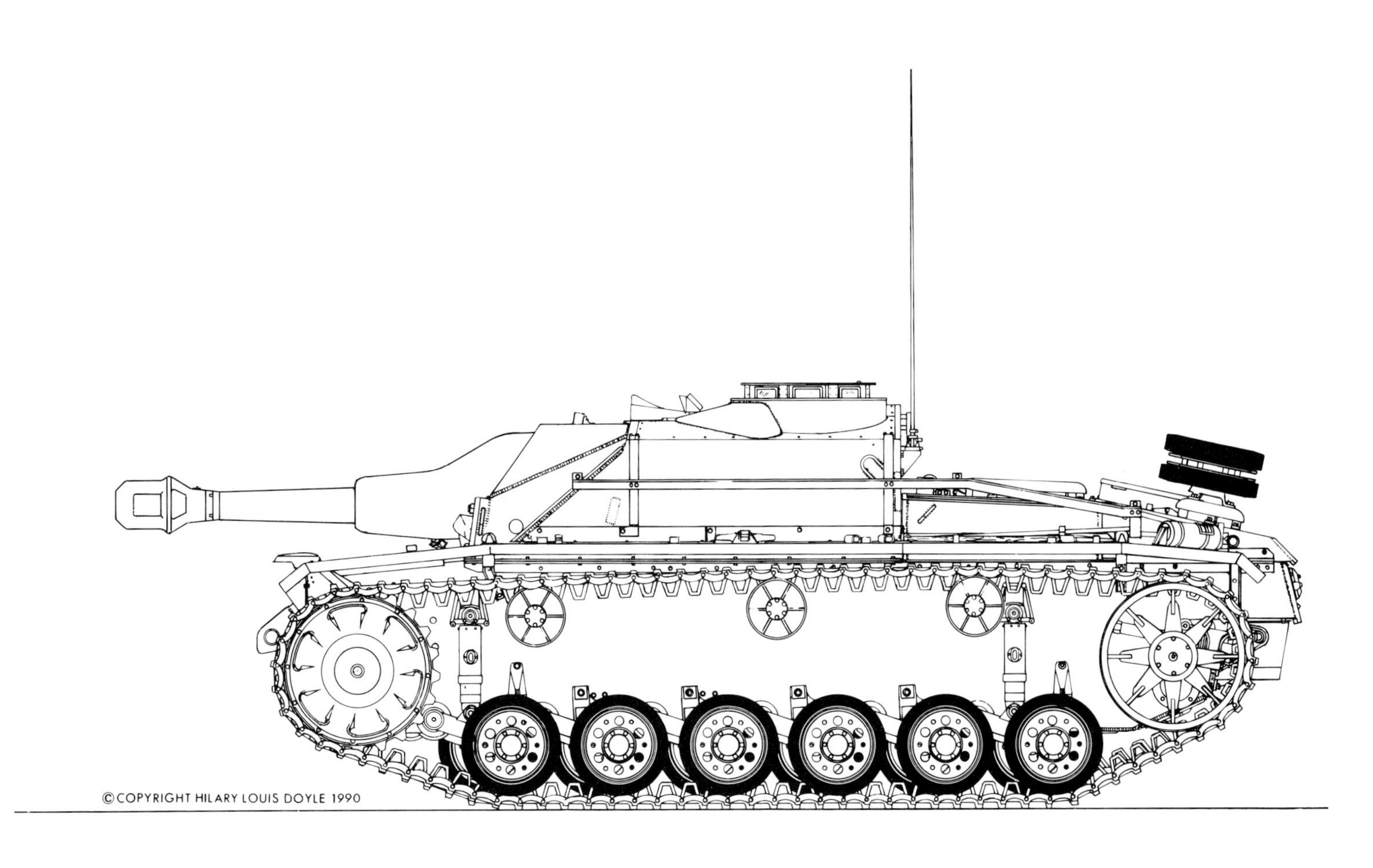

Sturmhaubitze, Ausf.G, with steel return rollers and Saukopf mantle.

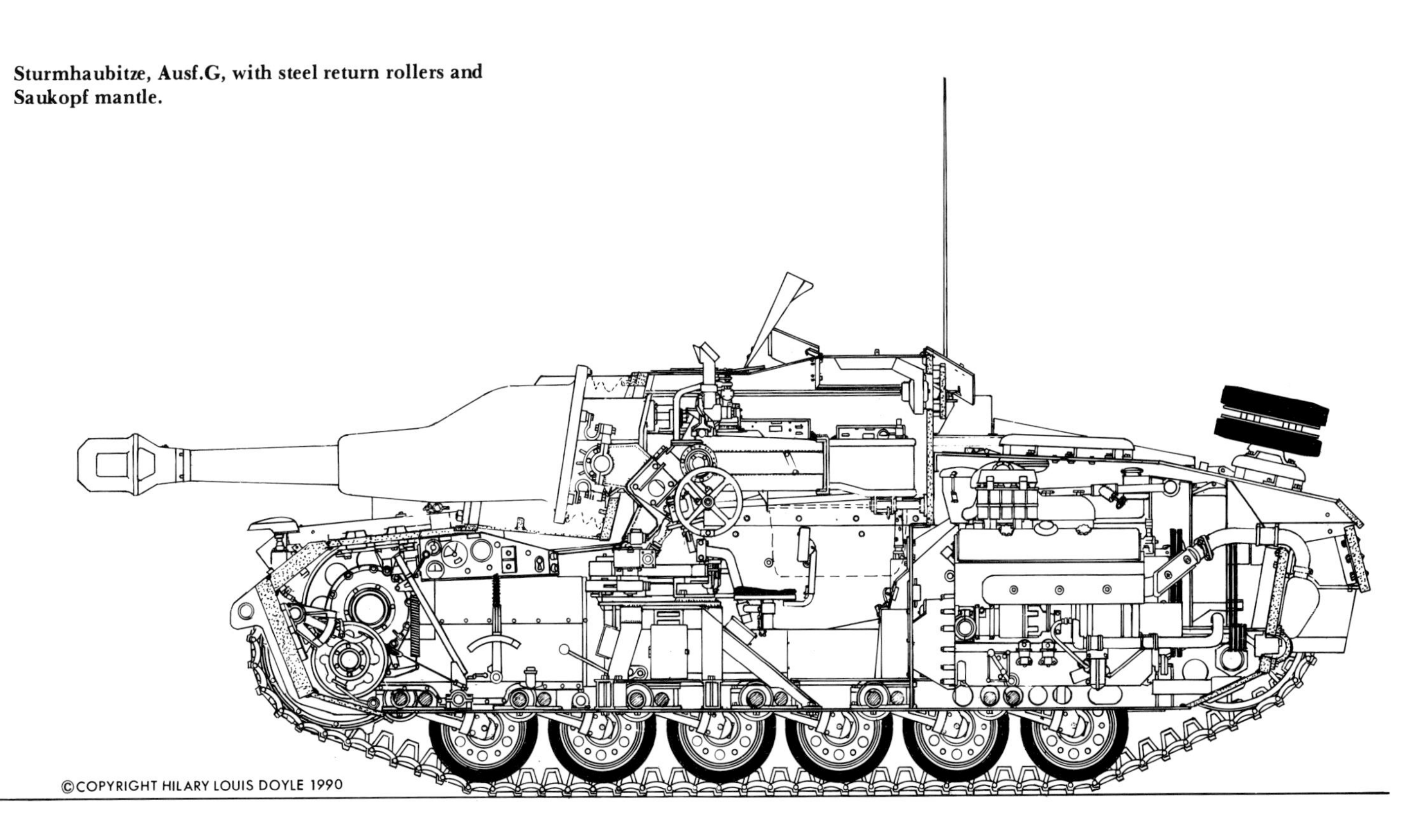

Sturmhaubitze, Ausf.G produced in April/May 1943 with smoke grenade launchers.

Sturmhaubitze, Ausf.G. Note the Saukopf mantle and side baffles on the muzzle brake.

Sturmhaubitze, Ausf.G, produced in May 1944, with a welded mantle for the main gun and 80mm basic armor plate for the right superstructure and hull front.

Sturmhaubitze, Ausf.G, with chassis number 105602, produced in June 1944. Triangular plates on the rails for mounting the Schürzen side skirts.

Sturmhaubitze, Ausf.G, from the final production run. A muzzle brake in no longer mounted on the Sturmhaubitze 42. Coaxial machine gun in welded mantle.

Sturmhaubitze, Ausf.G, with features that were present on those produced near the end.

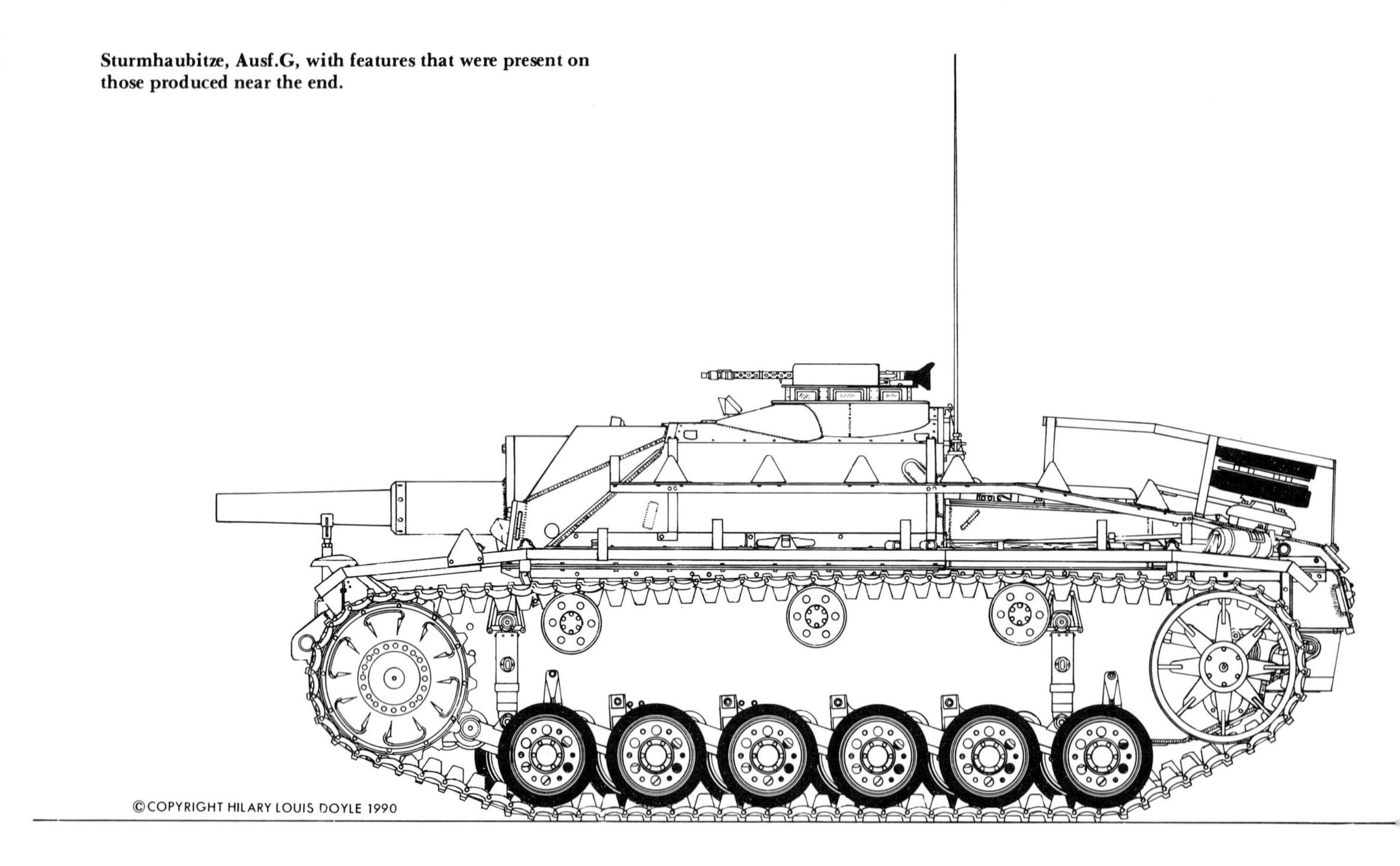

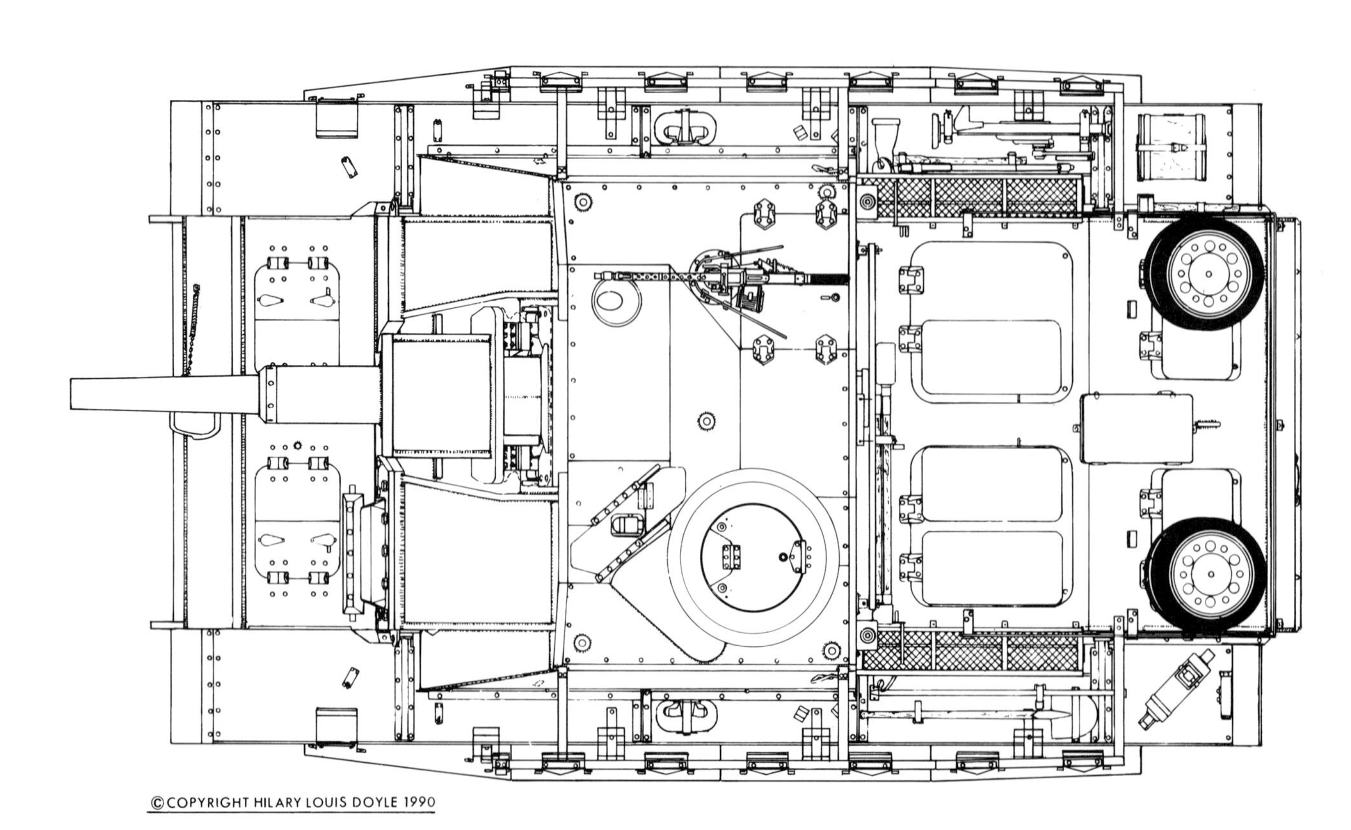

Sturmgeschütz IV

In order meet the Army's requirements for a medium battle tank with a 75mm gun, the firm of Fried.Krupp AG in Essen delivered their "Proposal for the B.W." to the Heereswaffenamt on April 13th 1935. The B.W., abbreviation for Begleitwagen (escort vehicle) was the cover name for the Panzerkampfwagen IV. Already in 1935 Krupp received contract number 41/2203 for one B.W. experimental vehicle, whose development status was reported on June 30th, 1936: "As promised, at the end of the second shift, the vehicle stands complete and ready for use." Krupp, as the engineering design firm, producer of the armored components, and also the assembly firm, was awarded the contracts for B.W. series production.

On November 15, 1937, Fried.Krupp Grusonwerk AG, Magdeburg-Buckau informed the artillery design department of its parent company, Fried.Krupp AG that the acceptance of the first two B.W. (complete with turrets and armament) would take place on November 29th, 1937 in Magdeburg. This event signaled the beginning of the long history of the Panzerkampfwagen IV, the only tank of the German Wehrmacht which remained in production and action for the entire duration of the war.

The two standard tanks of the German Wehrmacht, the Panzerkampfwagen III and IV were quite similar in external measurements and combat weight. The one essential difference between them could be found in the layout of the running gear. While the Panzerkampfwagen III, starting with the fourth production series, received a modern torsion bar suspension, the Panzerkampfwagen IV continued to be produced on the outdated but reliable and easily repairable, leaf spring suspension right through to the end of the war. Countless suggested and attempted improvements to the Panzer IV suspension never advanced past prototype tests.

The running gear consisted of eight road wheels on each side, set up in four pairs of two. Each pair was affixed to a hull bracket via suspension arms. A leaf spring was clamped to the leading axle arm of each road wheel pair. The free end of this leaf spring rested atop a shackle pin roller on the other axle. By pairing the roadwheels, the effective spring length was longer and the suspension considerably softer. Because of the dampening ability of the leaf springs, shock absorbers were not needed. Maximum deflection of the leaf springs was limited by bump stops. The chassis moved on two dry-pin tracks. Each track consisted of 99 individual track links, held together with pins. Located at the front of the chassis, the double sprocket, drive wheel propelled the track. The track center guides ran between the flanges of the roadwheels. At the rear end of the vehicle, the track was returned by the idler wheel. In order to avoid throwing a track, the idler wheel contact surface was made of steel. On its way back to the drive wheel, the track was supported by four return rollers which kept slack to a minimum. The idler wheel, mounted on an eccentric axle, was used to tension the track.

The armored hull served as the frame for the chassis. The hull consisted of armor plates of various thicknesses which were welded together. The aft engine compartment was separated from the crew compartment by a fire wall. The hull was reinforced by fore and aft crossbraces and in the center by the rectangular cover for the fuel tanks.

Just as in the Panzerkampfwagen III, the Maybach HL 120 TRM high-performance engine was installed.

The transmission selected for the Panzerkampfwagen IV was the proven and tested, technical state of the art, synchromesh transmission designed and produced by Zahnradfabrik Friedrichshafen (ZF).

Both steering gears, designed as planetary gears, were installed from the outside of the hull, together with the final drives. The final drive was a reduction gear which was linked to brakes. There were two sets of brakes, one for steering and for bringing the vehicle to a halt.

At the latest, after the Battle of Kursk in 1943, the role of the German Panzertruppe changed dramatically. There role was no longer lightning attacks by large tank units to rapidly advance and cut up massive armies, but rather defense through counter-attacks. The main battle tank lost out in favor of the Sturmgeschütz, Panzerjäger and Jagdpanzer as tank destroyers. The ready evidence is displayed in the production numbers.

The first known proposal for a Sturmgeschütz on a Panzer IV chassis is seen in Krupp drawing number W 1468 dated February 1943. These drawings reveal the fact that Krupp had based this design on a proposed 9./B.W. chassis, instead of the normal 8./B.W. production chassis. At this point in time, Krupp apparently did not have the final drawings for a Sturmgeschütz, Ausf.G superstructure from Daimler-Benz in their hands.

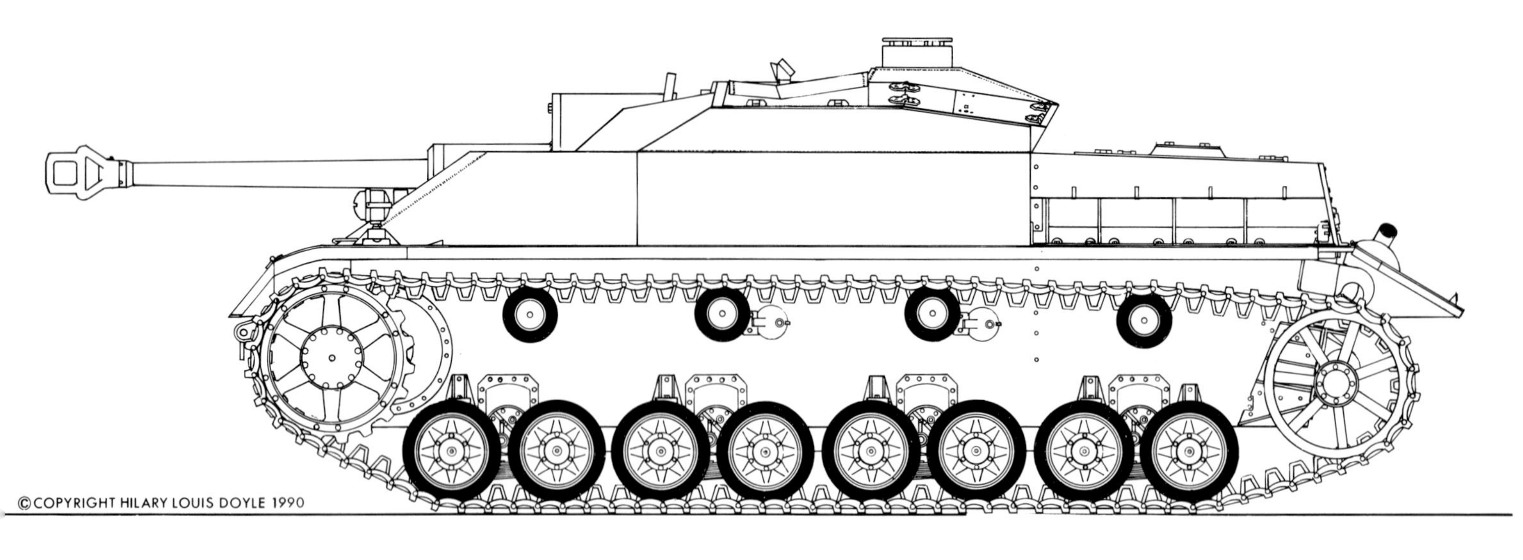

Krupp's proposal for a Sturmgeschütz on a Panzer IV chassis: Sturmgeschütz auf 9./B.W. Fahrgestell (Krupp drawing number W 1468).

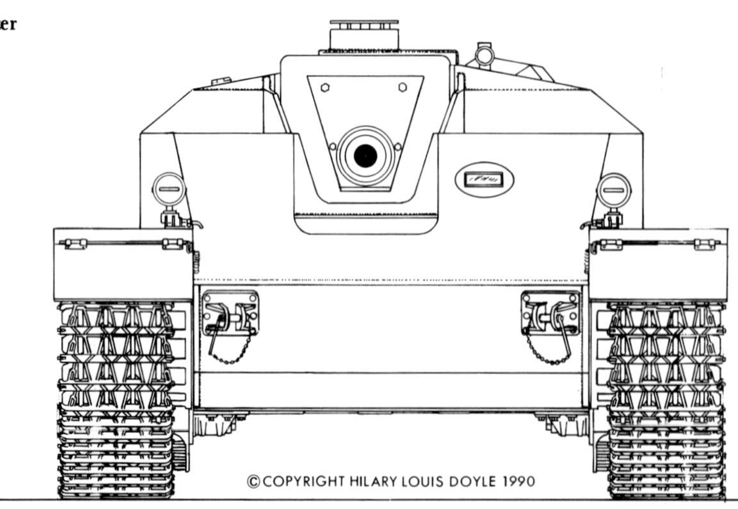

Instead, a drawing of the proposed vehicle was created using the outdated superstructure design from the Sturmgeschütz, Ausf.F.

This drawing contained the following details of the proposal:

Hull armor:

Hull, glacis plate	50mm	56 degrees
Hull, front	80mm	12 degrees
Hull, lower front	30mm	59 degrees
Hull, side	45mm	0 degrees
Hull, rear	45mm	10 degrees
Hull, below, rear	30mm	74 degrees
Forward belly plate	20mm	
Rear belly plate	10mm	
Deck over driver	16mm	

Superstructure armor:

Front	50mm
Sides	45mm
Rear	30mm

Provision was made for a driver's periscope in the glacis plate. The selected armament was the 7.5cm Sturmkanone 40 L/48 with a minimum elevation of -6 degrees and maximum of +20 degrees, andtraverse arc of approximately 20 degrees. The Panzerkampfwagen IV fuel tanks remained, as they had always been, under the floor of the fighting compartment. The engine was rated at 320 metric hp with a with a power-to-weight ratio of 11.3hp per ton. Maximum speed was 38km/h.

The overall length (including main gun) was 6350mm. Overall height (above the superstructure fan) 2180mm. Overall width, 3260mm. Ground clearance, 400mm. Width of the tracks, 560mm. Ground pressure, 0.76kg/cm². Combat weight, 28.26 tons: of which the chassis weight was 23.61 tons, the superstructure 2.4 tons, and the gun with mount 2.25 tons.

On February 5th 1943, Krupp wrote to the OKH, (Ch Rüst and BdE), the Heereswaffenamt, and Prüf. 6/II that according to their research, converting the Panzerkampfwagen IV into a Sturmgeschütz in accordance with the proposals made by the Munitionsministerium could not be considered because there was no expected savings in weight by removing the turret of the tank. The proposal for the modified 9./B.W. chassis with sloping and thicker side armor was rejected due to the anticipated disruption in production, deemed intolerable for the Spring and Summer of 1943.

Had Krupp used the normal (8./B.W.) chassis with 40mm wide tracks and rubber rimmed road wheels for this proposed Sturmgeschütz, the resulting combat weight of the vehicle would not have exceeded 23.95 tons. Krupp's statement, that the expected overall weight would be too heavy, was based on the decision to use a modified 9./B.W. chassis for the proposed Sturmgeschütz design. This decision resulted in the idea of a Sturmgeschütz produced on the Panzerkampfwagen IV chassis being dropped until the end of 1943.

At the end of November 1943, Sturmgeschütz production at Alkett was brought to a standstill by a bombing raid. Alternative production was urgently needed.

During the Führerkonferenz of 19-22 August 1943, reports from the front that had been presented to Hitler, showed him the value of the Sturmgeschütz. According to some reports, the Sturmgeschütz had proven itself superior to the Panzerkampfwagen IV within the restraints of how it had been deployed up to that time. It was therefore prudent to expect that — as soon as the new "Panzerjäger IV" was acceptable for field service — Panzerkampfwagen production would be converted over to the Panzerjäger IV. This was to occur as soon as possible without causing disruptions production and allowing production to steadily increase. (This meeting does not refer to the Sturmgeschütz IV, but rather to the Jagdpanzer IV (previously referred to as the Panzerjäger IV). However, it does relate the attitudes during this period, in which "Sturmgeschütz" production was being increased at the expense of Panzerkampfwagen production.)

In his conference from 6-7 December 1943, Hitler welcomed the suggestion that in quick compensation for the lost Sturmgeschütz production capacity, Sturmgeschütz components which had been mounted on the Panzer III chassis could be mounted on the Panzer IV chassis. He saw the possibility of issuing these vehicles to the Panzer Abteilungen. The advantage being that the same line of replacement parts could supply both the Panzerkampfwagen IV and the Sturmgeschütz on the Panzer IV chassis.

On December 11th 1943 the following memo was written by Herr Woelfert of Krupp A.K.):

Subject: B.W.-Sturmgeschütz

At Alkett, Borsigwalde, Sturmgeschütz on the Z.W. chassis can no longer be produced due to loss of one major production facility. At the instigation of the Munitionsministerium (Speer), as an emergency solution, by December 15th 1943, a B.W. chassis will be fitted with a Z.W. Sturmgeschütz superstructure at Daimler-Benz, Marienfelde. Initially, a design engineer from the Lanz department will be available for the creation of conceptual drawings, he will be assisted beginning December 13th 1943 by an additional design engineer from the Lanz department. It is yet to be decided if special production drawings are to be completed in the Krupp Artillerie

Sturmgeschütz IV. Modified Sturmgeschütz superstructures were mounted on thirty Panzerkampfwagen IV chassis supplied by Nibelungenwerk.

Sturmgeschütz IV, right side, with mounts for the Schürzen side skirts.

Sturmgeschütz IV, view of the right rear.

design office. Grusonwerk will still be required to deliver an additional 50 vehicles in December. This might be impossible, however, as production (of the Sturmgeschütz) cannot even begin until the middle of December.

In the first prototype and the following series production vehicles, the "normal" chassis along with the rear deck from the Panzerkampfwagen IV series were combined with a modified superstructure from the Sturmgeschütz, Ausf.G series.

The Sturmgeschütz superstructure was modified for use on the new chassis. Due to the longer chassis of the Panzer IV, a box shaped compartment (to provide protection for the driver) was added to the superstructure right front corner.

In the days that followed, two "B.W.-Panzerjäger", one from Vomag and one from Daimler-Benz, took part in a demonstration at the Führer's headquarters. During the Führerkonferenz of 16-17 December 1943, the Sturmgeschütz on the Panzer IV chassis received Hitler's complete approval. To make up for the large deficit in the Sturmgeschütz production as quickly as possible, full support was to be provided to attain a high production rate. The production goal for December was given as 350, while in January the goal of 500 was to be reached, inclusive of the Sturmgeschütz on the Panzer III chassis with the 7.5cm StuK 40, the Sturmhaubitze on the Panzer III chassis with the 10.5cm StuH 42, Panzerjäger IV from Vomag and the Sturmgeschütz on the Panzer IV chassis from Krupp Grusonwerk.

Krupp Grusonwerk wrote to Abteilung Apparatebau III of Fried.Krupp AG, Abt. in Essen on 18 December 1943:
As you already are aware, we are to begin assembly of 605 Panzerkampfwagen IV as Sturmgeschütz auf Panzer IV. Therefore, the hulls must be altered. From this point on, we will require about 100 hulls per month, the rest are to remain as B.W. hulls. We cannot give you the exact number we will require as this depends on when you can begin the delivery of the altered hulls. Until then we must make the alteration to the B.W. hulls here.

Alterations to the Sturmgeschütz IV

On January 24, 1944, the *Inspektor der Artillerie* determined that supplementing armor protection by adding concrete had been shown by test firing to be without benefit. It only increased the vehicle weight and fragmentation effect (on the armor piercing round striking the plate) was less favorable. The Waffenamt was not in favor of concrete for supplemental protection. Even when the assembly firms followed this order, troops in the field sometimes added concrete for extra protection to their armored vehicles.

Krupp-Grusonwerk in Magdeburg in a letter to Wa J Rü (WuG 6), dated March 10 1944, stated that according to information from WaPrüf 6 the DKW-motor-generator set (providing power for the electric turret drive in the Panzer IV) was to be dropped beginning in February. As a result, Krupp made provisions for an additional fuel tank in the rear of the vehicle in place of the DKW-motor-generator set. Krupp also requested that the suppliers of the hulls be informed accordingly.

A memorandum concerning a discussion on March 14th 1944 in Berlin addressed installation of a firewall in the Panzer IV hull between the engine and the new fuel tank, after discontinuation of the DKW-motor-generator set. This interfered with their plans to install a new ammunition bin in the engine compartment of the Sturmgeschütz IV. The number of hulls that were to be manufactured for the Sturmgeschütz IV had increased to 1500. Difficulties arose because the hull manufacturers only learned immediately prior to shipping whether the hulls were for Panzerkampfwagen IV tanks or for Sturmgeschütz IV. WaPrüf 6 was asked to clarify if the 12-round ammunition bin in the engine compartment of the Sturmgeschütz IV could be replaced with an additional fuel tank.
Krupp submitted a new proposal for an altered ammunition bin (for 8 rounds instead of 12) which could be installed in the new hull without further alterations.

A memo recording a meeting in Magdeburg on March 20th 1944 provides an insight into further modifications:
Subject: 7.5cm Sturmgeschütz 40 IV
1. Ammunition storage
Krupp presented a new proposal. The ammunition storage area (in the engine compartment) is to be protected from exhaust heat by a metal plate.
2. Gun mount
The simplified model based on drawing SKB 6124 was presented. A test model is to be made, installed in a vehicle and tested in Hillersleben. Results of this test firing must be received before it can be authorized for production.
3. Adjustable drivers' seat
Two experimental seats are to be prepared, of which one is to be sent to Kummersdorf for testing.

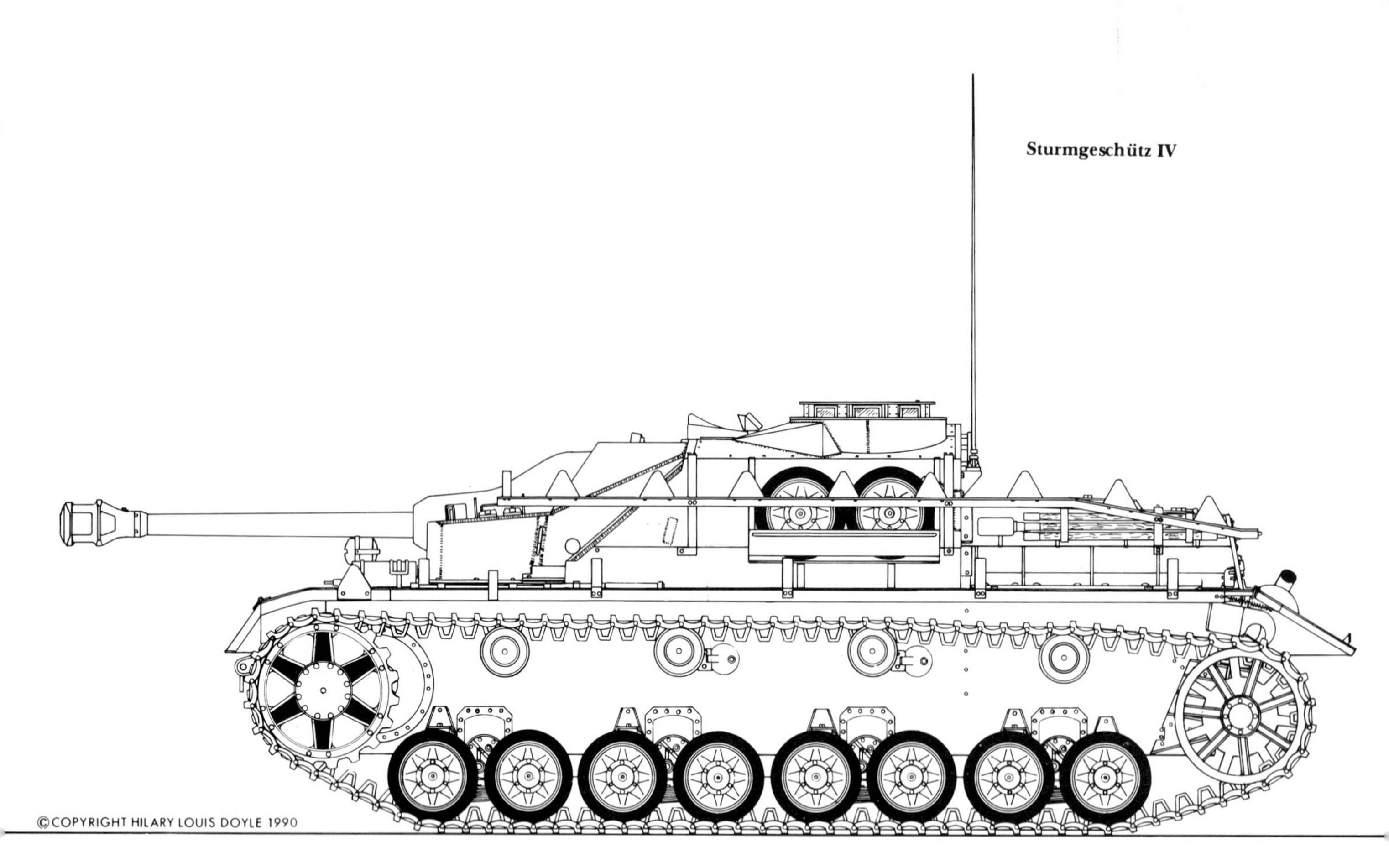
Sturmgeschütz IV
©COPYRIGHT HILARY LOUIS DOYLE 1990

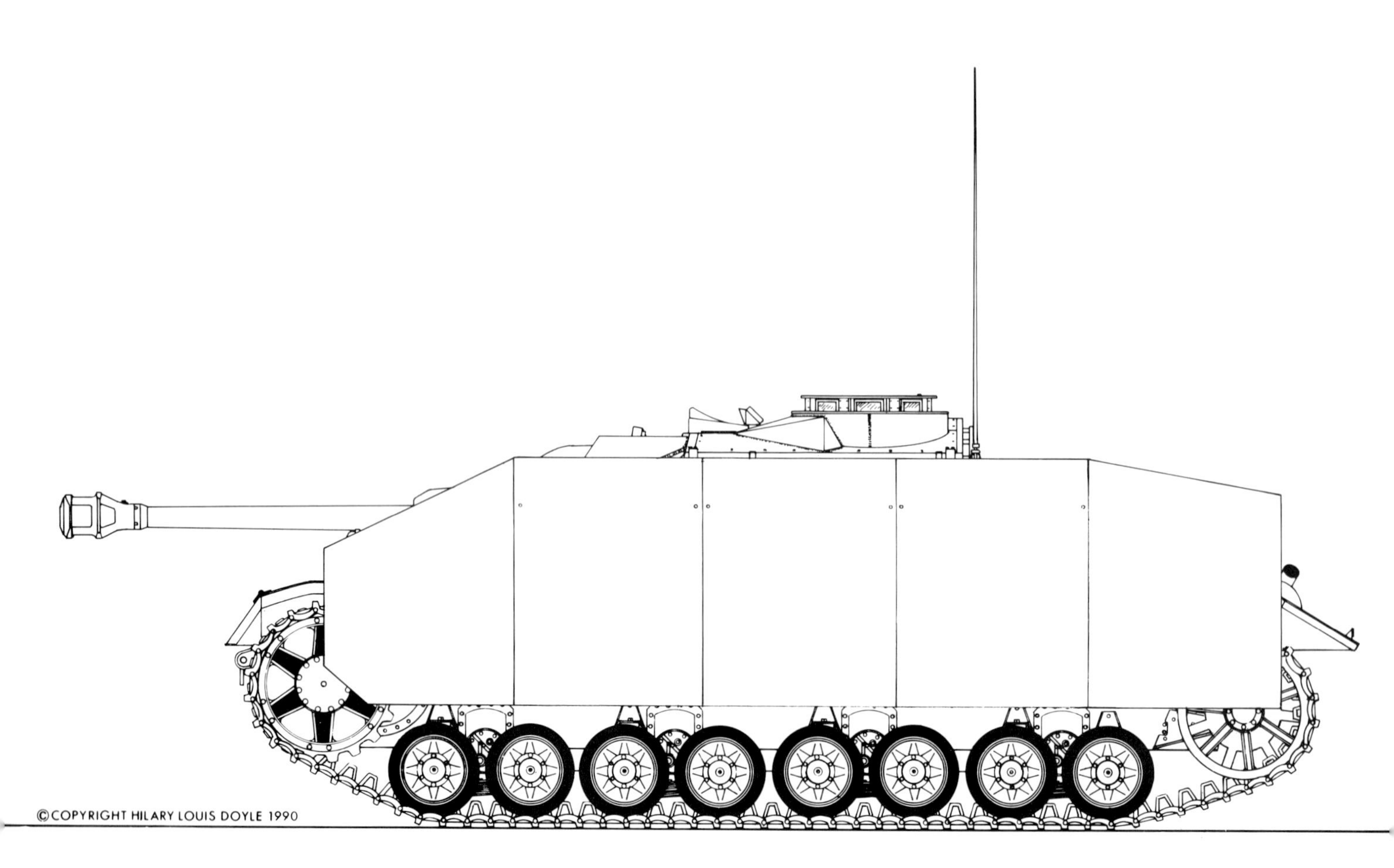
©COPYRIGHT HILARY LOUIS DOYLE 1990

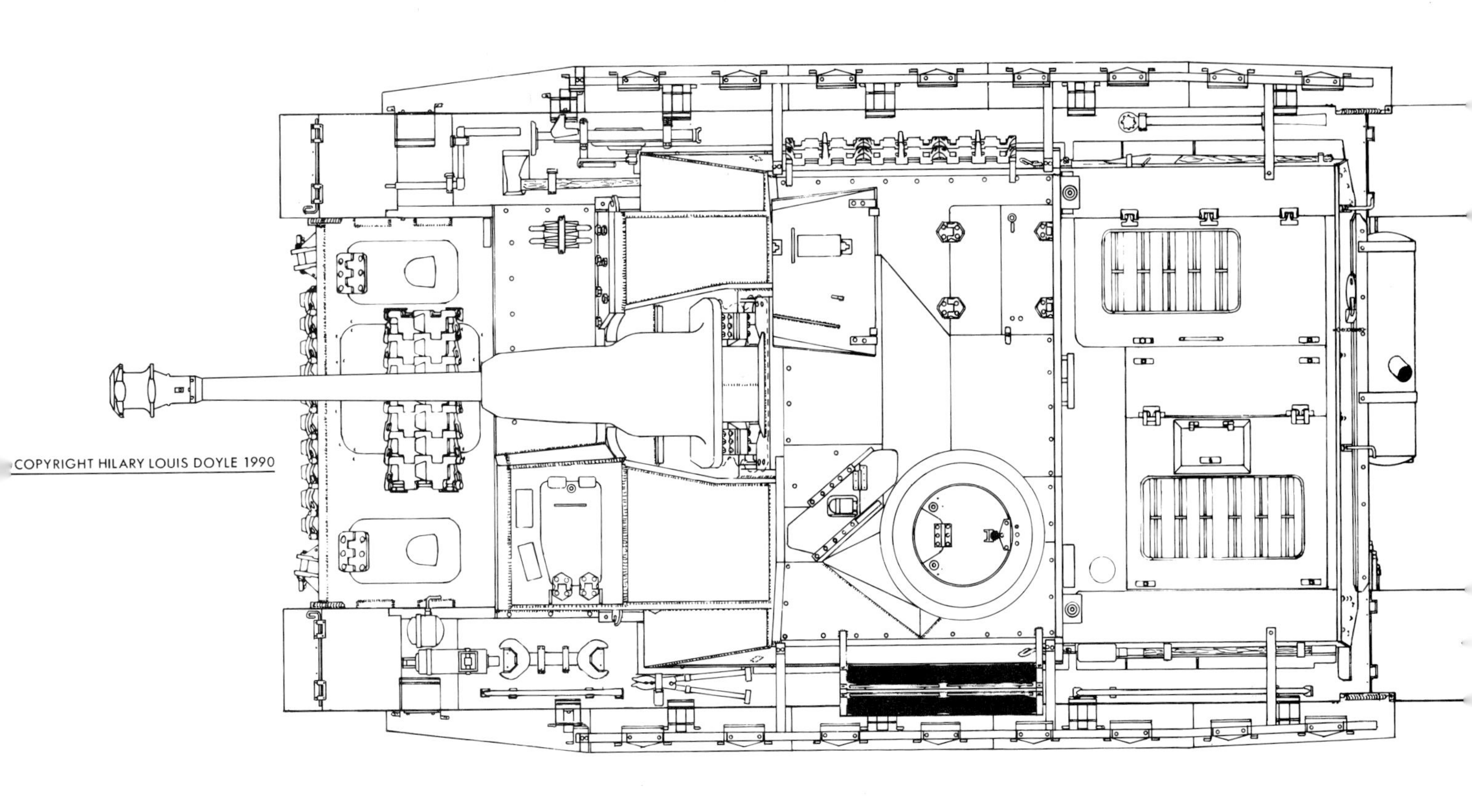

COPYRIGHT HILARY LOUIS DOYLE 1990

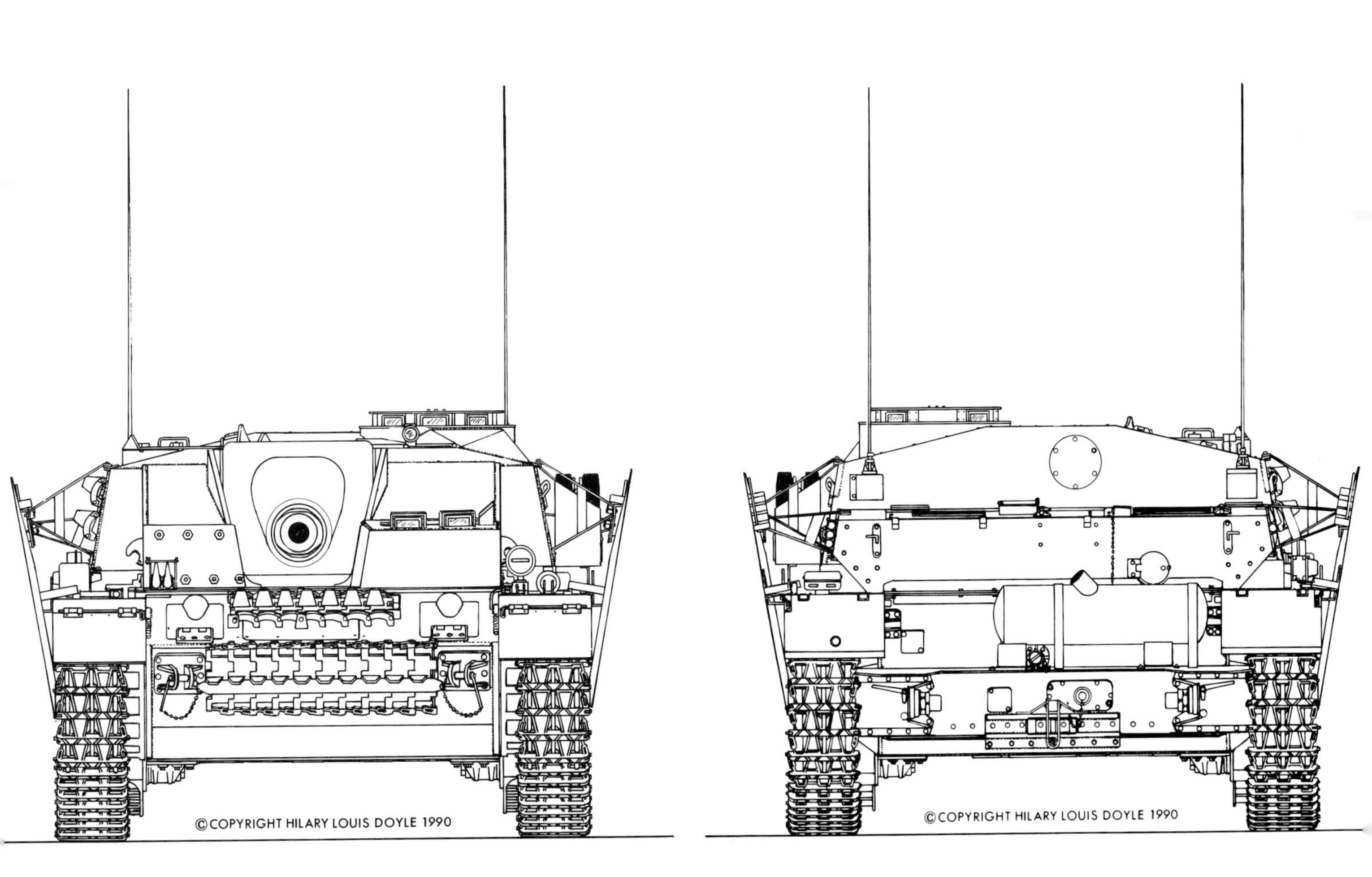

©COPYRIGHT HILARY LOUIS DOYLE 1990
©COPYRIGHT HILARY LOUIS DOYLE 1990

Sturmgeschütz IV, produced with an older hull before the modification to interlock the 80mm front hull plate with the hull side plates.

Sturmgeschütz IV with Schürzen side skirts.

4. Driver vision
Krupp is submitting a proposal for a better mount for the periscopes. The model in use is not acceptable due to the large tolerances for the periscope and the mounting components.
5. Close combat device, external machine gun and barrel support for the gun. These items are to be mounted on a prototype vehicle.

Abteilung Artillerie Konstruktion in Essen received a contract for design work amounting to 500 days of design engineer time to be applied to design of further modifications.

On March 21st 1944, Krupp-Grusonwerk reported to the OKH,
Subject: 7.5cm Sturmgeschütz 40 IV

The vehicle with the simplified gun mount has chassis number 100247 and hull number 86497. It was sent to Königsborn on March 20 1944 for further testing in Hillersleben. The modifications save 18% in production materials, 45% in welding electrodes and 35% in costs when compared with the Daimler-Benz design.

On March 27th 1944 Krupp reported that the floor escape hatch under the radio operator's seat in the Panzer IV was no longer needed. A radio operator did not occupy this position in the Sturmgeschütz IV. Permission was being sought for the assembly firms to weld the hatch closed.

In the Heerestechnische Verordnungsblatt (army technical bulletin) article Nr.256 dated 1 May 1944, mounting the Ostkette on the Sturmgeschütz IV was described, which closely corresponded to the directions for

Sturmgeschütz IV. The driver's compartment has been strengthened with reinforced concrete.

mounting this track on the Sturmgeschütz III.

On May 27th 1944, Krupp-Grusonwerk submitted a proposal for modifying the mounting brackets for the Schürzen side-skirts.

During a discussion on June 7th 1944 in Magdeburg the following changes to the Sturmgeschütz IV were addressed:

1. Protective rain shield over the driver's periscopes. In order to ensure that the driver can see when it rains, a protective rain shield is to be installed over both periscope brackets.
2. Driver's periscope
The alignment of the driver's right periscope has already been changed so that only a dead space of 5 to 6 meters remains in front of the vehicle.
3. Strengthening the front of the driver's compartment.
The thickness of the forward wall was 80mm. An additional 20mm armor plate at an angle of 15 to 20 degrees could still be installed. Krupp is to work out a proposal and submit it by June 24 1944 for WaPrüf 6 approval.
4. Removal of the gun
In order to insure easier removal of the gun, Grusonwerk recently started fastening the mantle with hexagonal bolts, which are simply secured with a center punch.
5. Ammunition storage
A total of 87 rounds (39 in bins, 48 in ammunition canisters) were stored in the vehicle. An increase in the ammunition stowage was not possible. Daimler-Benz will be called upon for advise on stowing the smoke grenade ammunition.
6. Smoke grenade launcher
Krupp recently received a delivery of 65 smoke grenade launchers. There were difficulties with the Freudenberg firm over the delivery of sealing rings designed to prevent water from entering the launcher.
7. Barrel support
The barrel support developed by Daimler-Benz for the Sturmgeschütz III could not be used on the Sturmgeschütz IV because securing the gun at 0 degrees elevation robbed the driver of his vision to the right. Krupp was called upon to propose a solution to this problem.
8. Adjustable driver's seat
The design of the adjustable driver's seat proposed by WaPrüf 6 was essentially completed. The back rest remained foldable, so that the driver can escape to the rear, if the Sturmgeschütz is knocked out.

On June 8th 1944, testing began with a Panzerkampfwagen IV with six instead of 8 return rollers. Permission was being sought to carry out this modification on the Panzerkampfwagen IV and all its variants. Steyr-Daimler-Puch (Nibelungenwerk) promised that this would result in saving 2000 "bottle-necked" ball-bearings. The savings in time were equivalent to that needed to produce an additional 24 Panzerkampfwagen IV annually.

Although the modification had been tested in June of 1944, Sturmgeschütz IV with six return rollers were not produced until December 1944, and from then only on part of the production run.

On June 6th 1944, an order specified that all C-Hooks were to be replaced with the S-Hooks. From June 10th 1944, only one S-Hook was to be carried on each Sturmgeschütz instead of two C-Hooks.

On June 13th 1944, a modified travel lock for the Sturmgeschütz was discussed. The current travel lock was for a gun positioned at 0 degrees elevation. A new travel lock at 12 degrees, was necessitated by the intended changes to the external barrel support.

On June 14th 1944, an order concerning the installation of the "Pilze" mounts for the 2-ton jib crane, was followed on the 30th of June by an Army order to Krupp-Grusonwerk to begin immediately equipping the current production series with the "Pilze" mounts.

A memorandum for a discussion held in Magdeburg on July 6th 1944 contained the following details:

1. Additional armor for the Driver's Compartment
The suggestion to weld the bracket for the additional armor directly to the front of the driver's front plate was rejected. The supplemental plate is to be held by six M 20 bolts. The supplemental armor plate is to be increased to a thickness of 30mm. It was not possible to drill holes through the 80mm-thick plate under the makeshift conditions in the field. Therefore, a new design for securing the additional armor by welding instead of bolting is to be implemented in the field.
2. Driver Seat
Krupp presented a new design for a driver's seat which can be raised and lowered (drawing number SKA 6300 b).
3. Barrel support for the gun
The support was developed for a gun elevation of 6 degrees. The original request for 10 degrees elevation was not achievable due to interference with the inertial starter and the hand-operated fuel pump.
4. Rain shields for the driver's periscopes
Wa Prüf agreed with the design as shown in drawing SKC 6286.
5. Guard rail on the rear deck
A slightly altered version of the guard rail developed by Daimler-Benz for the Panzer III was approved. The new version at 300mm above the deck and projecting 100mm over the right side, facilitates unincumbered use of the ventilation fans hatch.
6. 2-ton Jib crane
Simpler "Pilze" mounts are to be produced with 50mm

Sturmgeschütz IV with the 2-ton jib crane. The gun has been removed and the radiator cooling fans swung open.

Sturmgeschütz IV. Gun prior to installation.

Sturmgeschütz IV. The superstructure roof is being removed. The Rundumfeuer-MG machine gun mount and Nahverteidigungswaffe (close combat defense weapon) are visible.

Sturmgeschütz IV. The additional protection in front of the driver's compartment and on the opposite side of the superstructure are modifications made by the troops.

May 5, 1945. The war is over. The 34. Infanterie Division surrender their Sturmgeschütz IV intact. A Bergepanzer III is behind the three Sturmgeschütz vehicles. These Sturmgeschütz IV were produced by Krupp-Grusonwerk in August 1944. The additional steel plates and concrete were modifications made by the troops.

Sturmgeschütz IV produced by Krupp in September 1944.

diameter and a 10x12 drain channel. An attempt was made to avoid vertical installation of these mounts, as well as any special anti-dust closure. An investigation was to be made to determine if the three mounts could be arranged in such a way that removal of the main gun would be facilitated. A test crane was provided for Krupp.

7. Smoke grenade ammunition

Stowage could not be resolved because a vehicle was not available. It was agreed to use the mounting developed by Daimler-Benz for the "Sturmgeschütz III/IV."

By the end of July 1944, Krupp-Grusonwerk was to outfit a production series Sturmgeschütz IV with all the modifications described in items 1 through 6 above. WaPrüf 6 was to seek approval from In 6 and arrange for transport to Kummersdorf.

Beginning in August 1944, the rotatable commander's cupola was to re-introduced (not 100%).

By August 8th 1944, development of the flash suppressor on the exhaust pipes was almost completed. This modification was introduced in the production run in August of 1944.

In September 1944, the assembly firms received the order to stop applying Zimmerit protective coating to new

Sturmgeschütz IV, side and rear views. Flame suppressor stacks had replaced the exhaust muffler.

Sturmgeschütz IV. Typical layout for those produced near the end.

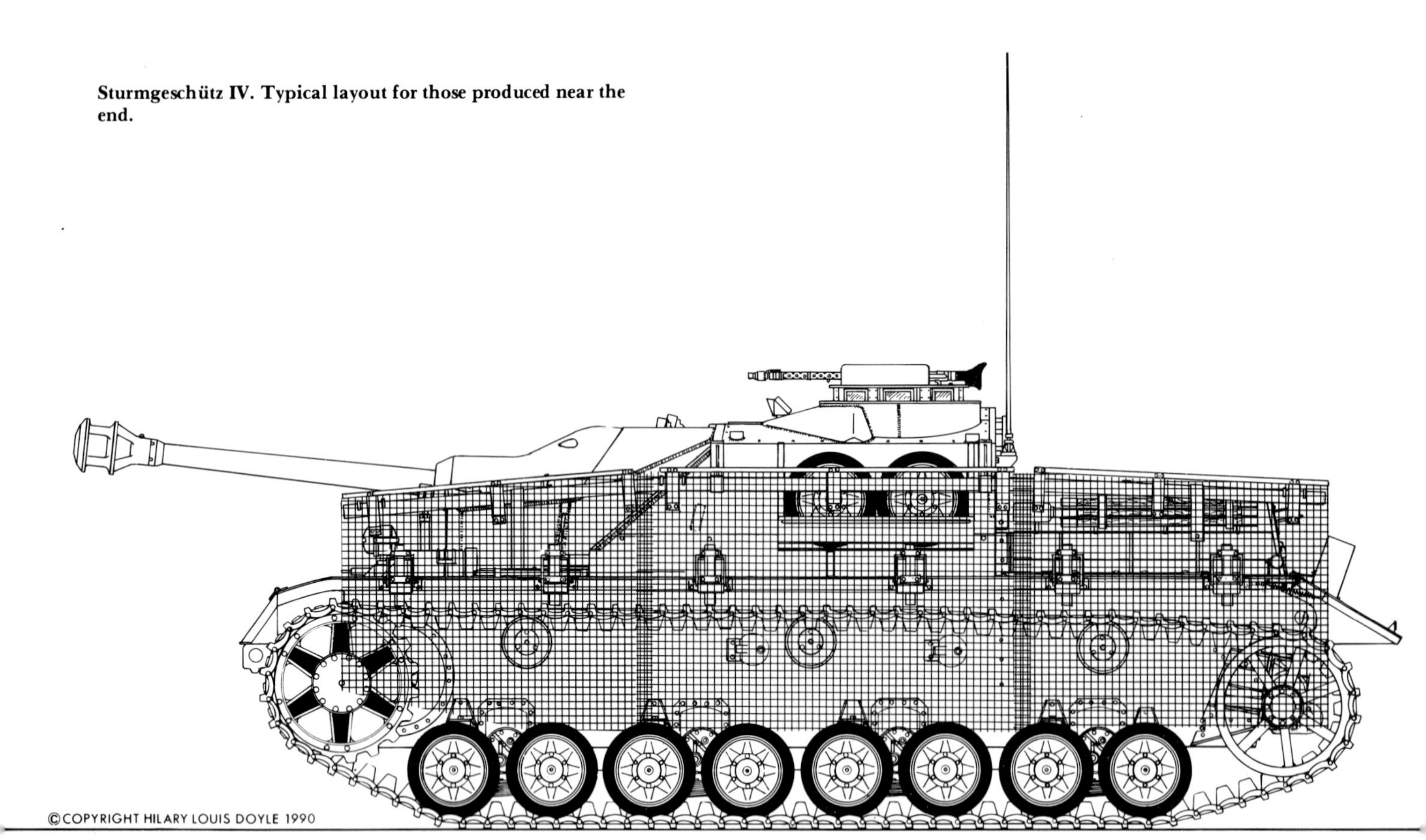

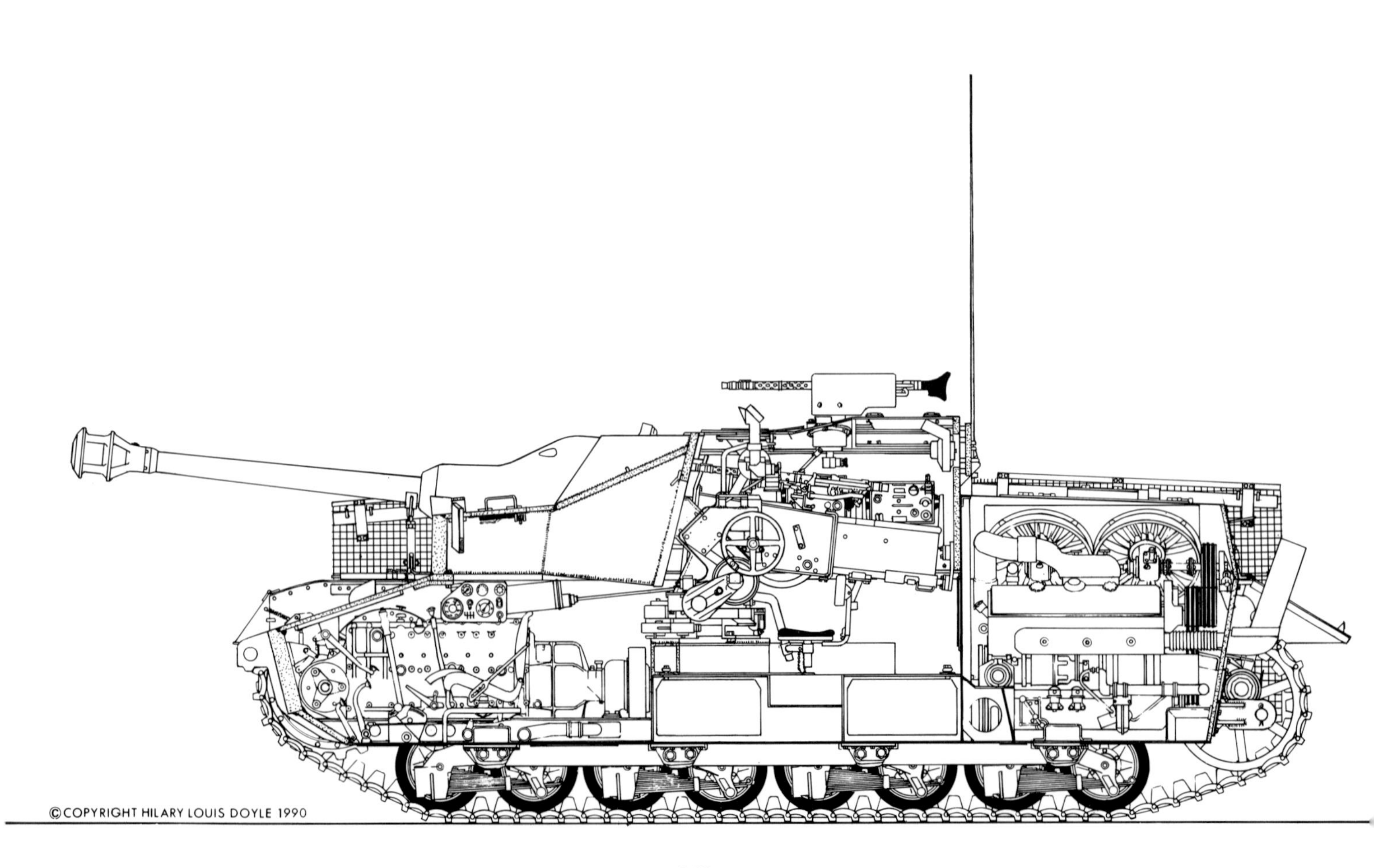

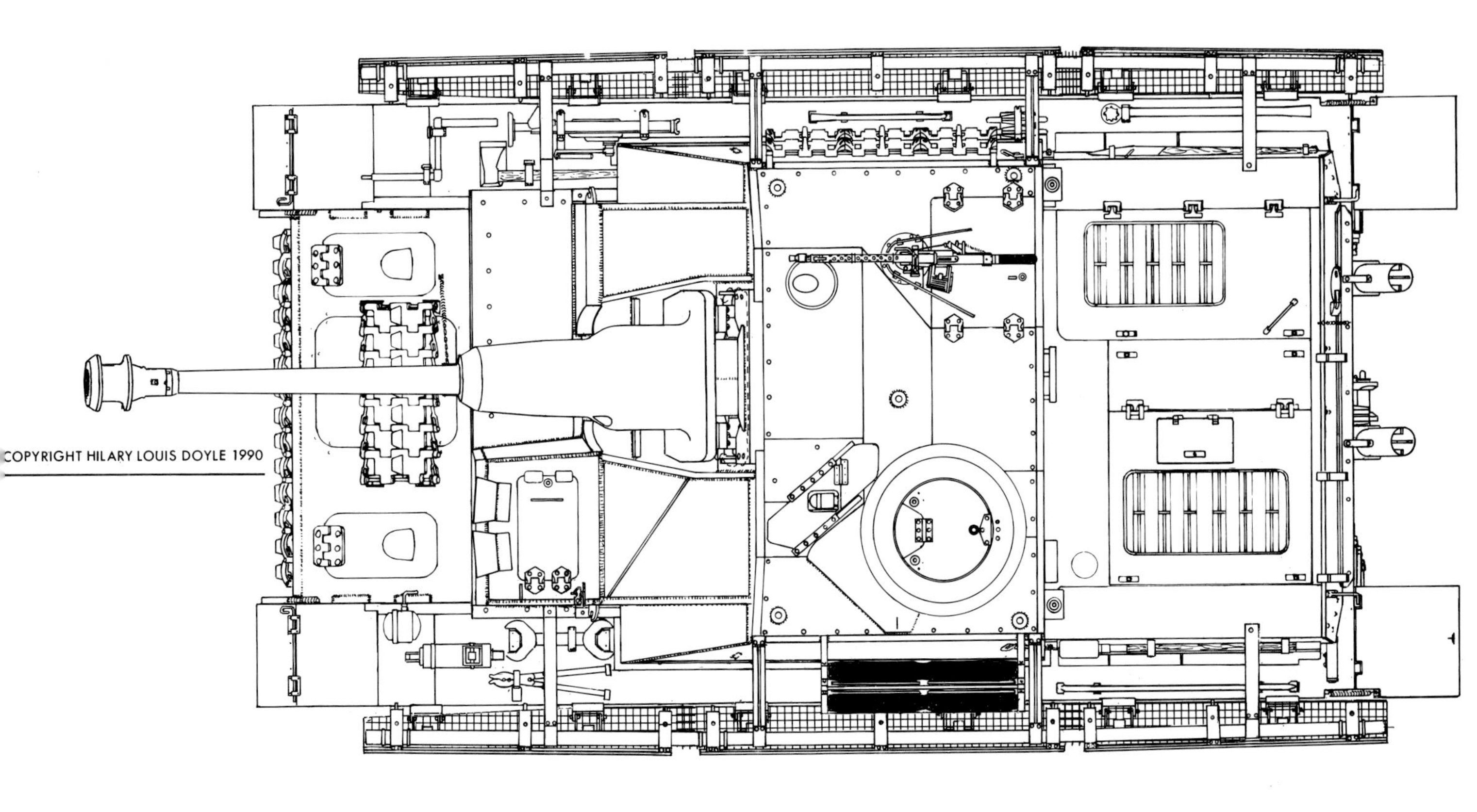

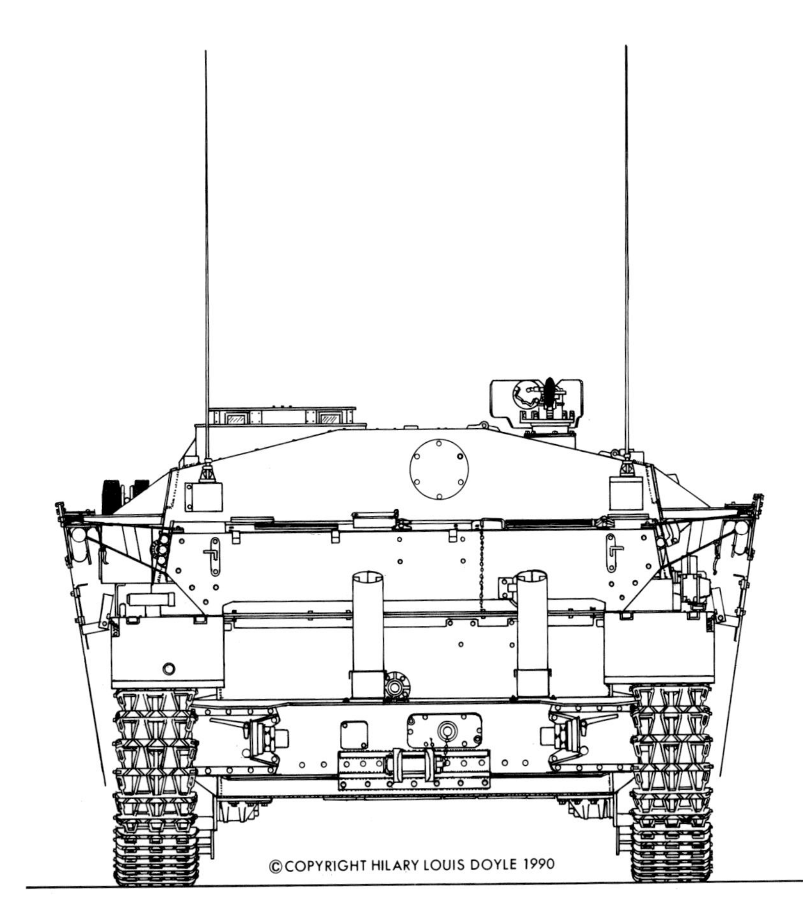

production armored vehicles.

Starting in November of 1944, a barrel support (supplementing the internal travel lock) was installed on the glacis plate of the Sturmgeschütz IV. The gun was held at an elevation of 6 degrees. Installation by the troops was to be done according to Krupp drawing SKD 6437. Those parts required for the installation could be obtained through supply channels.

Also beginning in November 1944, a new rain guard design was installed over the driver's periscope brackets. Up to this point, the driver had been practically blinded during rain and snowfall. Beginning in November 1944, to avoid damage, a protective sheet metal cover was installed over the open fuel line to fuel tank III. The field units were authorized to install both of these modifications.

In a attempt to halt the never-ending breakage, the base of the teeth of "A" gear in the final drives were strengthened. Standardized implementation of these changes in the Sturmgeschütz IV was not possible due to the non-uniform delivery of needed parts.

While driving the Sturmgeschütz IV, accidental closure of the driver's hatch was the cause of problems. This was prevented by installing a lock to secure the hatch in its open position. There were complaints about low power from the storage battery when it was cold. Beginning in November 1944, the battery was insulated by a wooden covering and warm air from the radiator was routed to it.

Beginning at the end of December 1944, a new horizontal, tow coupling was installed in the middle of the hull rear. Attaching tow bars had not previously been possible. The modifications for the driver's hatch, battery heater, and the tow coupling were authorized to be implemented by the field units.

Beginning in December 1944 changed to the vehicles were made to further improve towing. The old towing brackets bolted to the front and rear were dropped. In their place holes were drilled in extensions to the hull side walls. This modification was not fully implemented for all Sturmgeschütz IV by the end of the War.

In addition to these changes, constant improvements were made to the superstructure of the Sturmgeschütz IV. These were already discussed in detail in the chapter concerning the Sturmgeschütz, Ausf.G. The same armor supplier completed superstructures for both the Sturmgeschütz IV and the Sturmgeschütz III.

The Production Firms

This chapter tries to more closely define the role of the assembly firms. Similar to automobile production of today, numerous sub-contractors and suppliers were included in the process, whose products poured into the assembly firms.

The Sturmgeschütz consisted primarily of two main components: the chassis with the automotive components and the superstructure with the weapon. This distinction was obvious in the fact that the chassis was produced, tested and officially accepted apart from the superstructure. It then returned to the factory where the superstructure was mounted. A large part of the work of the assembly firms lay in the preparation of the many components for final assembly. After the hull was delivered by the steel industry, there was multiple drilling for the installation of the torsion bar suspension, track drive and track adjustment among other things. Only then could the hull be placed on the assembly line for final production.

Hull side plates being cut into shape. (All of the following photos are from Alkett.)

Welding apparatus.

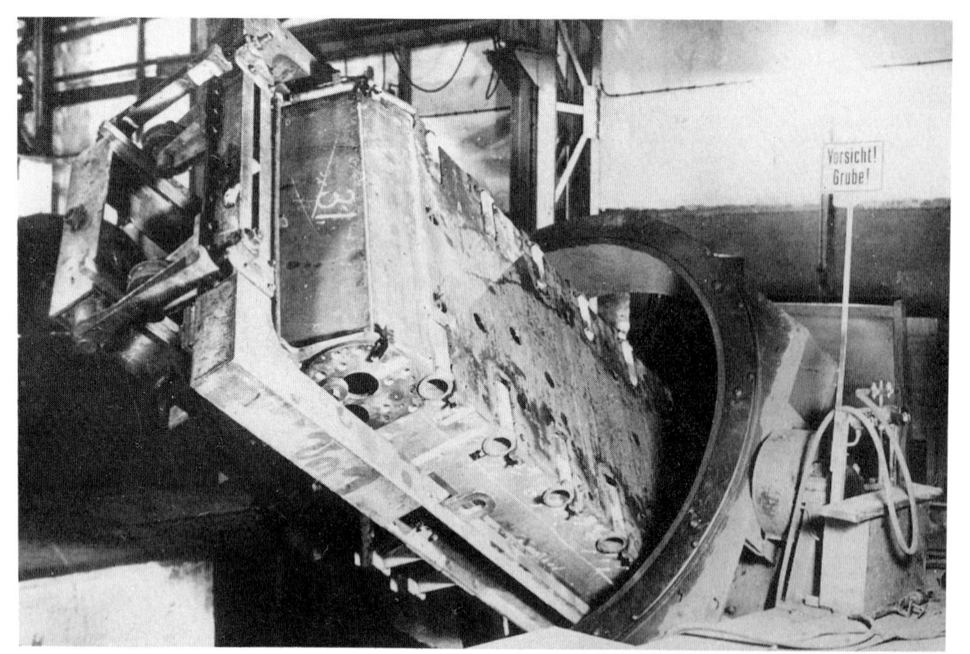

Welding apparatus.

Welding apparatus.

Twelve-spindle drilling machine.

Bolt hole boring machinery.

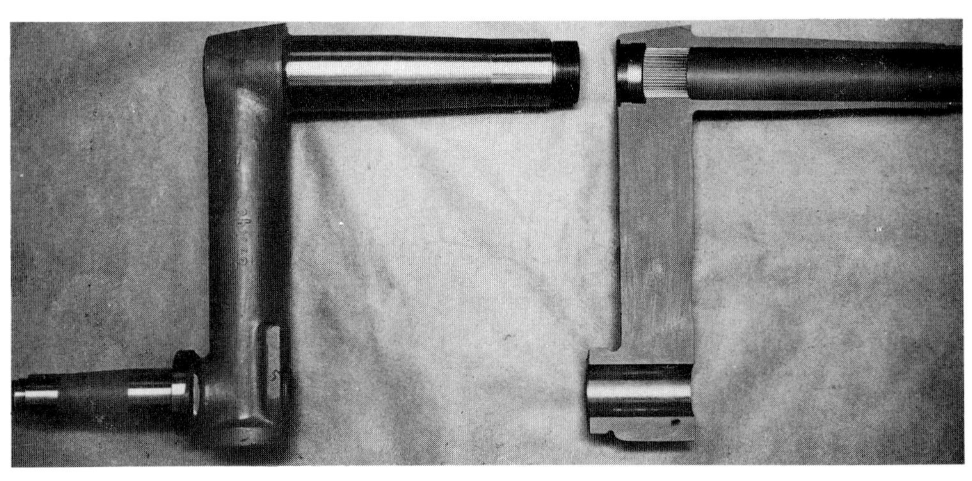

Suspension arms.

Boring the suspension arms.

Boring the suspension arms.

Machining the suspension arms.

Milling the suspension arm threads.

A Woerner multi-steel lathe.

Milling a pin.

A G.F. rigid lathe.

Milling a support axle.

De-burring bench.

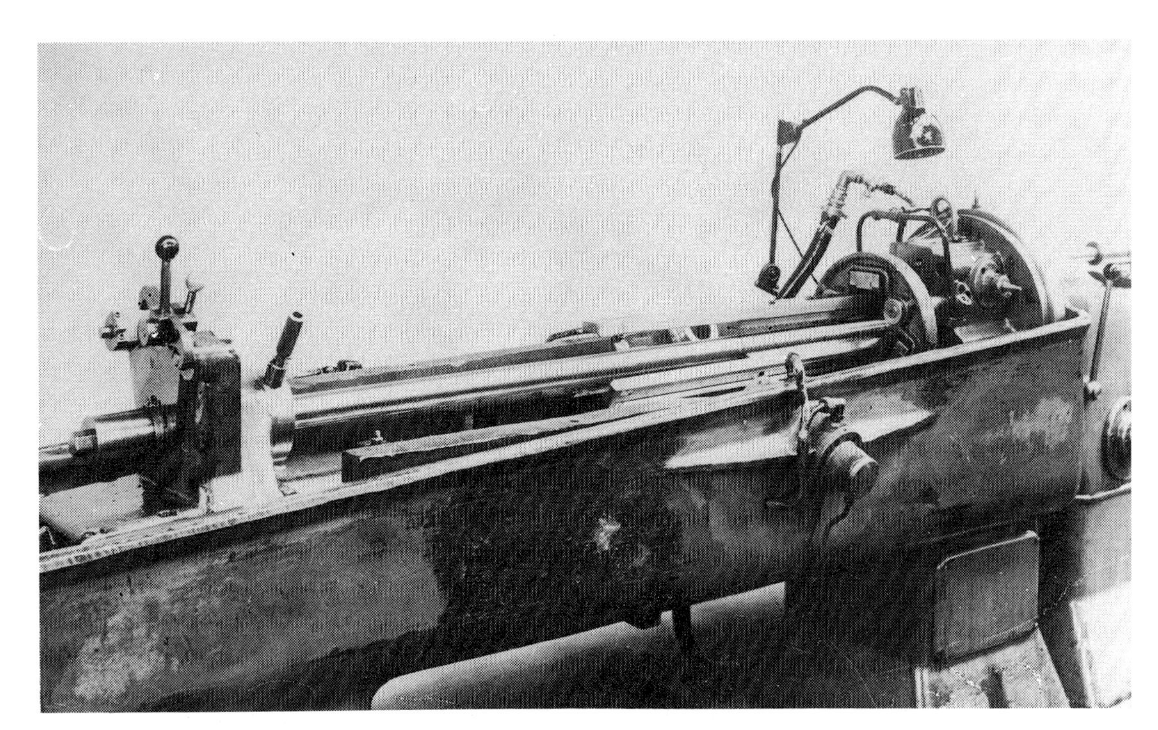

De-burring bench.

Torsion bar testing.

De-burring machine head.

Torsion bar shaft.

Hardening the drive sprocket.

Face-hardening drive sprockets.

Machining the drum brakes.

The assembly line at ALKETT for the production of the chassis was done in 18 individual steps in the following manner:

Step 1. Reinforced laminate collar installed and reamed, gaskets installed, swing arm guides bolted on, base for track adjustment mounted, and electrical wiring laid out in the engine compartment.
Step 2. Mounted bearing sleeves, installed swing arms, mounted idler wheel cranks, and bolted on the brackets for the shock absorbers.
Step 3. Installed swing arm limiters, attached track tensioner, bolted on return rollers, laid out various electrical wiring.
Step 4. Set torsion bar bolts, installed torsion bars, mounted idler wheels and continued electrical wiring.
Step 5. Attached shock absorbers. Prepared to mount road wheels and installed the regulator and solenoid.
Step 6. Mounted road wheels, installed final drives with drive wheels and sprockets.
Step 7. Mounted brake housings, placed the transmission tunnel and clutch housing.
Step 8. Completed installation of track brakes, installed fuse box and laid out electric cables.
Step 9. Installed exhaust fans, wall and supports for the fuel tank, and the base for the driver's seat; and fastened the support for the gas pedal.
Step 10. Installed steering gear and transmission. Drive shaft was installed as well as the fuel tank.
Step 11. Fuel pump, steering installed, instruments installed and connected, engine prepared.
Step 12. Engine installed and appropriate connections made.
Step 13. Radiators installed, fuel and vent linkage connected, and ammunition bin installed in the engine compartment.
Step 14. Engine connected to fuse box. Cooling fans installed, as is the battery box.
Step 15. Tubing to the radiators installed, floor covering fastened and air filter connected.
Step 16. Inertia starter mounted; lights, starter, and battery connected; and electrical system tested.
Step 17. Cooling air baffles mounted and exhaust system installed.
Step 18. Engine and transmission filled with oil, the radiator filled with water, chassis lubricated, test run conducted and track is connected and tensioned.

Location of production facilities.

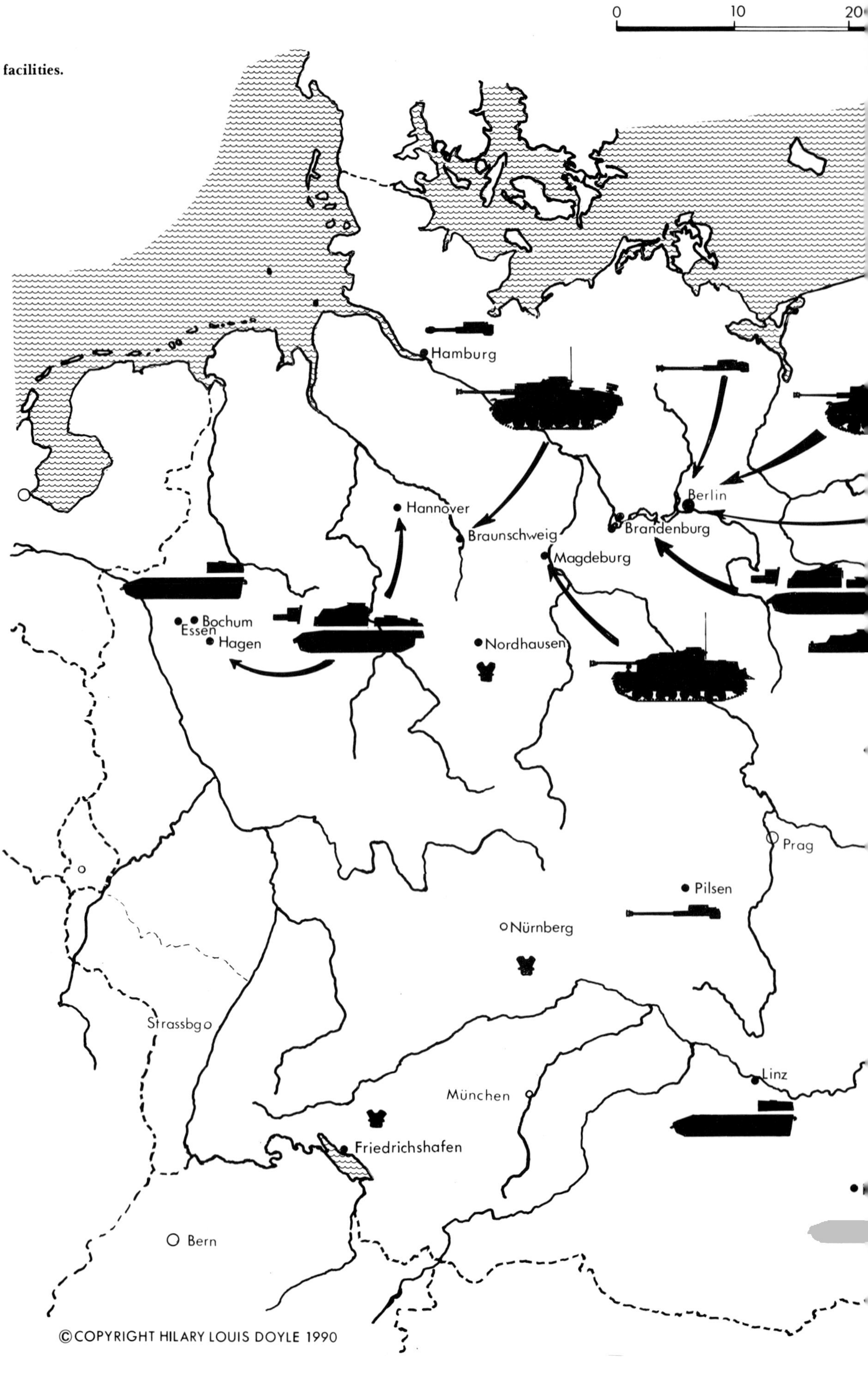

0
10
20
Hamburg
Berlin
Hannover
Braunschweig
Brandenburg
Magdeburg
Bochum
Essen
Hagen
Nordhausen
Prag
Pilsen
Nürnberg
Strassbg
München
Linz
Friedrichshafen
Bern
©COPYRIGHT HILARY LOUIS DOYLE 1990

The chassis was then driven a distance of 200km on a test course enroute to final inspection. Problems were corrected, and acceptance was done by a Heereswaffenamt Inspektor. Then the chassis was returned to an assembly plant.

The assembly of the superstructure was decidedly easier than that of the chassis. The assembly plant was required to make many drillings which were intended for mounting accessories. The superstructure was bolted to the chassis, the final assembly completed and the Sturmgeschütz was presented to the Heereswaffenamt Inspektor for acceptance. Then the Sturmgeschütz was transported by rail to the Heereszeugamt (ordnance depot). Here the vehicle was outfitted with tools, spare parts, ammunition, radios and light weapons. Because electrical impulses generated by the vehicle itself created interferences with the radio sets, the vehicle was fitted with a suppressor at the Heereszeugamt. After final testing at the Heereszeugamt, the Sturmgeschütz was issued and transported by rail to the receiving training or combat unit.

During the last year of the war, due to fuel shortages, the break-in period was limited to 50km. Beginning in April 1945, due to the utter collapse of the transportation system, the vehicles were no longer broken in or shipped to the Heereszeugamt offices —they were completed in the assembly plants and handed over to the crews who came to the assembly plant.

Assembly Plants

Daimler-Benz AG, Werk 40, Berlin-Marienfelde

The Motorfahrzeug- und Motorenfabrik AG in Marienfelde, Berlin emerged from an engine factory founded by Adolf Altmann (1850-1905). It existed under this name from 1898 to 1902 and from then until 1926 as Werk Marienfelde of the Daimler-Motoren-Gesellschaft (DMG), Cannstadt (which beginning in 1904 was known as Stuttgart-Untertükheim).

In the general structure of the firm, Cannstadt soon only produced personal automobiles, while in Berlin they concentrated on commercial vehicles. After the fusion of Daimler/Benz in 1926, Werk Marienfelde was threatened to be closed as an assembly plant and only utilized as a repair shop. Production commenced again in Berlin in 1934 with primarily military vehicles and aircraft motors being produced. Already in the 1920s the first prototypes of armored wheeled and tracked vehicles were completed

Daimler-Benz Werk 40 in Marienfelde. Group production of Panzerkampfwagen III chassis (June 1940).

in Marienfelde. Zugkraftwagen (half-tracked towing vehicles) were also developed and mass-produced. Finally in 1934 the design of the tank that later became known as the Panzerkampfwagen III was initiated. Its final form, development and production became the responsibility of the Marienfelde plant of the Daimler-Benz AG. Initially, production was accomplished in a so-called group assembly fashion, which means that the vehicle remained in one spot during the entire assembly; a production line was only planned later. Daimler-Benz was responsible for the development, testing and initial production of the Sturmgeschütz.

The contribution made by Daimler-Benz Werk 40 to the Sturmgeschütz production was 30 Sturmgeschütz, Ausf.A in 1940 and 1 Sturmgeschütz IV prototype in December 1943.

Production facilities were engaged from 1935 to 1938 with assembly of military trucks, Zugkraftwagen (half-tracked towing vehicles), the Panzerkampfwagen I Ausf.A and Ausf.B, and kleine Panzerbefehlswagen (small armored command vehicles).

From 1936 to 1939, Werk 40 assembled superstructures and turrets for the Panzerkampfwagen II, which were mounted on chassis delivered to them. Production of the first series of Panzerkampfwagen III's began in 1937. Werk 40 received an order for the production of a series of each model of the Panzerkampfwagen III. The last Panzerkampfwagen III, assembled at Werk 40, left Marienfelde in October 1942. Werk 40 was likewise the sole producer of the grosse Panzerbefehlswagen (on the Z.W.-chassis), beginning with the Ausf.D1 in 1938 and ending with the last Ausf.K in February 1943.

Werk 40 had to retool for the production of the Panther tank, whose delivery began in January 1943 and ended in April 1945, when the Red Army conquered Berlin.

Daimler-Benz Werk 40 was one of the main suppliers of special vehicles for the Heer as shown by the following production statistics:

Year	Zgkw.*	Pz.Kpfw.*	StuG*
1936	329	109	—
1937	282	129	—
1938	252	63	—
1939	157	90	—
1940	262	184	30
1941	329	387	—
1942	343	301	—
1943	204	545	1
1944	22	1175	—
1945	—	220	—

Zgkw. = Zugkraftwagen, Pz.Kpfw. = Panzerkampfwagen, StuG = Sturmgeschütz

Altmärkische Kettenwerk GmbH, Berlin-Tegel (Alkett)

As opposed to Daimler-Benz and other firms, which offered a large palette of products, the Alkett firm was almost completely occupied with arms orders. In 1940, at the start of the Sturmgeschütz production, 1500 employees worked in armor production. In the framework of orders from the Wehrmacht, the construction of armored fully-tracked vehicles took up the largest amount of room. In early 1936 the Oberkommando des Heeres demanded the establishment of a production site for armored vehicles by the Rheinmetall-Borsig AG firm. Among various projects, the choice went to the territory of a former boiler and pressure vessel factory in Berlin-Borsigwalde, Breitenbachstrasse.

To carry out the project, a GmbH (similar to an American corporation) was founded under the name of Altmärkisches Kettenwerk Berlin-Borsigwalde. Shareholders of the 50,000 Reichsmark capital were
— the Montan-Industriewerke GmbH with 40%
— the Rheinmetall-Borsig AG with 60%.

The existing factory halls were torn down, partly renovated and enlarged. Among other new buildings, a new boiler house was constructed.

Up to the end of 1938 the Panzerkampfwagen II was produced in slowly-increasing numbers. In the same year, preparations were begun for the production of the Panzerkampfwagen III. This vehicle was then built as a replacement for the Panzerkampfwagen II until October 1942. In 1940, assembly of the Sturmgeschütz (on the Panzer III chassis) was added. This vehicle was produced until the end of the war in 1945.

It was in this context that A. Wasmuth wrote on March 1, 1978 to *Oberst* (ret.) Icken: "The Alkett firm was licensed firm, meaning the Panzerkampfwagen II and III, including the Sturmgeschütz were produced under license according to someone else's designs. Only after the startup of the first series were three metal workers from the experienced qualified employees sent to Berlin along with the Meister Hahne. The first armored vehicles, the Panzerkampfwagen I Ausf.A were still assembled at Rheinmetall-Borsig in Berlin-Tegel.

Rheinmetall-Düsseldorf did not make a single designer available. These resources were needed for other programs, such as heavy mortars on self-propelled mounts, railroad guns, etc."

Complete (chassis and superstructure) were produced and assembled in Borsigwalde up to the Fall of 1943. Suppliers provided forgings, cast steel parts, armor casings and other important parts for installation, such as the engine and parts of the transmission among other things.

Production in the Alkett-Borsigwalde was interrupted from November 1943 to January 1944 due to extensive destruction caused by the air attack of November 26 1943. For this reason, the Reichsbahn-Ausbesserungswerk Falkensee near Spandau (Rheinmetall entered into the Demag-Fahrzeugwerke, Duisburg agreement) was included in the production at the end of 1943, as was the Montan-Industriewerke in the Fall of 1943. The Falkensee works only produced chassis for the Sturmgeschütz as a replacement for the Borsigwalde works, which were fitted with superstructures and completed in the Spandau works. Transmission and other sub-components were produced in Borsigwalde.

Transmission construction in Borsigwalde, set up for a maximum of 150 gear boxes per month, was rebuilt after the bombing for an output of 300 transmissions (150 gearbox and 150 steering gears) and equipped with the required machinery as well as tempering systems. The remaining gearbox and steering gears needed for maximum output were provided by other suppliers.

Alkett, Werk Borsigwalde (Werk I).

Alkett in March/April 1944.

The size of the individual plants and the number of their employees gives an idea of the size of these operations:

	building area (m²)	employees
Borsigwalde	ca.62,000	3,000
Spandau	ca.60,000	2,700
Falkensee	ca.40,000	5,500
	ca. 162,000	11,200

The Spandau plant had received two single-spindle boring machines simply in order to be able to fulfill special orders. In the Falkensee works there were four single-spindle boring machines of which two stood across from one another in order to be able to drill the steering mechanism and idler mount, on both the left and right side of the hull at the same time. Additionally, there were two 12-spindle boring machines on hand for the boring of the swing arm mounts.

Torsion bars were also processed in the Borsigwalde works after delivery of the blanks. Monthly output was between 800 and 1200 pieces.

Alkett, Werk Borsigwalde after its destruction by Allied bombers.

Alkett, Werk Falkensee.

Alkett, Werk Falkensee.

The suspension arm production line in the shop of the firm of Jachmann.

The torsion bar output always created a bottleneck. Suppliers were:
— Röchling, Wetzlar
— Hösch, Hohenlimburg
— Dittmann-Neuhaus, Herbede

A small factory in Berlin-Stralau was also set up.

After the bombing of the ball bearing manufacture in Schweinfurt, Panzer III chassis could still be produced for five weeks without requiring a single ball bearing from Schweinfurt. Armor manufacturers had so overstocked themselves with ball bearings (this was also true with tools, crowbars, hammers, and jacks) that continued manufacture was assured by the forced trading of individual bearing types. By means of moving it, ball bearing manufacture was so well under control within five weeks that they were temporarily able to do without the planned manufacture of the Panzer III and IV with roller bearings.

The suspension arm production line was not destroyed in the bombing attack on Borsigwalde because it was not located on the factory grounds. It was moved to the Jachmann firm in Berlin-Wittenau. The only production difficulties experienced there was a shortage of forging materials.

Four bombing raids were flown against Alkett, during which 3609 tons of bombs were dropped.

The Alkett firm assumed a special place in armor production compared to the other armor manufacturers. A test facility was established in 1941, as was a design office under the direction of Dipl-Ing. Michaels. Based on its long history of building armored vehicles as well as its first class skilled labor, the plant was constantly employed by the Reichsminister für Rüstung und Kriegsproduktion to carry out important developmental work and production of vehicles yet to be developed as well as already developed types. Prototypes were created for vehicles that were produced by other manufacturers. Alkett had a decisive influence on the creation of Panzerjäger and Artillerie self-propelled guns —also mounted on captured chassis — and the development of underwater and amphibious vehicles. Alkett was also preferred because besides the design office, it had a very flexible factory which time and again made it possible to translate new ideas into reality.

Legend for Amme-Werke of MIAG in Braunschweig

1. Offices, Door-keeper, Canteen, Sanitary
2. Workshops, Magazine Stores
3. Workshops, Tempering, Boiler House
4. Metal and Iron Foundry
5. Cast Steel Foundry, Track Casting
6. Track Manufacturing, Wash Rooms
7. Vehicle Assembly
8. Vehicle Assembly, Tube-Mill Production
9. Vehicle Assembly
10. Iron Cutting, Iron Stores

Mühlenbau und Industrie AG (MIAG)
Amme-Werke, Braunschweig

MIAG came into being in 1926 after the merging of the five most significant German milling concerns. Milling apparatus and other agricultural machinery were produced in Braunschweig, much of which went for export.

Armor production started in the Neupetrietor works in 1938 with the assembly of Panzerkampfwagen II's. Production of this vehicle continued until April 1939. The planned MIAG production of the Panzerkampfwagen III was delayed until the Fall of 1939 due to problems with the transmission. Beginning with two Panzerkampfwagen III in September 1939, monthly production was up to 15 units in June of 1940. The monthly output continued to climb steadily and reached its high point of 80 in December 1943.

The last Panzerkampfwagen III built by MIAG was accepted by the Heereswaffenamt in May 1943.

MIAG's Sturmgeschütz production started with ten vehicles in February of 1943 and they completed their Waffenamt contract in March 1945.

In 1943, MIAG was called upon to carry out the design of the Jagdpanther. The production series began in January 1944 and continued until the end of the war. In addition to the assembly of armored vehicles, MIAG produced steering apparatus for the Panzerkampfwagen III and the Jagdpanther and was the general manufacturer of tracks for the Panzer II and III chassis as well as the 3t, 8t and 12t Zugkraftwagen.

The Amme works in Braunschweig had an area of 180,00 square meters with 81,300 square meters of manufacturing area. The monthly production capacity at MIAG was rated at 150 Sturmgeschütz and 125 Jagdpanther units.

Diagram of the Amme-Werke of MIAG in Braunschweig.

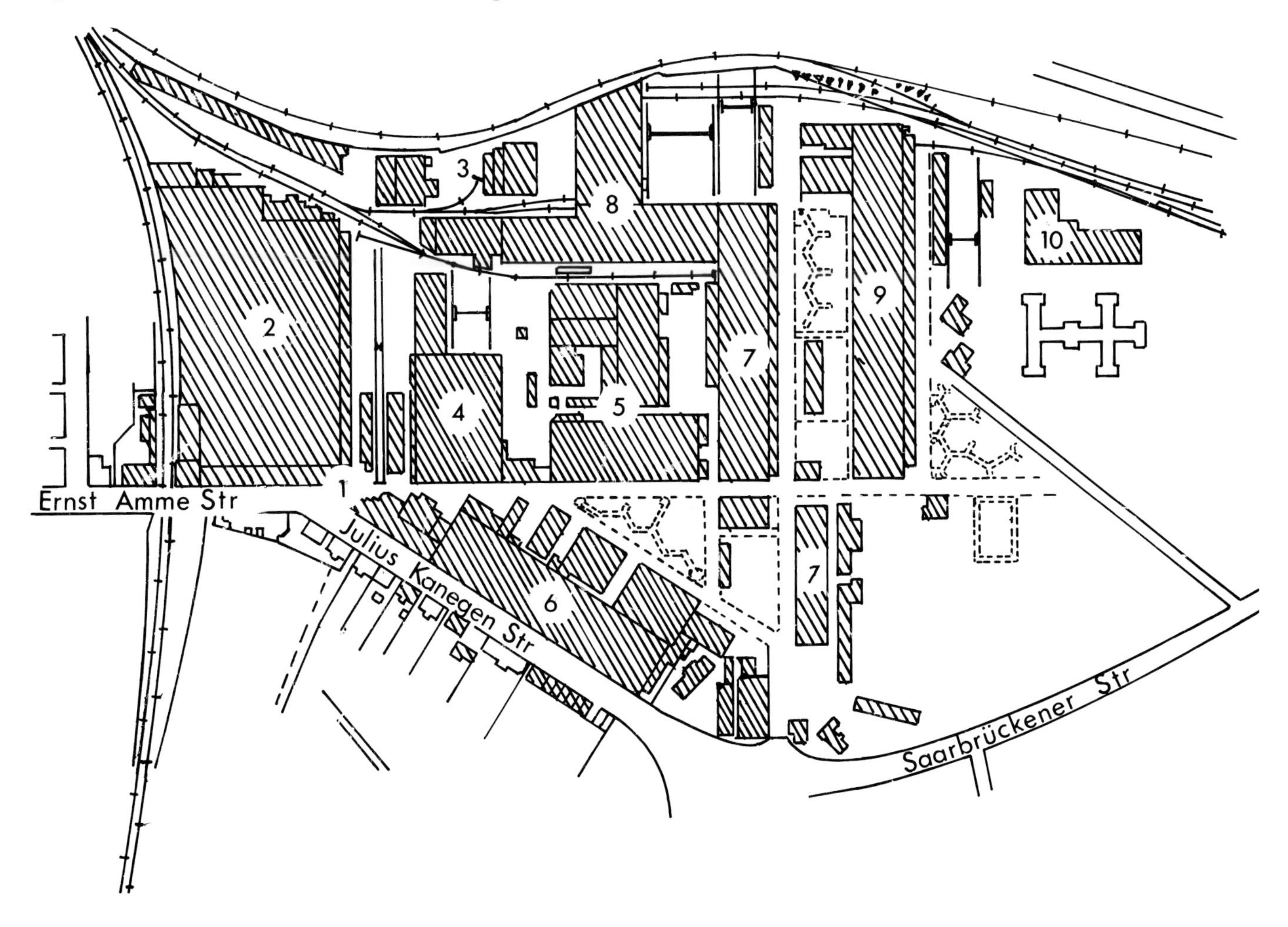

Average number of employees and the profit from Waffenamt T1 orders were given as follows:

year	production workers	office workers	earnings (RM)
1939	3745	1127	915,340.-
1940	5238	1419	8,935,200.-
1941	4973	1557	16,775,068.-
1942	5261	1420	33,572,275.-
1943	5791	1551	44,093,558.-
1944	5343	1465	60,178,545.-
1945	4478	1313	

A total of eight bombing raids were flown against MIAG which delivered a total of 4907 tons of bombs to their target. But only 283 bombs actually hit the target area of which 235 hit buildings. These hits destroyed approximately 60% of the built-up area, destroyed 49 machine tools and damaged 117 (of 701) machines for armored vehicle production. Intensive destruction in the residential areas of Braunschweig also had a notable impact on production.

At the Führerkonferenz of 1 and 2 September 1944, Hitler was briefed on the attacks on MIAG up to that point using the latest photographs and drawings, and was informed that despite the latest and heaviest damage done to the plant in the last month, the production of 100 Sturmgeschütz and 25 Jagdpanther was achievable. Hitler asked that his thanks be expressed to those workers who had achieved this deed, and at the same time that the plant foreman (Dr. Blaicher) be recognized by Hitler in his capacity as Chairman of the Panzer Committee for this determination in performance, as well as the performance of all the other Panzer plants which under similar circumstances had performed exceptionally.

Fried.Krupp-Grusonwerk AG, Magdeburg-Buckau

The Magdeburg plant of Krupp-Grusonwerk AG was founded in 1855 by Hermann Gruson and taken over in 1893 by Fried.Krupp AG in Essen. The firm developed world renown with its constant and successful broadening of its production program.

Before the Second World War, the firm was a world leader in the production of heavy machinery and in the area of structural steel fabrication.

Of the approximately 300,000 square meters available, only ten buildings with 93,000 square meters were devoted to armored vehicle production, which took up an area of 35,000 square meters. Up until 1938, no additions were made to the structures for armor production, and only then were the press-shop and a hall for the assembly of the Panzerkampfwagen IV added.

Assembly of Panzerkampfwagen began in the Krupp-Gruson plant in 1934 with the production of three type La.S. chassis, (code name for the tank later known as the Panzerkampfwagen I). Assembly of complete Panzerkampfwagen I's and chassis for other purposes began in 1934 and continued to 1937. The first two Panzerkampfwagen IV's were completed in November 1937, and Krupp-Grusonwerk remained the sole manufacturer of the Panzerkampfwagen IV until 1941 when Vomag and Nibelungenwerk were included in the assembly process. Production of the Panzerkampfwagen IV at Krupp-Gruson was converted to Sturmgeschütz IV production in December 1943 and continued that way until war's end.

In the years 1944 and 1945 Krupp-Gruson also manufactured Panzer IV chassis for the Flakpanzer IV, the Möbelwagen and Ostwind. In September 1944, Krupp-Gruson received the order to convert from production of the Sturmgeschütz IV to the Panzer IV (lang) E, with production of this "Jagdpanzer" to commence in January 1945. Only one prototype of the Panzer IV (lang) E was produced as the decision was made not to mass produce this Jagdpanzer. In November 1944 Krupp-Gruson was also called upon to convert their production to Panther Ausf.F tanks commencing in April 1945. Due to the pressing events in the war, this idea fell to the wayside in March 1945.

Production of the Sturmgeschütz IV, the chassis for the Flakpanzer and other tank components represented about 28% of the entire production by tonnage (about 47% of the sales) in 1944 and about 27% of the plants output in 1945.

Artillery pieces and shells made up 27% of sales and general machine production made up the remaining 26%. The highest rate of production was reached in July 1944 shortly before the bombing attack on August 5th. At that time Krupp-Gruson employed 1388 in tank production and utilized 460 machine tools for this purpose.

There were an additional 182 workers and 63 production machines for the manufacture of spare parts for tanks.

Krupp-Grusonwerk manufactured and processed numerous tanks components. The most important of these were the steering mechanisms, final drives and machining hulls and superstructures. About 75% of these components were completely manufactured at Krupp-Gruson, the rest came in semi-finished form from Vomag and Nibelungenwerke.

Despite the heavy bombing attacks in August, September and the beginning of October 1944, the plant recovered quickly in the months of October and November. This was possible in spite of heavy damage in the machine shops, because the plant was now in the position of being able to fall back on pre-built and warehoused components. Additional help was provided

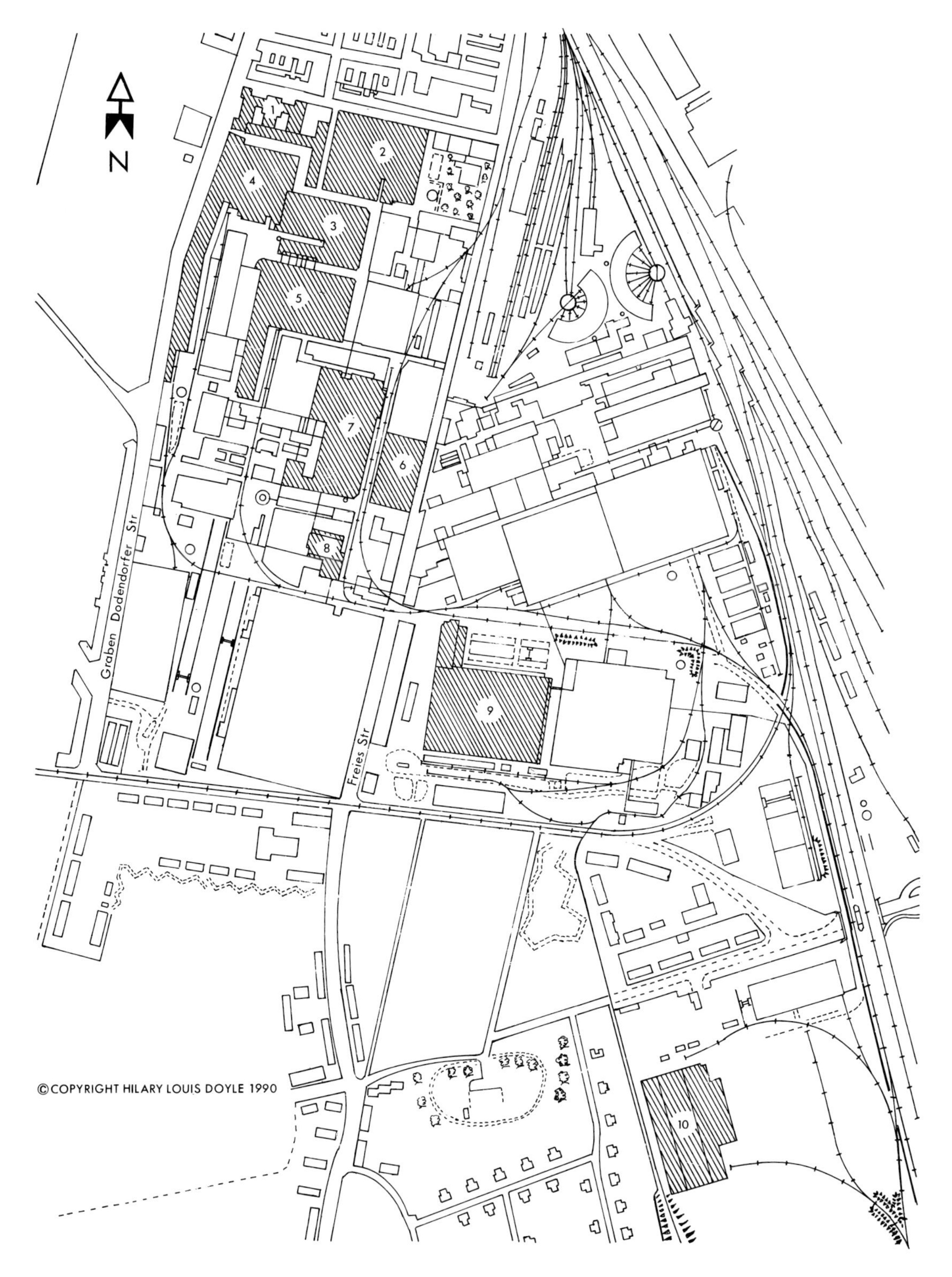

Diagram of the Krupp-Grusonwerke AG in Magdeburg.

Legend:
1. Administration Building
2. Machine Shop
3. Machine Shop
4. Tank Parts Storage
5. Forge Shop
6. Screws, Bolts and Nuts Shop
7. Steel Foundry
8. Central Power Plant
9. Iron Foundry
10. Tank Assembly Shop

Krupp-Grusonwerke in Magdeburg. Production line for Sturmgeschütz IV.

by the still undamaged Vomag plant in Plauen in the Nibelungenwerke plant in St. Valentin.

In addition, Krupp-Gruson had dispersed some of its tank parts production activities and was in position to fall back on them by October 1944.

Beginning in December 1944 the first signs of collapse appeared within the organization. Important parts at Vomag and Nibelungenwerk were used up, rail transport was brought to a standstill, gas and electric power was no longer guaranteed. In the first quarter of 1945 only 33% of the production capacity could be utilized.

In 1944 spare parts production had reached 4913 tons, independent of what Krupp-Gruson needed for their own production. These spare parts were shipped on to the Heereszeugamt in Magdeburg-Königsborn as finished products and stockpiled there for later dispersal to the troops. The bulk of those 4913 tons was produced in the second half on 1943.

The Krupp-Gruson plant's order situation at the end of 1944 is documented by the following:

Abschrift

Hauptausschuss Panzerwagen Berlin NWz7,
Der Geschäftsführer 3.11. 1944
FS-Nr. 01 1514
Geheime Reichssache Br. B. Nr. 216/44.g RSz
18 Ausfertigungen
8. Ausfertigung

Herrn Dir. Mehner o. V.
Krupp-Gruson AG
Magdeburg-Buckau

SUBJ: Armor Production

The following is for your plant's future production plans from November 1944 to December 1945:

type		Nov.	Dec.	Jan.	Feb.	Mar.	Apr.	May
StuG IV	100	130	130	130	130	100	50	
Fgst.IV	(30)	(30)	(30)	(30)	(30)	(30)	(30)	
Panther*	—		—	—	—	—	5	20

type		Jun.	Jul.	Aug.	Sep.	Oct.	Nov.	Dec.
StuG IV	—		—	—	—	—	—	—
Fgst.IV	—		—	—	—	—	—	—
Panther	40	60	70	70	70	70	70	

Signed, Kutscha

(*FOOTNOTE: Panther production never commenced at the Krupp-Gruson plant)

A copy of a document from March 2nd 1945 gives an idea of what contracts were in hand for the Krupp-Gruson plant toward the end of the war.

Abschrift

Fried.Krupp-Grusonwerk Magdeburg-Buckau 2.3.1945
Aktiengesellschaft
Magdeburg-Buckau
Hauptausschuss Panzerwagen Sonderausschuss PZ IIIm

Herrn Dir. Dr. Ing. Jansen
i. Fa. Miag, Braunschweig

SUBJ: Wehrmacht Contract Numbers

The requested information is as follows:
For 1945 we have under contract:
644 Sturmgeschütz on Pz.IV chassis
order SS 4902-0210-8917/43
120 Flapa-Fahrgestelle
order SS 4902-0210-8914/43 500 Panther
order SS 4900-0210-5940/44
monthly production of about 80 vehicles.

The gearboxes here, according to 021 St 48440/441, 021 St. 48436, 021 St 51156/157 and 021 St 51167, all have the priority of SS 4900 (priority group "Zi").

For replacement requirements, the following order outline is given for 1945:

Zu. 1. a)Pz.Kpfwg.IV and variants order
SS 4902-0212-48015/1
b)StuG IV order SS 4902-0212-4806/45/1
Zu. 2. a)Spare parts for the chassis and superstructure
b)Spare parts for the superstructure (various)
Zu. 3. a)Spare parts according to special supply list
and 4

The planetary gears as replacement parts have a priority of 4900.

FRIED.KRUPP-GRUSONWERK AG
(signed)

Further, the plant was pressed into further raising production rates; in the first three and one half months of 1945, 2400 tons of spare parts were made available to the troop units.

Suppliers of Main Components

While the assembly firms were directly tasked with carrying out the final assembly of steering gears, final drives and rubber-tired road wheels, many of the main components, such as the hull, superstructure, the engine, gearbox, weapons, and gun sights were manufactured by other firms and furnished to the assembly plants after acceptance by the Heereswaffenamt. These subcontractors played a predominant role in their responsibility to do justice to the equipment demands of the troops. Increased production rates of the assembly plants were only possible when the subcontractors were notified on time to increase their own production rates. These orders for the subcontractors needed to be awarded at least six-months in advance of the need for the component at the assembly plant in order to guarantee that increased production of the final product could be achieved.

Armor Industry

The production process at the firms engaged in (Panzerfertigung) armor component manufacturing was essentially separated into three main areas:
— mechanical processing, including gas welding
— annealing installation and tempering
— assembly, welding, quality control and shipping.

The welded hulls and superstructures for the Sturmgeschütz were delivered exclusively by the Brandenburger Eisenwerke GmbH of Brandenburg/Havel from 1939 to 1942. The Kirchmöser/Havel plant of the same firm was responsible for production. Production numbers were:

1939-41 836 hulls, 797 superstructures
1942 1049 hulls, 942 superstructures
1943 1910 hulls, 1907 superstructures
1944 approx. 2600 hulls and 2700 superstructures.

In January 1943 the Deutsche Edelstahlwerke AG in Hannover-Linden (DEW) joined this Sturmgeschütz construction program and delivered the following components for the firm:
1943 579 hulls, 630 superstructures
1944 728 hulls, 738 superstructures
1945 40 hulls, 40 superstructures.

Altogether, beginning in 1943, 1347 hulls and 1408 superstructures were delivered by the Deutsche Edelstahlwerke to the Sturmgeschütz III assembly plants.

The Harkort-Eicken Stahlwerke AG in Hagen/Westfalen, with plants in Wetter/Ruhr delivered the following approximate numbers for the Sturmgeschütz III:
1943 900 hulls, 850 superstructures
1944 1300 hulls, 1400 superstructures.

In mid-1944 the Königs and Bismarckhütte AG, Königshütte-Bismarkhütte O.-S. was added to the process. It manufactured about 200 superstructures for the Sturmgeschütz III between June 1944 and February 1945.

The Brandenburger Eisenwerke GmbH, Brandenburg/Havel, Kirchmöser plant manufactured all the superstructures for the Sturmgeschütz IV: 42 in 1943 and about 1400 in 1944.

The hulls for the Sturmgeschütz IV were produced by:
— Gebr. Böhler Co. AG in Kapfenberg-Deuchendorf
— Eisenwerke Oberdonau GmbH in Linz
— Fried.Krupp AG in Essen, and
— Eisen- und Hüttenwerk AG. (EHW) in Bochum.

Weapons Manufacturers

Only 14 of the 7.5cm Sturmkanone L/24 developed by Krupp/Essen were actually produced by Krupp. The remaining 800 Sturmkanone which were installed in the Sturmgeschütz, Ausf.A through Ausf.E models, were produced by the Wittenauer Machinenfabrik GmbH (Wimag).

The 7.5cm Sturmkanone 40 developed by Rheinmetall-Borsig was predominantly built by Wittenauer Maschinenfabrik GmbH (Wimag), Berlin-Borsigwalde, (about 60%) and Skoda, in Pilsen (about 40%). Price per gun was 13,500RM.

The 10.5cm Sturmhaubitze 42 developed by Rheinmetall-Borsig was built exclusively by the Manck & Hambrock GmbH firm in Hamburg-Altona.

Optical Industry

The periscopic gun sights were developed and produced by Carl Zeiss in Jena. The Sturmgeschütz, Ausf.A and Ausf.B utilized the Sfl.ZF scope produced between 1939 and 1941. Model Sfl.ZF1 produced from February 1941 to 1944 was installed into the Sturmgeschütz, Ausf.C through Ausf.E. The final model Sfl.ZF1a (for the Sturmgeschütz, Ausf.F through Ausf.G of the Sturmgeschütz) went into production in March of 1942. From October 1944 to March 1945 a total of 2250 Sfl.ZF gun sights were completed by Zeiss.

Manufacturers under license included in the production program were the Optische und Feinmechanishe Werke in Görlitz as well as the Ernst Leitz GmbH in Wetzlar. Leitz manufactured about 50% of these gun sights in the following quantities:

1941	504	Sfl. ZF 1
1942	1007	Sfl. ZF 1a
1943	4087	Sfl. ZF 1a
1944	6579	Sfl. ZF 1a
1945	1030	Sfl. ZF 1a

Transmission Manufacture

The 10-Gang-Variorex-Getriebe, Typ 328145, 10 speed transmission was installed in the Sturmgeschütz, Ausf.A. Although developed by the Maybach firm, this gearbox was in manufactured under license by Zahnradfabrik Friedrichshafen (ZF) and Alkett.

The transmission, which was used in the Sturmgeschütz, Ausf.B through Ausf.G, was the SSG 77, synchro-mesh 6-speed-transmission. This was a Zahradfabrik Friedrichshafen design which was produced in Werk I in Friedrichshafen, in Werk II in Berlin-Wittenau and in Werk III in Schwäbisch-Gmund.

Year	Werk I	Werk II	Werk III
1940	793	0	118
1941	1214	219	1393
1942	1364	585	1799
1943	1770	1456	700
1944	808	2625	0
1945	0	1076	0

Two bombing attacks on the 24th and 27/28th of April 1944 brought the entire output of finished transmissions in Werk I to a standstill. This did not cause a falter in the Sturmgeschütz production however, because the Alkett plant in Berlin-Borsigwalde had already taken up the production of the SSG 77 gearbox in 1943 and was manufacturing up to 100 such items per month. In 1944, the production rate increased to 200 to 300 per month.

The Arbeitsgemeinschaft Heilbronn (C. F. Weipert) was called upon to start making up for production shortcomings of the ZF Werk I in Friedrichshafen. Production started in September 1944 with a one-time delivery of about 120 SSG 76 and SSG 77 until it stopped in 1945 due to complete loss through bomb inflicted damage. The transmissions for the Sturmgeschütz IV and its forerunner the Panzerkampfwagen IV was a development of the Zahnradfabrik Friedrichshafen as well and was manufactured in the Werk I plant in Friedrichshafen and Werk III plant in Schwäbisch-Gmünd:

year	Werk I	Werk II
up to 1939	405	0
1940	297	0
1941	714	0
1942	1259	0
1943	1778	1561
1944	963	3645
1945	0	684

As was already emphasized, the bombing attacks on the ZF-Werk plant in Friedrichshafen completely interrupted its output. The Carl Hamel AG in Chemnitz assumed the production in 1944 and started with its first delivery in about September of 1944. Monthly production rates were about 200 SSG 76 transmissions by the end of 1944. A monthly output rate of about 300 transmissions can be assumed for the year 1945.

Engine manufacture

The one engine which almost without exception saw service with the Sturmgeschütz III and IV was the Maybach-developed HL 120TRM. Maybach/Friedrichshafen started with this production in 1938.

Unfortunately, almost all the Maybach files have been lost. The following annual production figures were estimated by plant employees after the war:

year	HL 120
1938	400
1939	900
1940	1400
1941	2600
1942	3500
1943	3600
1944	2229
1945	118

After the bombing attack on Friedrichshafen on April 25, 1944, 80% of the structures and 30% of the machinery was destroyed. There was no engine production in the months of May and June 1944. The production restart came about very slowly, and in the July to October 1944 time frame, a maximum output of about 400 HL 120 TRM engines was attained. In 1944/45 Maybach only delivered 107 of the HL 120 engines to Alkett, 45 to MIAG and 102 units to Krupp-Gruson.

In 1938 manufacture of the Maybach HLz120 TRM engine was conducted under contract with the Nordbau (Norddeutsche Motorenbau GmbH) in Berlin-Niederschöneweide. The fiscal year of this firm ran from April 1st to the 31st of March of the following year. The delivery figures, therefore, are given for this time period:

Fiscal Year	HL 120
1938/1939	160
1939/1940	370
1940/1941	544
1941/1942	1584
1942/1943	2743
1943/1944	6701
1944/1945	9474
1 to 20 Apr 1945	175

In the war, there was a concentration of the production capacity for tank engines in the Friedrichshafen area by the Maybach firm. On account of the Luftwaffe, a concentration of the production facilities within a 100 kilometer radius of Ludwigshafen evolved. Of the engine models HLz108 and HL 120, which came to be utilized in the Panzerkampfwagen III and IV together with all their variants, about 40,000 were produced from 1938 to the end of the war. Nordbau reached the highest output of HL 120 engines immediately after Maybach in Friedrichshafen was bombed out. Nordbau manufactured 1025 such items in June 1944 and a record 1100 HL 120 engines in July 1944. The first expansion of the HL 120 production was licensed under contract with the Maschinenfabrik Augsburg-Nürnberg (M.A.N.). The first completed engines came out of Nürnberg in January 1944. M.A.N. manufactured about 2284 HL 120 engines in 1944 and a further 131 in the first quarter of 1945.

Production of the HL 120 engines was further extended when reproduction licenses were granted to the MBA-Maschinenbau und Bahnbedarf firm in Nordhausen (formerly Orenstein & Koppel). Installation of the machinery was accomplished in October 1943, the production began in March 1944. A total of 955 HL 120 engines were produced by MBA between March 1944 and March 1945. The engines did not only go to the Heeres-Zeugämter for replacement stockpiles, but also directly to Krupp-Gruson, Vomag and Nibelungenwerke for installation in the armored vehicles manufactured there.

Other Suppliers

Material delivered	Supplier
Batteries	Accumulatorenfabrik Hoppecke
	Carl Boelker & Sohn Hoppecke
Fuel tanks	Gothar Metallwarenfabrik, Gotha
Torsion bars	Dittmann-Neuhaus, Herbede/Westfallen
	Gebr. Röchling, Wetzlar
Electrical components	Robert Bosch, Berlin
Fire extinguishing systems	Minimax, Berlin
Driveshafts	Rheinmetall, Sömmerda
Rubber tires	Continental, Corbach
Rubber gaskets	Hermann Wendt, Berlin
	Schoeps & Co., Mannheim
Cables and wiring	Valentin Klein, Freden/Leine
Track adjustors	Rudolf Sack, Leipzig
Fuel strainers/filters	Knecht, Stuttgart
Radiators	Hans Windhoff, Berlin-Schöneberg
	Längerer u. Reich, Stuttgart
Clutches	Stromag, Schlottmann & Co., Unna
Steering brakes	Süddeutsche Arguswerke, Karlsruhe
Air compressors	"Apag" Apollower, Gössnitz
Fans and fan wheels	Vereinigte Turbinenwerke, Meissen in Sachsen
Ammunition storage	Ankerwerke, Bielefeld
Smoke grenade launchers	Rohrleitungsbau, Unna
Pumps and fan drives	Fross-Büssing, Wien
Wheel rims and plates	Kronprinz, Solingen-Ohligs
Seeger-rings	Seeger & Co., Frankfurt/Main
Oil seal rings	Carl Freundenberg, Weinheim
	Goetze-Werke, Burscheid
Tachometers, oil pressure gauges	VDO-Tachometer, Frankfurt/Main
Telekin, hoses and tubing	Telekin, Baden-Baden

Sturmgeschütz Production

Even before the first contract for 30 Sturmgeschütz's of the first series had been incorporated into production planning, contracts for a second series consisting of 250 Sturmgeschütz had been awarded. Finally, in June 1939, the Heereswaffenamt worked out a production schedule for Daimler-Benz to deliver the first vehicle in December 1939. Daimler-Benz succeeded in producing the first chassis by this time. However, delays at the Krupp-Essen facility resulted in the first 7.5cm Sturmkanone L/24 not being delivered until January 1940. This one month delay between the original plan and the actual delivery continued to plague all 30 Sturmgeschütz assembled by Daimler-Benz and thereby delayed the timetable for the deployment of the first Sturmgeschütz units in the campaign in the West starting May 10, 1940.

Daimler-Benz Werk 40 in Berlin-Marienfelde was so overloaded at that time with contracts for the production of the Panzerkampfwagen III, the grosse Panzerbefehlswagen, the 12-ton Zugkraftwagen DB 10 (half-tracked towing vehicle) and the 6x4 cross-country truck LG 3000, that only 10% of the available capacity remained for the production of the Sturmgeschütz. For this reason, Sturmgeschütz production was transferred to the more spacious Borsigwalde plant of the Alkett firm. Their original production schedule, established in October of 1939, specified the month of April 1940 as a start date for the production of the second series of 250 Sturmgeschütz vehicles, with a monthly quota of 20 vehicles.

This plan was delayed, however, as Alkett was called upon to re-build 132 Panzerkampfwagen I tanks into self-propelled carriages for the 4.7cm Pak(t) anti-tank guns in the months of March, April and May 1940. Also contracted to mass-produce the Panzerkampfwagen III, Alkett completed their first 12 Sturmgeschütz in June of 1940. Production remained behind the planned schedule for the next three months, because of the need to fill higher-priority Heereswaffenamt contracts for the conversion of 133 Panzerkampfwagen III's and 48 Panzerkampfwagen IV's into submersible tanks for Operation "Sealion" (the invasion of England). At the same time demands were made to increase the production output of the Panzerkampfwagen III.

During the year 1941 production goals for the Sturmgeschütz were constantly increased from 36 in January to 50 in November 1940. Because a considerable number of HL 120 engines failed due to sucking in dust during the rapid advances in the Russian campaign. During June and July 1941, the Heereswaffenamt had diverted two months worth of engine production as emergency replacements to restore the combat effectiveness of the frontline units.

This potential engine shortage had its effect on the production goals. The goal for September 1941 was reduced to only 17 Sturmgeschütz. Alkett was nevertheless in a position in September 1941 to complete 38 Sturmgeschütz, convert production from the Ausf.D to the Ausf.E, and in addition achieved a record by completing 71 Sturmgeschütz in October 1941.

On December 4 1941 the following message was sent to *Generalfeldmarschall* Keitel concerning Sturmgeschütz production:

"In your memo of 1 Dec 1941, under point number 3, attention was drawn to the fact that Sturmgeschütz production should not be stifled."

"I draw attention to the fact that the production goal for 50 Sturmgeschütz, which up to now has been in accordance with schedules, has been reduced and the Panzerkampfwagen III increased. This was done with the justification that Sturmgeschütz are no longer to be used in Armor units, only in Infantry Division

"I ask you check into this matter so that the Führer does not one day complain about the shortage of Sturmgeschütz."

Enclosed with the message to *Generalfeldmarschall* Keitel was the following:
New production schedule for Sturmwagen (this name was again used instead of the official title of Sturmgeschütz) in conjunction with the increased production plan for the Panzerkampfwagen III at Alkett (Sturmgeschütz are only being produced at Alkett). Previous monthly goal: 50 Sturmgeschütz:

	Panzerkampfwagen III	Sturmgeschütz
December 1941	25	40
January 1942	28	40
February 1942	33	35
March 1942	37	30
April 1942	40	25
May 1942	42	25
June 1942	45	25

On the 7th of December 1941, the following message was sent to the Reichsminister for Armament and Munitions, Herr Dr. Todt:

Subject: Production of Sturmgeschütz
Reference: Your message of 4 December
Maintaining the monthly production rate of 50 Sturmgeschütz is of particular interest to the Führer. He has emphasized this point. The Führer considers it essential to equip Infantry Divisions with Sturmgeschütz to bolster the attacking forces in the East for the continuation of the campaign 1942. Accordingly, the OKH must determine the number of Sturmgeschütz that will be needed for the Ostheer (Eastern Army) to support an offensive, and furnish the estimated requirement figures to the OKW for presentation to the Führer. I do not consider a reduction possible.
Without regard to the above exchange of correspondence, the suggestion for reducing Sturmgeschütz production was never carried out, nor did the above communications have any effect on the Alkett production. In March 1942, the Alkett production practically came to a halt, due to the converting the Sturmgeschütz production from the 7.5cm Kanone L/24 to the 7.5cm Sturmkanone 40 L/43. In March 1942 only three Sturmgeschütz were produced, all with the 7.5cm Sturmkanone 40 L/43, the remaining assembled chassis were held back until additional longer guns became available for installation. Although this delayed final delivery, the sum of the production goals for the entire period of March to June was met. This resulted in the very effective Sturmgeschütz, Ausf.F becoming available for the Summer Offensive on the Eastern Front.

On May 13th 1942, Hitler concurred that 100 Sturmgeschütz should be produced per month at the expense of the Panzerkampfwagen III, whose output could fall to about 190.

This goal could not be reached immediately. It was specified as a monthly production goal that was to be achieved by December 1942. The monthly output at Alkett increased steadily, aided by phasing out the Panzerkampfwagen III production run at Alkett. The last ten Panzerkampfwagen III from Alkett were completed in October 1942.

One month ahead of schedule, the monthly production goal of 100 Sturmgeschütz was reached in November. Alkett was the first German manufacturer of medium-sized armored vehicles to achieve this monthly rate.

In Hitler's conference on 7 and 8 November, it was determined that the Panzerkampfwagen III with the 5cm Kw.K. L/60 could no longer meet the requirements for future combat. The question was to be examined, if expedited phasing out Panzerkampfwagen III production, would result in a corresponding increase in Sturmgeschütz production.

During the 1-3 December 1942 conference, after being presented with options, Hitler decided to accelerate the program for converting the Panzerkampfwagen III to Sturmgeschütz production in order to exceed the current goal of 120 vehicles in December. The long term monthly goal for June 1943 was increased to 220 vehicles.

During the 3-5 January 1943 conference, Hitler was informed that, based on an order, instead of Panzerkampfwagen III's with the 5cm Kw.K. L/60, a schedule had been established to complete 535 Panzerkampfwagen III between 1 January to 12 May as follows:
— 235 supplemental Sturmgeschütz with 7.5cm StuK L/48,
— 100 flamethrower tanks,
— 56 Panzerkampfwagen III with 5cm Kw.K. L/60 for Turkey, and the remaining
— 144 Panzerkampfwagen III with 7.5cm Kw.K. L/24.

The information recorded in these conferences were summaries of reports to Hitler relating the progress of programs that had already been set in motion as early as April 1942.

The MIAG Amme-Werk in Braunschweig had previously been ordered to phase out the Panzerkampfwagen III production and commence production of the Sturmgeschütz Ausf.G (using the chassis serial number series starting with 95001). The 235 "supplemental" Sturmgeschütz's (reported in the 3-5 January 1943 conference) were not conversions on the Panzer III chassis. These 235 simply reflected the schedule that MIAG was to comply with as they commenced Sturmgeschütz production (10 in February, 50 in March, 80 in April and 95 in May 1943 equaling a total of 235).

Additionally, on September 1, 1942, the Maschinenfabrik Augsburg-Nürnberg AG (M.A.N.) submitted their end-of-production-run plan for their Z.W. production run. M.A.N. was to deliver an additional 143 Panzerkampfwagen III's complete with superstructures and turrets, followed by 142 Panzer III (8./Z.W.) chassis. These 142 Ausf.M chassis (in the serial number ranges of 76111-76210 and 77351-77450) were apparently sent to MIAG or Alkett for the final assembly stage, mounting the Sturmgeschütz superstructures. In contrast to the Sturmgeschütz, Ausf.G in production at Alkett and MIAG, these 142 chassis were still produced by M.A.N. in accordance with the specifications for the Panzerkampfwagen III, Ausf.M. These Ausf.M chassis were completed by M.A.N. in 1943 as follows: 3 in January, 23 in February, 21 in March, 20 in April, 9 in May, 10 in June, 10 in July, 11 in August, 19 in September, and the last 6 chassis in October 1943.

During the Hitler conference of 6 March 1943, among other items was mentioned, that the previous order to convert a certain number of Panzer III's to Sturmgeschütz would in effect be lifted. Instead of this conversion, they were to be re-armed with the 7.5cm Kw.K. L/24. This order did not effect the Sturmgeschütz on the Panzer III chassis ordered from MIAG, or the chassis which M.A.N. was delivering for conversion to Sturmgeschütz. Instead, it clarified that the severely damaged Panzerkampfwagen III returned for major overhaul, after being rebuilt, were to be equipped with 7.5cm Kw.K. L/24, instead of being used for conversions to Sturmgeschütz.

By including MIAG, expanding the production facilities at Alkett, and obtaining additional chassis from M.A.N, Sturmgeschütz production grew from 130 vehicles in January to 395 in October. This production rate would have continued to grow to 500 per month (including the Sturmhaubitze) and even further, had the production not been interrupted by bombing attacks. The attack, in November 1943 on the Alkett plant in Borsigwalde, virtually eliminated the facility as an assembly plant. The immediate solution was to move chassis production to the available spaces in the Reichsbahnausbesserungswerk (rail improvement facility) in Berlin-Falkensee. The Demag firm had plans to assemble Panther tanks in this facility. Production of the Sturmgeschütz superstructures and final assembly were relocated to the Alkett plant in Berlin-Spandau.

As a further step, Krupp-Grusonwerk in Magdeburg was ordered to halt production of the Panzerkampfwagen IV and immediately commence assembly of the Sturmgeschütz IV in December 1943. An adapted Sturmgeschütz superstructure was fitted to the unaltered Panzer IV chassis. In order to expedite production, in December 1944, 30 Panzer IV chassis were confiscated from the Panzerkampfwagen IV production run at Nibelungenwerk. These 30 chassis were to be used for Sturmgeschütz IV production. Despite the bombing attacks on Magdeburg, in January Krupp-Grusonwerk managed to manufacture 108 Sturmgeschütz IV during the first full month of production.

Alkett was the target of another bombing raid in January 1944, followed by another attack in February on MIAG. Despite the extensive damage inflicted, it was considered certain that the following delivery schedule for the month of March 1944 could be relied upon:

— 250 Sturmgeschütz III from Alkett
— 120 Sturmgeschütz III from MIAG
— 90 Sturmgeschütz IV from Krupp-Grusonwerk.

In order to increase the number of Sturmgeschütz's being delivered to the troops, the idea of using Panzer III chassis for Sturmgeschütz was reactivated. Panzerkampfwagen III from the Ersatzheer (that had been used for crew training) were to be released for conversion to Sturmgeschütz. An equal number of Panzerkampfwagen IV (7.5cm Kw.K. L/24 and L/43) coming out of overhaul were to be issued to the Ersatzheer as replacements for their Panzerkampfwagen III.

At this time the inventory of Panzerkampfwagen III possessed by the Ersatzheer included: 55 with 3.7cm Kw.K. L/45, 105 with 7.5cm Kw.K. L/24, 357 with 5cm Kw.K. L/42 and 189 with 5cm Kw.K. L/60. About 150 to 200 Sturmgeschütz could be gained from this source. Conversions of Panzerkampfwagen III to Sturmgeschütz in 1944 occurred as follows:

	April	May	June	July	Total
gr.Pz.Bef.Wg.	0	0	2	0	2
Pz.Kpfw.III 50mm L/42	15	13	17	13	58
Pz.Kpfw.III 50mm L/60	28	28	33	11	100
Pz.Kpfw.III 75mm L/24	0	4	4	1	9
Pz.Kpfw.III Flamm	1	1	0	2	4
Total	44	46	56	27	173

The conversions were conducted under the control of:
Sonder-Ausschuss Pz III, Arbeitsgruppe Instandsetzung (Dr. Brender) MIAG
Sonder-Ausschuss Pz III, Arbeitsgruppe Instandsetzung (Niemoller) Alkett
In June of 1944 the Organisation-Abteilung of the Oberkommando des Heeres calculated the number of Sturmgeschütz that would be needed to completely outfit all of the current and planned operational combat units. The Sturmartillerie were only to be permitted to expand to a maximum strength of 45 Heeres-Sturmgeschütz Brigaden. The strength of each brigade was to be increased from 31 up to 45 Sturmgeschütz and Sturmhaubitze, for a permissible total of 2025.

In view of the estimated strength of Sturmgeschütz brigades at approximately 1200 Sturmgeschütz in June 1944 and considering the historical monthly loss rate of about 13%, a monthly quota of 250 Sturmgeschütz III and 125 Sturmhaubitze was authorized for the Sturmartillerie.

The initial goals for outfitting other units were:
— 693 Jagdpanzer IV for 33 Panzer Divisions (including the Waffen SS) at 21 per Panzerjäger-Abteilung
— 630 Sturmgeschütz III or Sturmgeschütz IV for 15 Panzer-grenadier Divisions (including the Waffen SS) at 42 per Panzer-Abteilung
— 465 Jagdpanzer IV for 15 Panzer-Grenadier Divisions (including the Waffen SS) at 31 per Panzerjäger-Abteilung
— 2200 Sturmgeschütz III, Sturmgeschütz IV or Jagdpanzer 38 for 220 Infanterie, Gebirgs and Jäger Divisions at 10 per Panzerjäger-Abteilung.

With an on-hand strength of about 1400 tank destroyers of this type in June 1944, taking into consideration the empirical value of the 13% losses per month, including the planned production of Sturmgeschütz IV, Jagdpanzer IV and Jagdpanzer 38 as well as that of the Sturmgeschütz III (minus the 375 units per month for allocated to the Sturmartillerie), it was determined that in the future a sufficient number of these vehicles would be on-hand to fill and maintain the combat strength of the units at their authorized organizational strength beginning in December 1944.
The assembly plants made great efforts to minimize the negative impacts from the bombing raids. Alkett set new production records with the assembly of 401 Sturmgeschütz III and Sturmhaubitze in December 1944. However, the overall effects of the bombing attacks on MIAG and Alkett reduced their output from a potential 6000 units down to an actual production of 4743 (903 of which were Sturmhaubitze). This production was supplemented by the Krupp-Grusonwerk output of 1006 out of a scheduled 1145 Sturmgeschütz IV. This was achieved in spite of two bombing attacks on Magdeburg and three further attacks aimed directly at the plant itself. The attacks in September of 1944 inflicted appreciable damage on the chassis production. This was partially compensated by acquiring major components from Vomag and Nibelungenwerk.

Production in 1945 declined due to the combined effects of bombing raids. These raids not only affected final assembly but also sub-contractors, the cities, the road and rail networks, and the utilities. MIAG ceased production of Sturmgeschütz III in March 1945 after their Heereswaffenamt contract had been fulfilled. Krupp-Grusonwerk suffered from heavy bombing attacks in January, February and March, greatly reducing their production until Magdeburg fell in April of 1945. The end for Alkett came when the Russian troops conquered Berlin. The last deliveries to the troops from Alkett in the closing days are documented as follows:
On 20 April 1945, 50 immobile 10.5cm Sturmhaubitze 42 were located at Alkett, consisting of:

19 10.5cm Sturmhaubitze 42 already mounted in Sturmhaubitze superstructures (owing to transportation difficulties these were earmarked for deployment in the Berlin area), and
31 10.5cm Sturmhaubitze 42 mounted on carriages (A daily output of 10 was earmarked for Festungs-Pak-Riegel in Heeres Gruppe Weichsel)
Tank production in the Berlin area was reported on 20 April 1945 as follows:

From Alkett: at 0200 on 21 April 7 Sturmgeschütz III, 1 Sturmhaubitze at 2000 on 21 April 6 Sturmgeschütz III
These vehicles were to be taken over by Sturmartillerie-Brigade 249. The prognosis after April 22 cannot be determined at this time. Alkett reported a 90% power failure.
On 23 April 1945, it was reported that the 1.Batterie/Heeres-Sturmartillerie-Brigade 249 with 10 Sturmgeschütz III up to 23 April had knocked out 29 enemy tanks, against their own loss of 4 Sturmgeschütz III.
The Brigade-Stab and 2.Batterie/Sturmartillerie Brigade 249 had departed with 10 Sturmgeschütz III on April 21 at 2000hrs.
The 3.Batterie/Sturmartillerie Brigade 249 departed on 22 April 1945 from Alkett in Spandau with 10 Sturmgeschütz

III.
The 8.Kompanie/II.Abteilung/Panzer-Regiment 2 with 10 Sturmgeschütz III at Alkett was ready for deployment at 2200hrs on 22 April 1945.
The following production was were achieved by the various assembly firms:

Daimler-Benz produced the first series of 30 Sturmgeschütz with the chassis numbers 90001-90030 and one Sturmgeschütz IV prototype

Alkett produced approximately 7734 Sturmgeschütz and Sturmhaubitze in chassis serial number series 90101-94250, followed by a second chassis serial number series beginning with 105001.

MIAG produced approximately 2643 Sturmgeschütz's in the serial number series 95001-100000.

M.A.N. delivered 142 Panzerkampfwagen III Ausf.M chassis in serial number series from 76111-76210 and 77351-77450 for final assembly as Sturmgeschütz between February and October 1943.

Krupp-Grusonwerk produced 1111 Sturmgeschütz IV on their own chassis in serial number series 100001-100650 and a second chassis serial numbers beginning with 110001. In addition, 30 were completed using chassis provided by Nibelungenwerk in the chassis serial number range of 89234-89383.

Monthly Sturmgeschütz and Sturmhaubitze production is contained on the following pages. Monthly production files from MIAG, Krupp-Grusonwerk and the Heereswaffenamt have survived. Unfortunately, the Alkett files were reported as having been burned directly after the War. In order to estimate the monthly production for Alkett Sturmgeschütz from June 1941 until the end of the war, the monthly figures for MIAG production were subtracted from the total Waffenamt acceptance figures.

Production of the Sturmgeschütz (Sd.Kfz.142 with 75mm L/24)

Date	Prod. Goals	Accepted by Waffenamt	Produced by: Daimler-Benz	Produced by: Alkett	Remarks:
1939					
December	1	0	0		
1940					
January	3	1	1		Start of Ausf.A
February	6	3	3		
March	10	6	6		
April	10	10	10		
May	10	10	10	0	
June	8	12	0	12	Start of Ausf.B
July	22	12		22	
August	32	10		20	Due to priority orders
September	36	29		29	
October	30	35		35	
November	36	35		35	
December	34	21		29	Goal exceeded previous month
1940 Total		184	30	182	
1941					
January	36	44		36	Delayed report for 8 from Dec.40
February	34	30		30	Goal exceeded previous month
March	38	30		30	Start of 3.Series
April	40	47		47	Start of Ausf.C
May	40	48		48	Start of Ausf.D
June	47	56		56	
July	47	34		34	Goal exceeded in June
August	50	50		50	
September	17	38		38	Start of Ausf.E
October	50	71		71	
November	50	46		46	Additional delivered in October
December	40	46		46	
1941 Total		540		532	
1942					
January	45	45		45	
February	45	45		45	

Sub-total	90		90
Overall Total	814	30	804

Production of the Sturmgeschütz (Sd.Kfz.142/1 mit 7.5cm StuK 40)

Date	Prod. Goals	Accepted by Waffenamt	Produced by: Alkett	MIAG	Remarks:
1942					
March	40	3	3		Start of Ausf.F
April	45	36	36		Converting to 7,5 cm StuK40 L/43
May	50	79	79		
June	50	70	70		Start of StuK40 L/48 and extra 30mm armor plate
July	55	60	60		
August	70	80	80		
September	80	70	70		Start of Ausf.F/8
October	80	84	84		
November	100	100	100		
December	100	120	120		Start of Ausf.G
1942 Total		702	702		
1943					
January	130	130	130	0	
February	140	140	130	10	
March	185	197	144	53	
April	218	228	151	77	
May	230	260	140	120	
June	235	275	155	120	
July	235	281	161	120	
August	240	291	171	120	
September	260	345	205	140	
October	280	395	255	140	
November	380	295	145	150	Bombing attack on Alkett
December	215	174	24	150	
1943 Total		3011	1811	1200	

Production of the Sturmgeschütz (Sd.Kfz.142/1 mit 7.5cm StuK 40)

Date	Prod. Goals	Accepted by Waffenamt	Produced by: Alkett	MIAG	Remarks:
1944					
January	320	227	77	150	Bombing attack on Alkett
February	320	196	59	137	Bombing attack on MIAG
March	310	264	144	120	Bombing attack on MIAG
April	365	294	194	100	Bombing attack on MIAG
May	380	335	255	80	
June	385	341	196	145	
July	395	377	242	135	Instead more StuH at Alkett
August	355	312	232	80	Bombing attack on MIAG
September	370	356	256	100	
October	375	325	253	72	Bombing attack on MIAG
November	375	361	251	110	Bombing attack on MIAG
December	355	452	361	91	
1944 Total		3850	2520	1320	
1945					
January	370	391	320	71	
February	220	189	152	37	
March	170	235	220	15	Bombing attack on MIAG
April	220	48	48	0	
1945 Total		863	740	123	
Overall Total		8416	5773	2643	

Production of the Sturmhaubitze (Sd.Kfz.142/3 mit 10.5cm StuH 42)

Date	Prod. Goals	Accepted by Waffenamt	Produced by Alkett	Remarks
1942				
October			9	produced using rebuilt chassis
November			0	
December			0	
1942 Total			9	
1943				
January	0	0	3	produced using rebuilt chassis
February	0	0	0	
March	20	10	10	Original contract for a series of 200
April	24	34	34	
May	30	45	45	
June	30	30	30	
July	30	25	25	Priority on 7.5cm Pak auf 38(t)
August	30	5	5	Delivery problems for howitzers
September	20	10	10	Delivery problems for howitzers
October	20	11	11	Delivery problems for howitzers
November	20	4	4	Bombing attack on Alkett
December	25	30	30	
1943 Total		204	204	
1944				
January	30	26	26	Bombing attack on Alkett
February	60	54	54	late delivery of chassis and hulls
March	60	56	56	late delivery of chassis and hulls
April	65	58	58	

Production of the Sturmhaubitze (Sd.Kfz.142/2 with 10.5cm StuH 42)

Date	Prod. Goals	Accepted by Waffenamt	Produced by Alkett	Remarks
May	70	46	46	Bombing raid hit engine production
June	75	100	100	
July	75	92	92	
August	125	110	110	
September	125	119	119	Priority on 7.5cm StuG
October	125	100	100	Bombing attack on Alkett
November	125	102	102	
December	125	40	40	
1944 Total		903	903	
1945				
January	60	71	71	
February	80	24	24	
March	80	49	49	
April	80	48	48	Additionally 19 StuH without engines were still at Alkett on 20 April 1945
1945 Total		192	192	
Overall Total		1299	1299	

Production of the Sturmgeschütz IV (Sd.Kfz.166 mit 7.5cm StuK 40)

Date	Prod. Goals	Accepted by Waffenamt	Produced by Krupp	Remarks
1943				
December	10	30	30	
1943 Total		30	30	
1944				
January	100	78	108	Bombing of Magdeburg
February	110	136	106	
March	90	87	87	
April	90	91	91	
May	90	95	95	
June	95	90	90	Bombing of Magdeburg
July	90	90	90	
August	90	70	70	Bombing of the Grusonwerke
September	90	56	56	Bombing of the Grusonwerke
October	90	84	84	Bombing of the Grusonwerke
November	100	80	80	
December	110	49	49	
1944 Total		1006	1006	
1945				
January	100	46	46	Bombing of the Grusonwerke
February	70	18	18	Bombing of Magdeburg
March	50	38	38	Bombing of the Grusonwerke
April	60	3	3	
1945 Total		105	105	
Overall Total		1141	1141	

15cm Sturm-Infanteriegeschütz 33

Developed by Rheinmetall, the heavy Infanteriegeschütz 33 (150mm heavy infantry support gun, model 33) was first introduced in 1927. It was the most powerful support weapon of the Infanterie. With a barrel length of 1748mm, a maximum firing range of 4.7km was achieved with a muzzle velocity of 240 m/s (with Charge 6). The rifled barrel had 44 lands and grooves with a constant twist of 1/21.

The first attempt to mount the sIG33 on a self-propelled tracked vehicle was completed in January 1940 when 38 of these weapons were mounted on the chassis of the Panzerkampfwagen I, Ausf.B. Six sIG Kompanien (numbered 701 through 706) were hurriedly established, each company receiving six of these self-propelled guns. Subordinated to six Panzer Divisions, they were first deployed in the Western Campaign in May and June of 1940. Most of these six companies stayed active in the field until 1943 when the sIG33 on the Panzer I chassis were replaced by sIG33 on the Panzer 38(t) chassis. Although the Panzer I chassis was overloaded, the troops preferred the self-propelled version to the towed weapon.

The entire gun, including wheels and trail spade, (weight of 1.7t) with light armor protection had been loaded onto the chassis of the Panzerkampfwagen I, Ausf.B.

Already in 1940, a second type of self-propelled chassis was being developed for the sIG33. The prototype completed in October 1940, consisted of an sIG33 mounted low in a normal Panzerkampfwagen II chassis.

Tests demonstrated that the normal Panzerkampfwagen II chassis was too narrow and too short and the engine lacked sufficient power. Major modifications were required in the design to create the Versuchsserie (trial series). Alkett was awarded contracts to design and assemble all 12 test vehicles. These were given the official title of "15cm sIG33B Selbstfahrlafette auf veränderten Panzer II Fahrgestell" (150mm heavy infantry gun model 33B self-propelled mount on the modified Panzer II chassis). The "veränderten Panzer II Fahrgestell was created by widening and lengthening the previous Panzer II chassis design and upgrading to a new engine and drive train.

A message from July 1941 stated that assembly of the 12 test vehicles was expected to occur in August/September 1941. As a result of production delays, the 12 test vehicles were completed and delivered in December 1941 and

150mm sIG 33, loaded on a Panzerkampfwagen I, Ausf.B chassis. The superstructure armor has been removed.

This photo shows the vehicle with armor superstructure mounted.

15cm sIG 33 auf verändertem Panzer II Fahrgestell. In North Africa in Spring of 1942.

Side view of the same vehicle.

January 1942. Six vehicles apiece were assigned to two newly-formed sIG-Kompanien (numbered 707 and 708)). The 708.sIG Kompanie was shipped to North Africa and arrived in Tripoli on the 28th of February 1942. The 707.sIG Kompanie arrived in Tripoli on April 4th 1942.

On 20 May 1942, directly before going into action for the first time in support of the major offensive against the British Gazala-Line, Pz. AOK Afrika sent the following teletype message to the OKH:

Multiple failures of the outer, rear coupling on the intermediate gear in the transfer case have occurred in the "gep.Sfl.f.sIG33 veränderten Fahrgestell Pz.II." Engine temperature is very high. Through QQu/V, urgently request that a specialist from Alkett be sent.

A report on the weapons and armored vehicles employed in North Afrika in Pz.AOK Afrika, dated 30 August 1942 stated:

The sIG33 was considered excellent from a ballistic and effectiveness standpoint. However, greater firepower is required. When mounted on the artillery carriage, it was infrequently used in the attack. This resulted from its short firing range and the poor towing characteristics of the carriage.

This Panzer II chassis, specifically converted to create a self-propelled chassis for the sIG, was not proven to be successful. The heavily overloaded vehicle rapidly overheated and thereby became immobilized. The suggestion of the troops to create a more fitting, armored self-propelled gun mount was not accepted based on the following reasons:

— The effective range of this gun could not be appreciably increased. This meant an expenditure of effort was not worthwhile. As a minimum, it was recommended to wait for experience reports about the "15cm sFH13 Sfl. auf Lorraine Schlepper (150mm heavy field howitzer, model 13, self-propelled, mounted on the captured French Lorraine tractor) which with the same caliber had a much greater range.

— In defense, the gun was subjected to massed artillery fire as soon as it opened fire. Therefore, the deployment of such a valuable self-propelled gun in defensive positions was too risky.

In response to an inquiry by OKH Gen. StdH Org. Abt (IIIb) dated 16 October 1942, the Pz.AOK Afrika Ia answered as follows in a teletype dated 17 October 1942:

1. The weapon has proven itself to be quite effective, especially when purposefully firing shells so that they ricochet.

2. The modified Panzer II chassis does not meet requirements and has not proven itself.

Difficulties noted include:

a) Engine (Büssing-NAG Typ L 8 V, engine numbers 95648 to 95659), gearboxes, and running gear are too weak for the total weight of the vehicle.

b) Specific problems include: Overheating of coolant and oil, caused by excessive demands on the engine. The ventilation is insufficient and the radiator surfaces too small.

c) Frequent damage occurs to the spur gears (the power train mounted on the right side of the chassis, transfers power from the engine to the transmission via two beveled gears). The gear teeth frequently break.

d) Cracks occur in the track links, track pins and drive wheels.

3. For any further production, it is considered necessary that the Panzer III chassis be selected.

The eight "sIG33 Sfl. auf ver.Pz.II" which had survived the Spring and Summer battles in North Africa, were all lost during the British offensive of El Alamein between 23 October and 2 December 1942.

The decision to mount the next series of sIG33 as self-propelled guns on the Panzer III chassis had already occurred prior to the recommendation from Pz.AOK Afrika. In the Führerkonferenz of 10-22 September 1942, the following was determined:

During the fighting in Stalingrad, it became quite clear that a heavy gun, mounted in a very heavily armored vehicle, firing mine-type ammunition (thin-cased high explosive shells) capable of destroying houses with a few rounds, was urgently needed. It was not necessary for the gun to be able to fire at long ranges or for the self-propelled chassis to have a large radius of action between refuelings. Instead, emphasis should be placed on effective armor protection. Everything possible was to be done to ensure that 12, a minimum of 6 were produced within 14 days. Should the installation of a sIG in the turret of a Panzer III of IV not be possible, then an attempt should be made to install it in a Sturmgeschütz.

Hitler was extremely satisfied with the report that by October 7th 1942, six sIG and by October 10th 6 additional sIG were to be installed into superstructures on Sturmgeschütz chassis. A further twelve of these weapon systems were to be produced each month. On 13 October 1942, Hitler was informed that in addition to the 12 that had already been completed, a further 12 "sIG als Sturmgeschütz auf Panzer III" with 80mm thick armor were to immediately enter production.

Waffenamt production files for the "sIG33 auf Fgst.Panzer III (Sfl) show:

	Delivered to Waffenamt	Available for issue by Fz.In.	Remarks
October 1942	24	12	one-time trial
November 1942	0	12	series mounted on rebuilt chassis

In the Führerkonferenz of 7-8 November 1942, it was revealed that the first 12 Sturmgeschütz's with sIG were missing their scissors-periscopes. On 9 December, it was reported that the chassis that had previously been used for mounts for the sIG Sfl. had been rebuilt from severely damaged chassis returned from the front. In consideration of the value of this weapon only new production chassis were to be used for further assembly.

With a five-man crew, the "15cm Sturm-Infanterie-geschütz 33" had a combat weight of 21 tons. The superstructure, which was closed on all sides, had 80mm thick frontal armor, 50mm on the sides, 15mm in the rear and the roof was protected with 10mm armor plate.

In addition to the main gun, the radio operator was provided with an M.G.34 machine gun in a ball mount. The main gun was manufactured by the firms of:
— AEG-Fabriken in Berlin-Henningsdorf and — Böhmisch Waffenfabrik in Strakonitz.

The price of the weapon was 20,450RM

Thirty rounds of partitioned ammunition, shells and shell cartridges, were carried in the vehicle.

The first 12 vehicles were issued to Sturmgeschütz Abteilung 177, which reached the area of Stalingrad on 8 November 1942. All 12 vehicles were lost in and around Stalingrad.

15cm Sturm-Infanteriegeschütz 33.

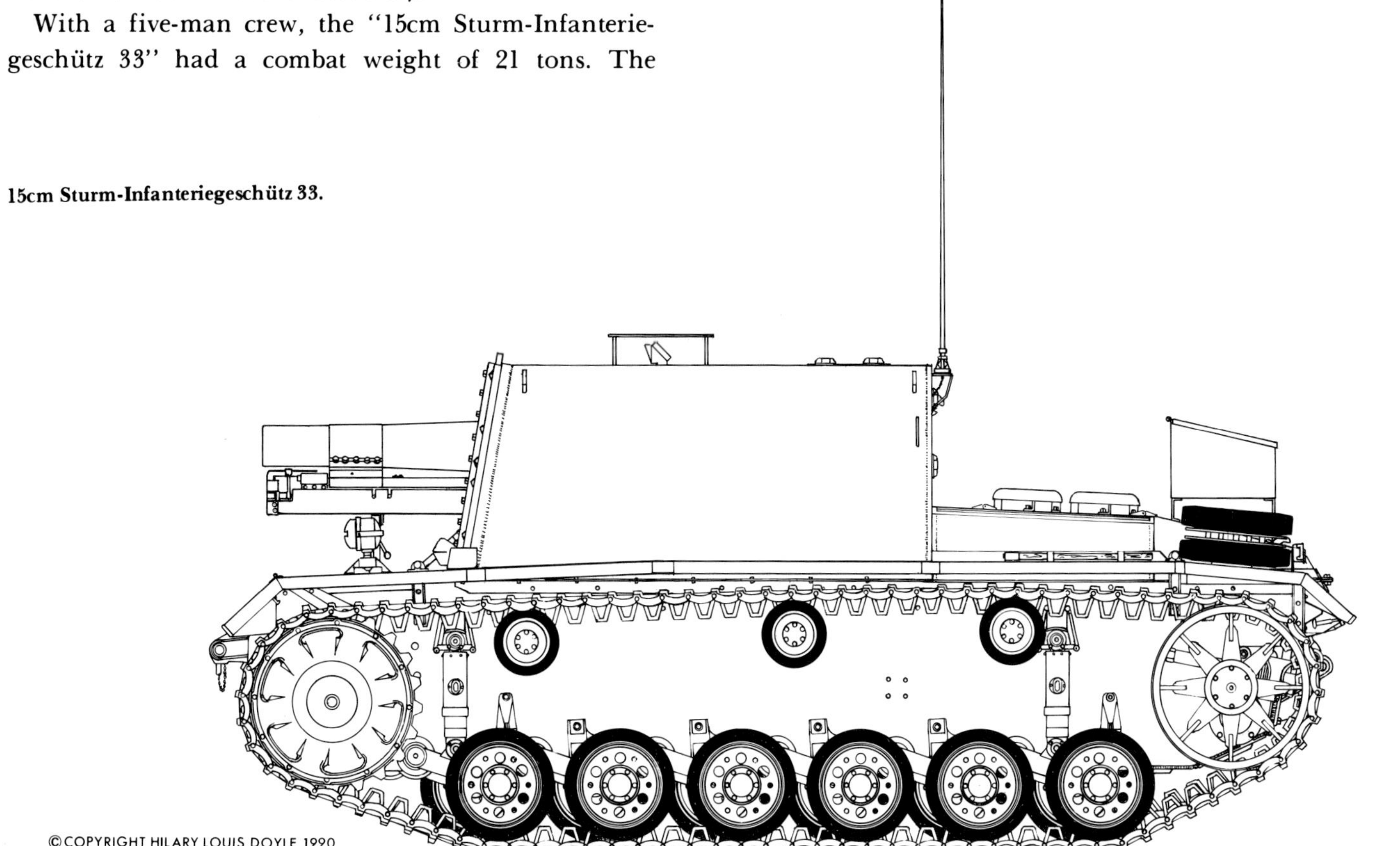

Sturm-Infanteriegeschütz converted from an overhauled Sturmgeschütz chassis.

Side view "Taktische Nr. G2" in the same unit.

Rear view of the 15cm Sturm-Infanteriegeschütz.

The second 12 vehicles were issued to the Sturm-IG Batterie attached to the Lehr Battalion XVII Armee Korps and later to the 22.Panzer Division. On April 11th 1943, the unit (with 7 Sturm-Infanteriegeschütz 33) was incorporated into Panzer-Regiment 201 of the 23.Panzer Division as "StuIG Battr./Pz.Regt.201." The 23. Panzer Division reported that the total loss of their last Sturm-Infanteriegeschütz had occurred in October 1943.

Technical Data for "15cm sIG33 Sfl.auf veränderten Pz.II Fgst."

Weight	12t
length	5480mm
width	2600mm
height	1830mm
crew	4 men
armor	
front	30mm
side	14.5mm
deck	10mm

engine: Büssing-NAG "L 8 V"
150hp
45 km/h
100 km
ammunition: 10 rounds, 15cm sIG33
elevation: -3 degrees to +75 degrees
traverse: 7 degrees to the left and right

Sturmgeschütz, Flammenwerfer

In the 1-3 December conference with Hitler, it was concluded that a number of the currently produced series of Sturmgeschütz would be put aside for use as flame-thrower vehicles. A single series of ten vehicles was planned.

Tests with flame-throwing tanks had already been conducted with the Panzer II and Panzer B2(f). On 18 January 1943, a meeting at was held at Wegmann in Kassel in which a representative of Koebe, Luckenwalde took part. Both firms had experience in this area.

On 23 January 1943, cold weather testing took place in Wünsdorf with a flamethrower installed in a vehicle. Wegmann had guaranteed that the flamethrower would ignite at temperatures down to -22 degrees Centigrade. A re-loadable powder device was in development at WaPrüf. An additional fuel pump was installed. The engine, a DKW 1100cc two-stoke motor which drove the pump, only sprang to life after being warmed with water. After the Maybach HL-120 engine had been running for only 5 minutes, the first flame oil ignition test was conducted. The equipment functioned adequately without problems and reached a range of 53 to 55 meters.

Sturmgeschütz (Flammenwerfer).

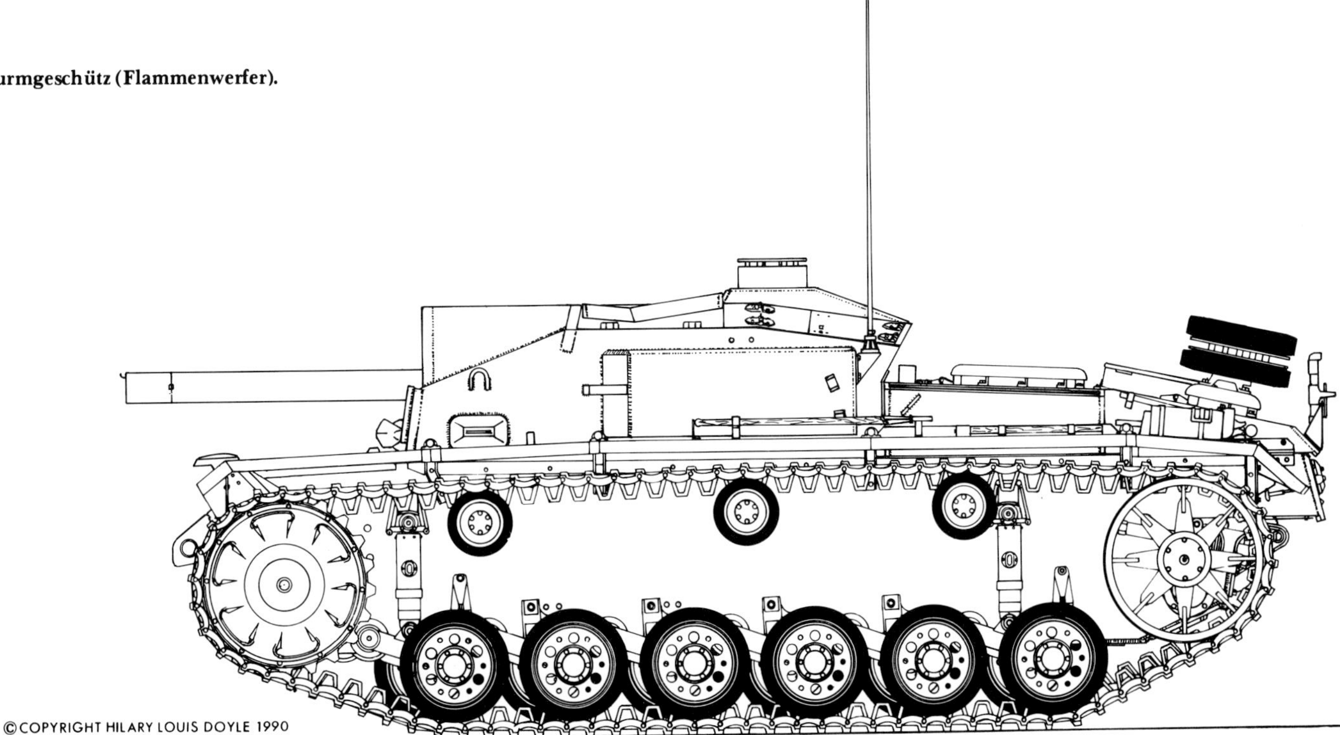

Flamethrower installation in a converted Sturmgeschütz, Ausf.F/8.

The oil was propelled out of a 14mm jet (mounted in the old location for the 75mm Sturmkanone 40). When checking the pressure, the device was to indicate a pressure of at least 15 Atü (15 atmospheres of excess pressure). The minimum acceptable range (with the wind or in calm weather) was specified at 50 meters.

During the first attempts, actual ranges of 53 to 59 meters were achieved. On several occasions, the oil did not immediately ignite and sometimes the oil did not ignite at all. Wegmann again guaranteed that these mistakes would be corrected.

These ten Sturmgeschütz-Flammenwerfer were not taken from newly produced vehicles, but rather were converted from chassis rebuilt in the overhaul facilities. Records reveal that nine StuG (Fl) in May 1943 and one StuG (Fl) in June 1943 were assembled from rebuilt chassis and delivered to the Heereszeugamt (HZA).

The ten finished Sturmgeschütz-Flammenwerfer were assigned to Panzertruppenschule 1 and transported by rail to them on 29 June 1943. A report indicates that one of these vehicles burned out. This StuG (Fl) was sent back to the HZA, overhauled and returned to the unit in September 1943.

The only photograph of the StuG (Fl), which has surfaced to date, shows an Ausf.F/8 chassis.

None of the ten vehicles were used in combat. All ten were returned to the Heereszeugamt in January 1944. Seven in February, one in March and two in April were converted back to normal Sturmgeschütz mounting 7.5cm StuK 40 guns.

Flakpanzer for Sturmgeschütz Units

In a presentation to Hitler on 22 December 1943, the *Generalinspekteur der Panzertruppen* dictated that armored 20mm Flak were to be provided for Sturmgeschütz units. The best approach appeared to be vehicles based on converted Panzer II chassis. These offered the crew a certain amount of protection and had a similar appearance as the Sturmgeschütz, at least not as different, as for example, an armored self-propelled anti-aircraft gun (on a half-tracked vehicle) might be.

Meanwhile, the *Inspektor der Artillerie* demanded that Sturmgeschütz units be equipped with anti-aircraft artillery for protection against strafing aircraft. Sturmgeschütz issued to the Sturmartillerie had Panzer III chassis while the Flakpanzer superstructures designed up to this point were mounted on Panzer IV chassis. At the direction of the Inspektor der Artillerie (In 4), research was to be conducted to determine if the new anti-aircraft turrets could fit on the Panzer III chassis. The diameter of the turret ring on the Panzer III was 1520mm. On the Panzer IV it was 1680mm.

In October 1944, officers from the Sturmgeschütz-Schule Burg near Magdeburg visited the Deutsche Eisenwerke plant in Duisburg. They were to inspect the Flakpanzer types produced there and select the most appropriate for the Sturmartillerie. There were two

Flakpanzer "Wirbelwind" on the Panzer IV chassis.

concepts to choose from. Those already designed for the Panzer IV chassis were:

— WIRBELWIND with a 2cm Flakvierling 38. The weapon weighing 1509 kg had four barrels each with a length of 1300mm (L/65). They were fed with 20 round magazines. The engagement ceiling was at an altitude of 3700 meters. Theoretically, the weapon fired 450 rounds per minute per barrel. Price was about 20,000 Reichsmarks. The weapon was produced by the firms Ostmark-Werke in Vienna, Auto-Union AG in Chemnitz and Bentler-Werke in Bielefeld.

— OSTWIND with the 3.7cm Flak 43. This gun, weighing about 1392kg, had a barrel length of 2112mm (L/57). It was fed by shell cartridges containing 8 shells each. Its engagement ceiling was at an altitude of 4800 meters. The theoretical rate of fire was 230 to 250 rounds per minute. The weapon was manufactured by the firms Dürkopp in Bielefeld and Weserhütte in Bad Oeynhausen.

Flakpanzer "Ostwind" on the Panzer IV chassis.

In accordance with Heereswaffenamt, memo dated 3 January 1945, the Thyssen plant (in Mülheim/Ruhr) of the Deutsche Röhrenwerke was to deliver the armor components for an OSTWIND turret to the Heereswaffenamt and a 3.7cm Flak 43 to the Sturmgeschütz-Schule Burg. A Panzer III chassis and two 2cm Flakvierling weapons were to be sent to the firm OSTBAU in Sagan, Schlesien.

Purpose: Trial construction of Flakpanzer III.

Regierungs-Baurat Becker received a contract to go to the Ostbau firm and start work. Ostbau had emerged as a shop in which a number of rebuilt Panzer IV chassis were used to assemble Flakpanzer under the leadership of a Lieutenant from the Panzerwaffe (Hans C. Graf von Seherr-Thoss) using relatively modest means.

There were differences of opinion between Becker and the leader from Ostbau. The leader of Ostbau considered it impossible to convert the Panzer III for this purpose. Becker was certain that it would work. No agreement was reached. Becker, having achieved nothing, returned.

The *Inspektor der Artillerie* demanded a rapid procurement of 90 Flakpanzer turrets. A comparatively low weight (about 2t for each superstructure) was deemed necessary and agreed upon. The necessary 16mm armor plates were available. A contract was soon placed for their manufacture. The chassis were to be diverted from the contingent of Sturmgeschütz allocated to the Artillerie, with some rebuilt chassis provided from overhaul facilities. The necessary superstructures, including ball bearing turret races, were to be provided from the stockpiles at the overhaul facilities.

In mid-March 1945, the trials were finalized. They confirmed that the turret called KEKSDOSE (cookie tin, possibly the same design as the "Ostwind" turret) could be mounted on the Panzer III chassis.

For this effort the following items were sent to Burg:

— a Panzer III hull from Sagan
— a second experimental hull
— two 2cm Flakvierling weapons
— two 3.7cm Flak weapons
— a complete superstructure with turret for the 3.7cm Flak
— a complete superstructure with turret for the 2cm Flakvierling

The technical and logistical preparations at the Sturmartillerie-Schule had been so thoroughly carried out that the Flakpanzer could have been assembled within short order after receiving the superstructure and turret components. However, as a result of a decision by the Panzer-Kommission, the allotment for these Flakpanzer superstructures and turrets was stricken from the Reichsminister für Rüstung und Kriegsproduktion "Emergency Program for Armament Production" dated 23 January 1945.

The *Inspektor der Artillerie* established the following proposal for the Sturmgeschütz-Schule Burg:

A shop is to be established at Burg, like those at Ostbau in Sagan. The means are available. We will conduct the design tests of the Flakpanzer III ourselves. Only in this manner can we guarantee the fastest solution and only in this manner would the Sturmartillerie come to possess the needed Flakpanzer.

The Inspektor der Artillerie, on 17 March 1945, once again requested that those superstructures already in production as well as the material for another 72 superstructures, to be delivered at a rate of 12 per month, be completed.

The *Generalinspekteur der Panzertruppen* responded to this succinctly:

". . . resolved by changes in the situation."

Nevertheless several Sturmgeschütz units reported that they were in possession of Flakpanzer:

— 2 Flakpanzer on 15 March 1945 with StuG Brig.244 in H.Gr.B
— 3 Flakpanzer on 15 March 1945 with StuG Brig.341 in H.Gr.B
— 4 Flakpanzer on 5 April 1945 with StuG Brig.667 in Ob West

Unfortunately, these reports do not mention the type of Flakpanzer or the type of Flak weapon. The issue records account for all allocations of the Flakpanzer IV, none of which were issued to the Sturmgeschütz-Brigaden. Therefore, one could assume that Burg was successful in producing and delivering at least nine "Flakpanzer III."

Sturmgeschütz in Radio-Controlled Units

In the German Wehrmacht a requirement was established for the design of remote-controlled vehicles with mine clearing rollers. This ushered in the development of an entire series of remote controlled vehicles. At first the Borgward B I. Followed by additional series of mine clearing vehicles, including the B II, the B III and finally the "schwere Ladungsträger" B IV (Sd.Kfz.301).

While en route, the fully-tracked vehicle was operated with normal steering levers by the driver. When deployed without the driver, the vehicle was switched over to remote guidance, which operated via radio signals from a command vehicle.

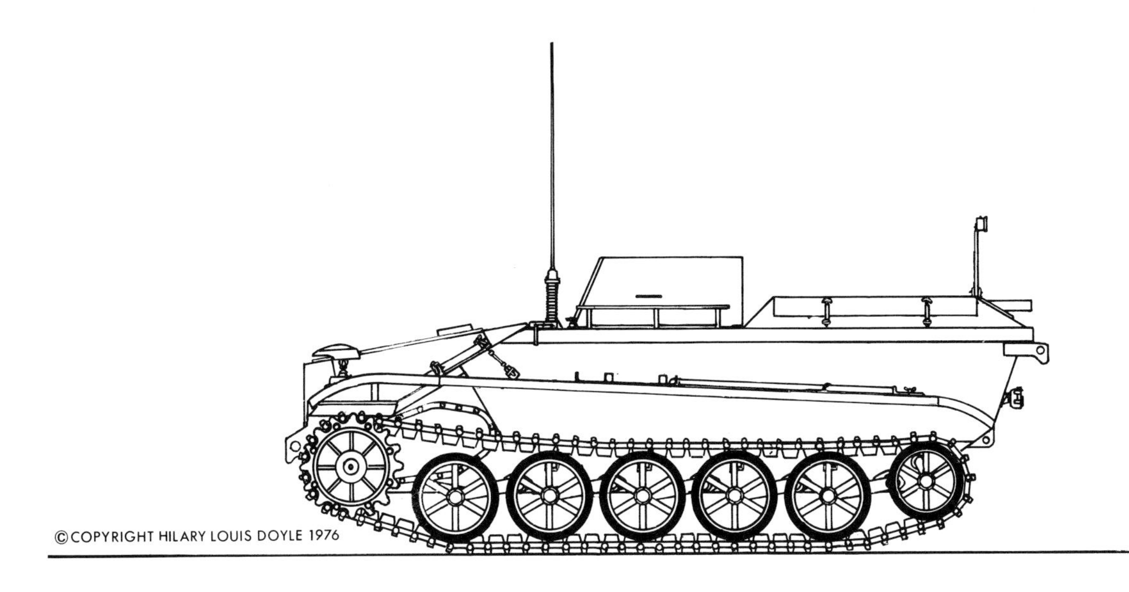

Schwerer Ladungsträger, Ausf.B (Sd.Kfz.301).

In the forward part of the vehicle there was an explosive charge. This 500 kilogram charge was mounted on the front of the vehicle. It was released by the use of explosive bolts. When released, the charge could slide down the 47 degrees incline by virtue of its own weight. The **B IV** could drop the charge and then retire to a safe distance. It was also possible to steer the vehicle right onto the target and set-off the charge.

Production of the **B IV** was terminated in September 1944. It was succeeded by the "mittlere Ladungsleger" (Sd.Kfz.304) Springer. The Springer could also be steered by a driver. It carried a smaller explosive charge of 330 kilograms. This charge was designed to remained with the expendable vehicle. A total of 50 production series Springer was completed before the end of the War.

Schwerer Ladungsträger B IV, Ausf.C (Sd.Kfz.301).

Demonstration of the B IV (1942)

A Ladungsträger B IV being remotely controlled by a Sturmgeschütz. (BA)

Sturmgeschütz, Ausf.G and five Ladungsträger B IV with the 1.Zug/Pz.Kp.(FKL) 314 at the Genzrode training area near Neuruppin.

When first formed on June 1st 1940, Minenräum Kompanie 1 (1st mine-clearing company) was issued kleine Panzerbefehlswagen (Sd.Kfz.265) (small armored command tanks on Pz.Kpfw.I chassis) as control vehicles for their Minenräumwagen (mine-clearing vehicles) (Sd.Kfz.300). Increased to the size of an Abteilung by the end of 1940, Minenräum-Abteilung 1 saw action for a short time in June and July of 1941 before it was withdrawn, reorganized and given a new designation. In Mid 1942, the unit was re-deployed to the Eastern Front. Its new unit designation was Panzer-Abteilung 300 (F.L.). It had two companies which were equipped with the Panzerkampfwagen III as control vehicles for the "B IV" Sprengstoffträger (Sd.Kfz.301). The unit was transferred to Heeres-Gruppe Nord in September 1942 and renamed as Panzer-Abteilung (FKL) 301. It remained on the Eastern Front until early 1943 and then moved back to Neuruppin for rest and refitting.

In early 1943, the number of Funklenk (radio remote control) units was expanded resulting in the Panzer-Abteilung (FKL) 301 and four independent companies (FKL 311 through 314). The companies organized in accordance with K.St.N.1171f still retained the Panzer III as the control vehicles. However, because the armor industry had begun to switch over to the production of Sturmgeschütz and Panzerkampfwagen Panther at the end of 1942 and beginning of 1943, it was only a matter of time before the Panzer III would no longer be available for these units. K.St.N.1171f Ausf.B, permitted the ten Sturmgeschütz in each company. From April until June 1943 a total of 61 Sturmgeschütz were issued to the Funklenk units. Panzer-Kompanie (FKL) 313 was the sole operational unit to retain the Panzer III as a control vehicle.

The following Funklenk units with the Sturmgeschütz as control vehicle were sent to the Eastern Front in 1943: To support the offensive at Kursk: Panzer-Kompanie (FKL) 312 (with 10 Sturmgeschütz and 36 B IV) was assigned to support the Tiger tanks of s.Pz.Abt.505 and Panzer-Kompanie (FKL) 314 (with the same equipment as the 312th) was assigned to support the 8.8 cm Pak 43/2 "Ferdinand" tank destroyers of s.Pz.Jg.Abt.653.
Panzer-Kompanie (FKL) 311 (with 10 Sturmgeschütz and 36 B IV) was assigned to Panzer Grenadier Division "Grossdeutschland" and sent to the Eastern Front in August 1943, remaining deployed until May 1944.

On 31 December 1943 all Funklenk units — with the exception of Panzer-Kompanie (FKL) 311 — were back at training bases and reported the following strengths:

Pz.Abt.(FKL) 301: 31 Sturmgeschütz
Pz.Kp.(FKL) 312: 2 Sturmgeschütz
Pz.Kp.(FKL) 314: 4 Sturmgeschütz
Pz.Kp.(FKL) 315: 10 Sturmgeschütz
Pz.Kp.(FKL) 316: 10 Sturmgeschütz

In 1944/45, further deployment of Funklenk units for frontline service included:

In Italy near Anzio, February to March 1944:
Pz.Abt.(FKL) 301 with its 2nd, 3rd, and 4th companies with 31 Sturmgeschütz and 108 B IV.

In the West after June 6th 1944:
Pz.Kp.(FKL) 315 attached to the 21.Panzer Division with 10 Sturmgeschütz and 36 B IV's (June to July 1944)
Pz.Kp.(FKL) 316 in the Panzer-Lehr Division with 9 Sturmgeschütz, 3 Tiger and 36 B IV (June to July 1944)
4.Kp./Pz.Abt. 301 attached to the 2.Panzer Division with 10 Sturmgeschütz and 36 B IV (June to September 1944)
Pz.Kp.(FKL) 319 with 10 Sturmgeschütz and 36 B IV (September 1944 to January 1945)

In the East, 1944-1945:
Pz.Abt.(FKL) 301 under Heeres Gruppe Nordukraine with 30 Sturmgeschütz and 108 B IV (June to August 1944)
Pz.Abt.(FKL) 302 at first under Heeres Gruppe Mitte with 40 Sturmgeschütz and 144 B IV (August 1944 until war's end)
Pz.Zug (FKL) 303 under Heeres Gruppe Weichsel with 4 Sturmgeschütz and 12 B IV (February until war's end)

Organization of the Funklenk Companies
"leichte Panzer Kompanie f" K.St.N.1171f (Ausf.B) dated 1 January 1943

Kompanie Trupp	2 gp.Sfl.f.Sturmgeschütz
1. Zug	4 gp.Sfl.f.Sturmgeschütz
	12 Sprengstoffträger (Sd.Kfz.301)
2. Zug	4 gp.Sfl.f.Sturmgeschütz
	12 Sprengstoffträger (Sd.Kfz.301)
Sondergerät Reserve	12 Sprengstoffträger (Sd.Kfz.301)

"leichte Panzer Kompanie f" K.St.N.1171f dated 1 June 1944

Kompanie Trupp	2 Sturmgeschütz III für 7.5cm StuK 40 (Sd.Kfz.142/1)
1. Zug	4 Sturmgeschütz III für 7.5cm StuK 40 (Sd.Kfz.142/1) 12 Sprengstoffträger (Sd.Kfz.301) 1 mittlerer Schützenpanzerwagen (Sd.Kfz.251/1)
2. Zug	4 Sturmgeschütz III für 7.5cm StuK 40 (Sd.Kfz.142/1) 12 Sprengstoffträger (Sd.Kfz.301) 1 mittlerer Schützenpanzerwagen (Sd.Kfz.251/1)
Sondergerät Reserve	12 Sprengstoffträger (Sd.Kfz.301)

"leichte Panzer Kompanie f" K.St.N.1171f dated 1 October 1944

Kompanie Trupp	2 Sturmgeschütz III für 7.5cm StuK 40 (Sd.Kfz.142/1)
1. Zug	4 Sturmgeschütz III für 7.5cm StuK 40 (Sd.Kfz.142/1) 9 Sprengstoffträger (Sd.Kfz.301 or 304) 1 mittlerer Schützenpanzerwagen (Sd.Kfz.251/1)

Leichte Panzerkompanie f, K.St.N.1171f dated 1 January 1943 and 1 June 1944.

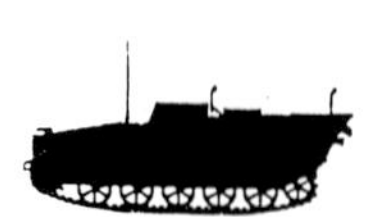

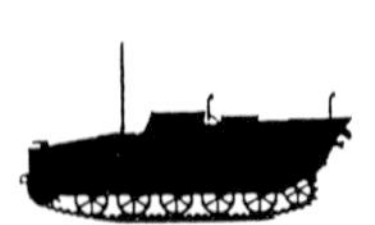

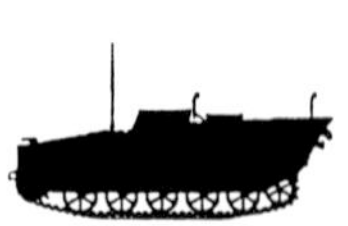

2. Zug	4 Sturmgeschütz III für 7.5cm StuK 40 (Sd.Kfz.142/1) 9 Sprengstoffträger (Sd.Kfz.301 or 304) 1 mittlerer Schützenpanzerwagen (Sd.Kfz.251/1)

Sturmgeschütz issued and delivered to Funklenk units:

Month	Unit	number from HZA	date transported
1943			
April	Pz.Abt.300	21	
May	Pz.Abt.300	20	12 May 1943
June	2 FKL Kp.	20	28 June - 1 July 1943
August	FKL Kp.31510	18 August 1943	
	FKL Kp.316	10	8 September 1943
December	FKL Kp.312	1	4 January 1944
1944			
May	Pz.Abt.(FKL) 3018		17 June 1944 (rebuilt)
June	Pz.Abt.(FKL) 30211		1 July 1944
July	Pz.Abt.(FKL) 3022		3 August 1944
August	Pz.Kp.(FKL) 311 10		13 August 1944
October	Pz.Kp.(FKL) 319 4		1 November 1944

The remote control steering units for these vehicles were delivered by the firms Hagenuk, Kiel and Dr. Hell, Berlin. the radio receiver for the remote control device carried the nomenclature "EP[3] mit UKE 6."
The following are after-action-accounts from Funklenk units on the Eastern Front.

Pz.Abt.(FKL) 301Ia/Kdr. 380/43

Memorandum concerning the further use of the radio controlled weapons system based on lessons from experience gained in action from 5-8 July 1943 during Operation "Zitadelle"

During the attack beginning 5 July to the south of Orel, three independent radio controlled companies were deployed in the 9th Armee sector. Two companies were assigned to the Pz.Jäg.Rgt. 656 and one to the s.Pz.Abt. 505. The companies were deployed as cohesive units under command of the company commander, the platoons closely cooperating with the forward-deployed companies of the supported unit. The combat mission was the same for all companies, namely, advancing battlefield reconnaissance in combat strength, marking mine fields and clearing lanes through mine fields, and destroying difficult targets, such as dug-in anti-tank weapons and heavy enemy tanks.

Reports from the troops create the following picture:

1) Deployment of the Pz.Kp.(FKL) 314 with the I./Pz.Jg.Rgt.656.

A very dense and deeply-laid mine field barred access to the Russian main line of defense, which, at the same time was being protected by artillery barrage fire. In accordance with orders, the company began to blast three lanes through the mine fields. As a result of the sheer depth of the mine field, a total of 12 B IV were needed. The control vehicles passed through the cleared lanes without any damage. The plans to utilize engineers for marking the mine fields did not materialize. The engineers did not advance due to the heavy artillery fire. This artillery fire caused the attack to halt. Because of the numerous artillery impacts on the battlefield, the s.Pz.Jäger (Ferdinand) were not able to identify the lanes cleared (by the B IV's) through the minefields. Because the lanes had not been marked and were no longer clearly identifiable — the B IV track marks could not be seen on the firm turf. Even though lanes had been cleared, Ferdinands were immobilized when they struck mines.

During further action, in total, seven B IV's were detonated. One of the seven fell into a trench and detonated killing the occupants of the trench who had attacked the B IV with hand grenades and close combat weapons.

Two B IV's were directed against a heavily occupied section of forest, that had stalled our infantry advance, then detonated. Afterward, no further resistance came from this area. During the entire attack, four B IV's were lost to artillery hits. One of which was "live," that is to say it had detonators installed, and exploded. The other three, without detonators set, simply burned.

2) **Deployment of the Pz.Kp.(FKL) 313** with the II./Pz.Jg.Rgt.656 was accomplished under similar circumstances. Four B IV's were damaged when one platoon of the company, advancing to attack, encountered one of our own unidentified mine fields. The other platoon only blew one lane through the Russian mine field. Four B IV's were expended in this effort.

One B IV in the assembly area was hit and detonated by artillery fire, causing an additional two B IV vehicles to catch fire and detonate as well. The undersigned could not satisfactorily determine the cause of this incident because the driver and engineer troops belonging to those vehicles were killed. It is assumed that the detonators were already installed in the charge and were detonated by the heat of the burning charge. An additional B IV was hit by artillery fire while it was being guided into position and also exploded.

In the further course of the attack, dug-in anti-tank positions and one bunker were destroyed by 3 B IV's, thereby achieving a tactical and moral success.

3) **Deployment of the Pz.Kp.FKL 312** with the s.Pz.Abt. 505.

The deployment of the company in front of the Tiger tanks as combat reconnaissance reflected the tactical guidelines and achieved the following good results.

A B IV was deployed at a range of 800 meters against an anti-tank nest consisting of 2-3 anti-tank guns. This position was completely destroyed along with the infantry which had massed there.

A B IV was deployed at 400 meters against a T-34 tank. The T-34 was totally destroyed when it tried to ram the B IV, which completely detonated.

Three B IV's were guided to three heavy gun bunkers at a range of 400-600 meters and destroyed all bunkers through total detonation. One B IV was guided against an anti-tank bunker. Although it caught fire ten meters before reaching the bunker, its detonation destroyed the bunker.

Two B IV's employed at a range of 800 meters against an anti-tank crew and one infantry gun position, destroyed both.

One B IV reached a russian position, was hit with Molotov cocktails and caught fire. Its total detonation had a devastating effect on the enemy position.

In four cases, B IV's were knocked out by defensive weapons while they were being remotely controlled. In these instances, two of the FKL devices could be retrieved, the other two burned out. A total of 20 B IV's were deployed during four days of combat.

signed Major Reinel,
Kommandeur, Pz.Abt.(FKL) 301

Pz.Kp.(Fkl.) 311O.U. 25 October 1943

Report on the deployment of a B IV during the forcible reconnaissance of Belsk on 14 August and about the blowing of a bridge by using a B IV on 14 August 1943.

1.) **Mission:** Forcible reconnaissance of Belsk with a Fkl-platoon and a Grenadier platoon. After scouting the area, both platoons assembled in a hollow west of the town. The Fkl platoon took up hull-down positions. The Grenadier platoon went forward 600 meters toward the edge of town under protective fire from the Sturmgeschütz. At the same time, three B IV's were deployed for reconnaissance, one of which reached the edge of town. Enemy machine gun and rifle fire was markedly reduced as the B IV advanced. The enemy seemed baffled. As they advanced further, the Grenadiers received heavy mortar fire and artillery fire coming from the northern edge of Kotelva. After retiring a short distance from the town, the B IV was once again guided toward the edge of the town and the charge was dropped. The B IV itself was guided back. The pause in the enemy's fire as a result of the explosion was utilized by the Grenadiers to retire. The pause lasted about 5 to 7 minutes.

Results: The reconnaissance was conducted, the enemy lost countless dead, several machine guns and three mortars were destroyed.

Losses: none

Lessons: The deployment of the B IV was possible based on the fact that the controlling vehicle and the B IV could be calibrated and tested together while still in the rest area. The time needed to accomplish this task prior to each deployment is about 5 to 6 hours. If a road march (of 10 kilometers) occurs or a time span of 12 hours lapses, the devices must be recalibrated.

The terrain also favored the **B IV** deployment. In the hollow west of the town, the necessary equipment check could be accomplished and the **B IV**'s prepared for action (which took about 10 minutes) without being observed by the enemy prior to the attack. The steadily increasing slope permitted vehicle guidance out to a range of 1200 meters. There were no hindering craters, ravines or wooded areas. The breakdown of two **B IV**'s can be traced to defects which came to light during the deployment. A relay jammed in one of the **B IV**'s.

2.) **Mission:** Blowing a bridge near Kholovdevchiya by means of a **B IV**.

The bridge to be blown was under enemy observation. Blowing the bridge with a **B IV** was therefore necessary.

The **B IV** was guided onto the bridge from a hull-down position. The enemy acted cautiously and restrained, allowing the **B IV** reach the bridge. The command to drop the charge failed, so that the entire vehicle had to be detonated.

Results: Mission accomplished

Losses: None

Lessons: The **B IV** was guided without any problems onto the bridge at a range of 500 meters from a hull-down position. The enemy seemed surprised and remained quiet. The 18-ton wooden bridge was completely destroyed. Entire supports were destroyed down to below the surface of the water. The length of the bridge was 25 meters.

signed Bachmann,
Oberleutnant und Kompanie-Chef

Battlefield Support Vehicles

In a Generalstab des Heeres memorandum Nr.355/36 gKdos 8.Abt.(III) sent to the AHA dated 15 December 1936, it was reported that the Oberbefehlshaber considered the first trials with the sPak to be most promising. He therefore directed that detailed testing of the Pak (Sfl.) be accelerated so decisions concerning the Pak (Sfl.)'s acceptance could be made in the Fall of 1937.

But for purpose of the trials, it was deemed necessary for the test platoon to possess not only the Pak (Sfl.), but also support vehicles.

The following were deemed to be the necessary support vehicles:

A) **Munitionswagen**

(These requirements initiated the design for the LEICHTER, GEPANZERTER MUNITIONSTRANSPORTWAGEN (SD.KFZ.252))

I. The Munitionsfahrzeug was to supply ammunition to the Pak (Sfl.) on the battlefield, that is, it travelled between the Pak (Sfl.) and the ammunition trucks.

As a rule, to reload, the Pak would be driven back to the nearest available cover and there take on ammunition. Driving the ammunition trucks forward into an open firing position would be an exception.

II. Requirements:

1. Good cross-country mobility, similar to that obtained by a half-tracked vehicle
2. Speed from 4 to 60 km/h
3. All-round armor, effective against armor piercing, small arms fire
4. Ability to tow a trailer of up to 2 tons weight (loaded)
5. Load capacity, for the tractor and trailer each, 1000kg useful load = 200 rounds of 75mm shells with packing
6. Crew: one driver, one ammunition handler
7. Armament: 1 machine gun which, mounted on a boom, for engaging attacking aircraft

III. For the near term, consideration was to be given to a fast-moving half-track vehicle.

B) **Beobachtungswagen**

(These requirements initiated the design for the LEICHTER, GEPANZERTER BEOBACHTUNGSWAGEN (SD.KFZ.253)

I. The Beobachtungswagen should be used during:

1. Target reconnaissance
2. Directing fire for a small unit, for example a platoon or company conducting direct fire (target assignment)
3. Fire direction for indirect fire when the entire company is deployed together directly behind the front line. In this case it would act as a forward, armored observation post. The Beobachtungswagen serving the Pak (Sfl.) has earned this designation in the earlier sense of the Beobachtungswagen belonging to the Artillerie. Nowadays, the Beobachtungswagen of the Artillerie is primarily just a transport carrier for all sorts of technical equipment. It is no longer an observation vehicle. Earlier, for example when it was deployed to an observation point with an observer, it had earned its name.

II. Requirements based on the expected use:

1. Vehicle with similar cross-country maneuverability as the sPak (Sfl.), which means a fully-tracked vehicle
2. Speed: faster and more maneuverable than the Pak (Sfl.), walking speed, that is about 4 to 40 km/h
3. All-round armor, effective against armor-piercing small arms fire, effective against 20mm projectiles in the front
4. Crew: one commander (observation officer), one observation NCO, one driver

1. Observation devices: Scissors periscopes with 360 degree traverse. The scissors periscopes must be extendable, upward and downward as a periscope. The height of the periscope head must be such that observation can still be conducted, in most cases, without the vehicle having to leave its covered position. The scissors periscope must be able to be dismounted from the vehicle for use on the ground.
2. Drivers' vision device
3. View ports

Means of communication:
1. Wireless telephone with a range of about 3 to 4 km to communicate with the company in closed deployment and in covered positions. The device must be dismountable. Serviceable by the driver or observation NCO (remote station as a mobile station on the ground).
2. Semaphore flags or signals for communication with the guns in open firing positions.
3. Other equipment: A sub-machine gun for close-combat self defense, an R.K., a flare pistol and 4 smoke grenades.
The chassis of the Panzer I should be considered for the vehicle.

The Generalstab des Heeres asked if two ammunition vehicles and one observation vehicle could be complete and operational for the tests with the Pak (Sfl.) beginning in May 1937. Based on the test results, the following was to be decided in the Fall of 1937:

— If vehicles of the type were necessary and suitable
— In what direction should the final design and procurement be pursued.

For purposes of testing, it was not necessary for the vehicles to be fitted with full armor plating. They only needed to represent the final measurements and weights. In the same respect, a slow-moving towing vehicle could be used for testing instead of a fast "Lilliput" Zugkraftwagen (1-ton "Lilliput" half-track).
The Generalstab des Heeres asked for confirmation of the fact that the vehicles would be ready on time for testing. For these tests, prototypes were envisioned of the light half-tracked Zugmachine, type "D ll 3" with an armored superstructure. It was quickly evident that the longer, and therefore heavier, armor plating severely limited the vehicle's useful load. As a result of this revelation, the running gear on the armored models of the light Zugkraftwagen were shortened by one road wheel.

The first armored experimental model with the running gear of the 1-ton (Sd.Kfz.10), Typ D ll 3 designed by Demag.

Development of the armored model of the leichten Zugkraftwagen

When the development of the Schützenpanzer began, the demand was not only for squad-level vehicles, but also for team vehicles. The half-track chassis of the 1 ton Zugkraftwagen designed by Demag, Wetter/Ruhr was available for this purpose. Essentially, the following modifications were made to the non-armored, "Typ D7" to create the basic armored version, specified as the "Typ D7p":

— armored superstructure replaced the sheetmetal superstructure

— running gear shortened by one road wheel

— altered radiator, steering wheel, fuel tank and exhaust system

The vehicles intended for use by Sturmgeschütz units were fitted with the Maybach 6 cylinder in-line engine, type HL 42. The engine, first available in 1939, had a 90mm bore and 110mm stroke, with a swept volume of 4.2 liters. Maximum output was 140 metric hp at 4000 rpm. Originally, this engine was intended for installation in the Maybach SW 42 automobile, the military version of which, with came to be installed in the D7, HL kl 6, D7p, H Kl 6 p and the schwere Wehrmachtschlepper half-tracks. Production of this engine was permitted under license beginning in 1938, even before the motor went into mass production. In 1939, Maybach offered 100 HL 42 engines for the D 7, capable of burning fuels with a 60 octane rating, 93 HP at 1800 rpm, at the price of 3575 RM per unit.

The transmission was a semi-automatic Maybach VG 102128 H. It had seven forward and three reverse speeds. The gears were selected by hand, while the actual gear changes were accomplished by a vacuum system that was activated by depressing the clutch. This transmission was also produced by other licensed domestic and foreign

The chassis for the 1-ton Zugkraftwagen for armored superstructures with shortened suspension.

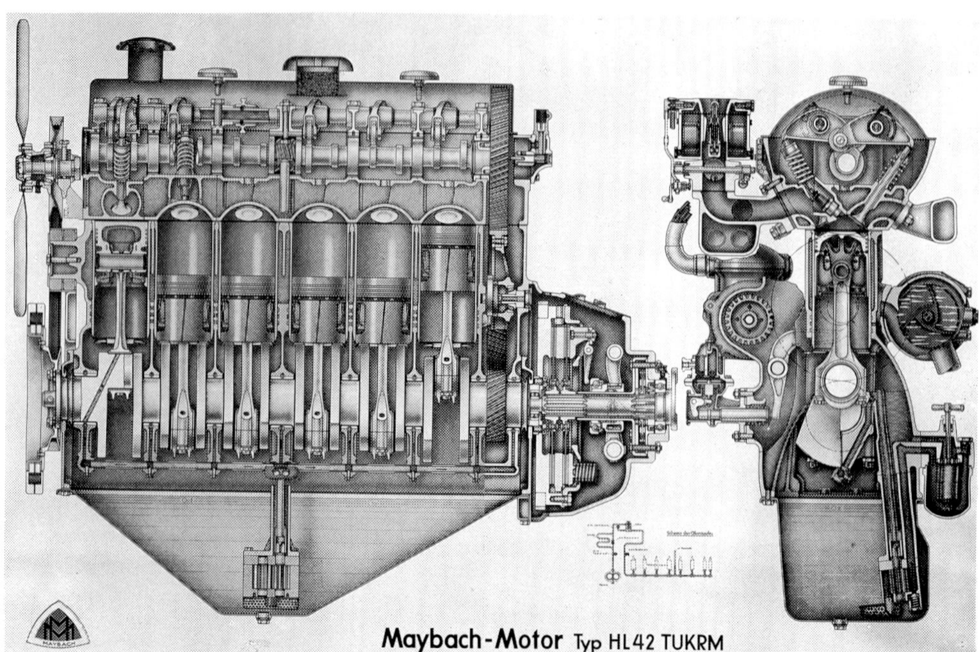

Cross-section and plan view of the Maybach HL 42 TUKRM high-performance engine.

firms.

Leaving the transmission, a pair of beveled gears drove each of the steering gears. Turning the steering wheel, moved the front wheels first, before the track steering engaged.

In the case of heavier steering, the steering brakes were actuated. The forward mounted, drive wheels were driven by the steering gears. The drive wheel with 12 rollers engaged the track. The brake drums for the hydraulic brakes were located in the drive wheels. The road wheels were hung on suspension arms sprung by means of torsion bars. Overlapping one another, alternatively inboard and outboard, the roadwheels were constructed as inter-changeable steel disk wheels. Drive wheels, inner road wheels and idler wheels guide the track center guides. Each of the two Zpw.51/240/160 tracks consisted of 38 links, fastened together with pins. The drive teeth on the track were designed as lubrication chambers. Each track link featured a rubber pad secured with four screws which were easily replaced. The forward axle was designed as a floating axle, supported at the side of the chassis by cross-mounted leaf springs. The tubular axle was supported in the middle of the chassis by a triangular brace, which rotated to absorb impact forces.

Leichter, Gepanzerter Munitionstransportwagen (Sd.Kfz.252)

The self-supporting superstructure was composed of three welded pieces (front section, front shield, and rear section) with riveted cross braces. The front and rear armor sections were fastened together with bolts. They consisted of welded armor plates effective against armor-piercing, small arms fire. The superstructure was bolted to the chassis. The fuel tank, located in the rear, had a capacity of 140 liters. The front armor with its removable shield protected the engine compartment. The front

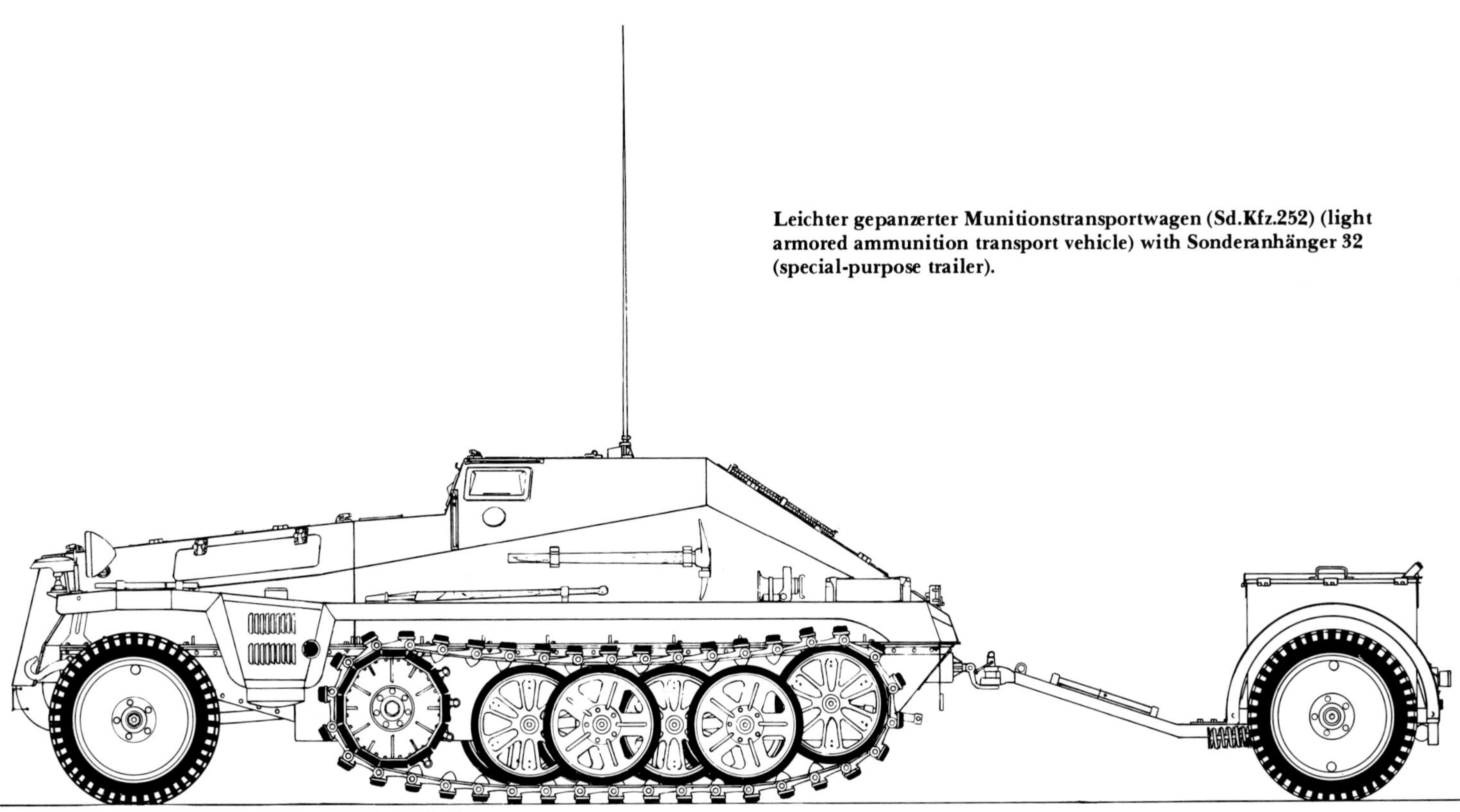

Leichter gepanzerter Munitionstransportwagen (Sd.Kfz.252) (light armored ammunition transport vehicle) with Sonderanhänger 32 (special-purpose trailer).

Leichter gepanzerter Munitionstransportwagen (Sd.Kfz.252) (light armored ammunition transport vehicle) . Side view of the first model. Early design for the road wheels.

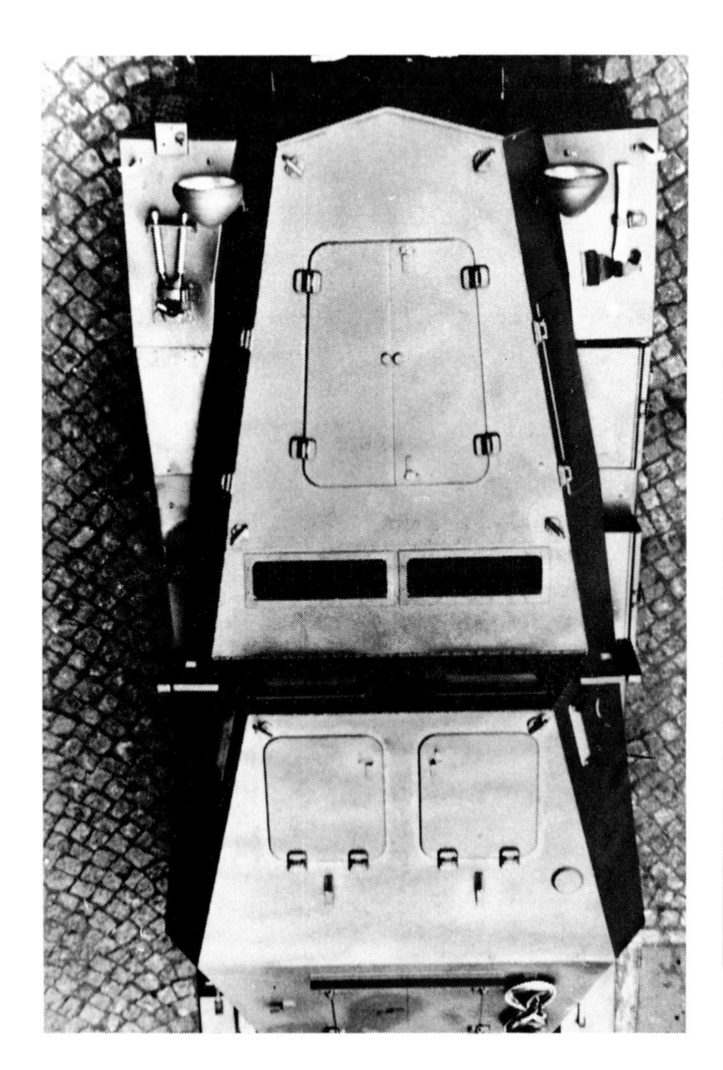

This overhead view of Sd.Kfz.252 shows the enclosed superstructure.

Interior view with open hatches of a Sd.Kfz.252 belonging to StuG-Abt.184.

Sd.Kfz.252 with Sd.Anh.32 during testing by Wegmann in Kassel.

shield also protected the steering rods and shock absorbers. The rear armor formed the cargo area, which was separated from the engine compartment by a firewall. On the forward part of the superstructure roof, two access hatches were located next to one another, one for each member of the two-man crew. A two-section access door, installed in the inclined rear, could be locked from the inside. A towing hitch installed on the rear of the hull, could move to each side and rotate 360 degrees.

The chassis was assembled in automotive running order at either the Demag Wetter/Ruhr plant (chassis number in series 96001-103000) or Büssing-NAG, Berlin-Oberschönewalde (chassis numbers in series 310001-311000).

Wegmann & Co. in Kassel developed the superstructure. The armor components were manufactured and welded by armor sub-contractors. Up to December 1939 there were a total of 55 Beobachtungs- and Munitionskraftwagen in production for the sPak, 20 of which were to be completed by 31 December 1939, a further eight were expected in January of 1940. There were, however, constant delays in the delivery of the armor components. After Austria was incorporated into the German Reich in 1938, introduction of austrian firms into the arms industry was immediately initiated. Gebr. Böhler & Co. AG in Kapfenberg-Deuchendorf constructed a new armor plate facility by the summer of 1939.

The new Böhler facility in Kapfenberg-Hafendorf for welding armored components was 40% by December of 1939. After the new plant was completed, in 1940 Böhler moved the production of armor superstructures to Hafendorf. In the early years, the firm manufactured armored superstructures for armored reconnaissance vehicles and armored personnel carriers. Their own Army acceptance office for armored superstructures was set up at the Böhler plant in Deuchendorf.

After delays, the production of armor components for the Sd.Kfz.252 from Gebr. Böhler began in September 1940 with 20 superstructures. This was followed by:
58 superstructures in October to December 1940
123 superstructures in January to March 1941
113 superstructures in April to June 1941
99 superstructures were shipped in July to September 1941.

The contract was completed. In sum, Böhler produced 413 Sd.Kfz.252 superstructures
Beginning in January 1941, in addition to Wegmann, the Deutschen-Werke in Kiel was added as an assembly plant.

Production of the Sd.Kfz. 252

Month	Produced	Remarks
1940		
March	0	
April	0	
May	0	
June	10	
July	17	
August	3	
September	0	delayed deliveries from Böhler
October	3	startup problems at Wegmann due to new series
November	16	
December	5	
1941		
January	21	startup problems at Deutsche Werke, Kiel
February	54	
March	50	
April	44	
May	35	loss in production rate due to enemy action
June	30	loss in production rate due to enemy action
July	44	
August	52	
September	29	
October	0	
total	413	

Production of the Sd.Kfz.253

Month	Produced	Remarks
1940		
March	1	
April	9	
May	10	
June	5	
July	0	difficulties with sub-contractors
August	10	
September	0	delayed deliveries from Böhler
October	31	
November	19	shortage of parts for radio equipment
December	0	shortage of parts for radio equipment
1941		
January	50	
February	40	
March	40	
April	30	
May	39	
June	1	
total	285	

Leichter, Gepanzerter Beobachtungskraftwagen (Sd.Kfz.253)

The final superstructure design, conceived by Wegmann had a closed crew compartment. A circular, flat cupola somewhat offset to the left of middle was mounted in the superstructure roof. The twin cupola hatches opened to the sides. There was a smaller opening in each of the cupola hatches to accommodate the heads of the extended scissors periscopes. Directly behind the cupola, on the left side of the superstructure roof, a rectangular hatch was hinged on one side. On the right side of the superstructure roof, a wood trough was mounted to protect the 2 meter-antenna in its rest position. This guard extended beyond the forward edge of the superstructure roof. Mounted in the superstructure rear was a one-piece access hatch with a view port.

The firm Gebr. Böhler & Co. AG in Kapfenberg/Steiermark participated in the production of the armored superstructures for the Sd.Kfz. 253. Delivery of the superstructures began in August of 1940 and up to 1 October 1940, 40 superstructures had been accepted. From October to December 1940, an additional 133 superstructures followed. In the reporting period from January to March 1941, 60 superstructures were produced, which were followed by an additional 17 from April to June 1941.

Böhler received an additional order for 150 Sd.Kfz.253-superstructures in August of 1940, of which they only completed 35.

The leichter, gepanzerter Beobachtungskraftwagen (Sd.Kfz.253) weighed 5.73 tons. For radios, it carried a 10-watt transmitter, two VHF receivers as well as a removable backpack radio.

After the production of the Sd.Kfz.253 had been completed in June of 1941 and startup of the leichte Schützenpanzerwagen (Sd.Kfz.250) had been assured, Sturmgeschütz units were also issued the leichten Beobachtungs-Panzerwagen (Sd.Kfz.250/5). Beginning in 1943, a simplified version of the Sd.Kfz.250/5 was produced.

Leichter gepanzerter Beobachtungskraftwagen (Sd.Kfz.253). (light armored observation vehicle)

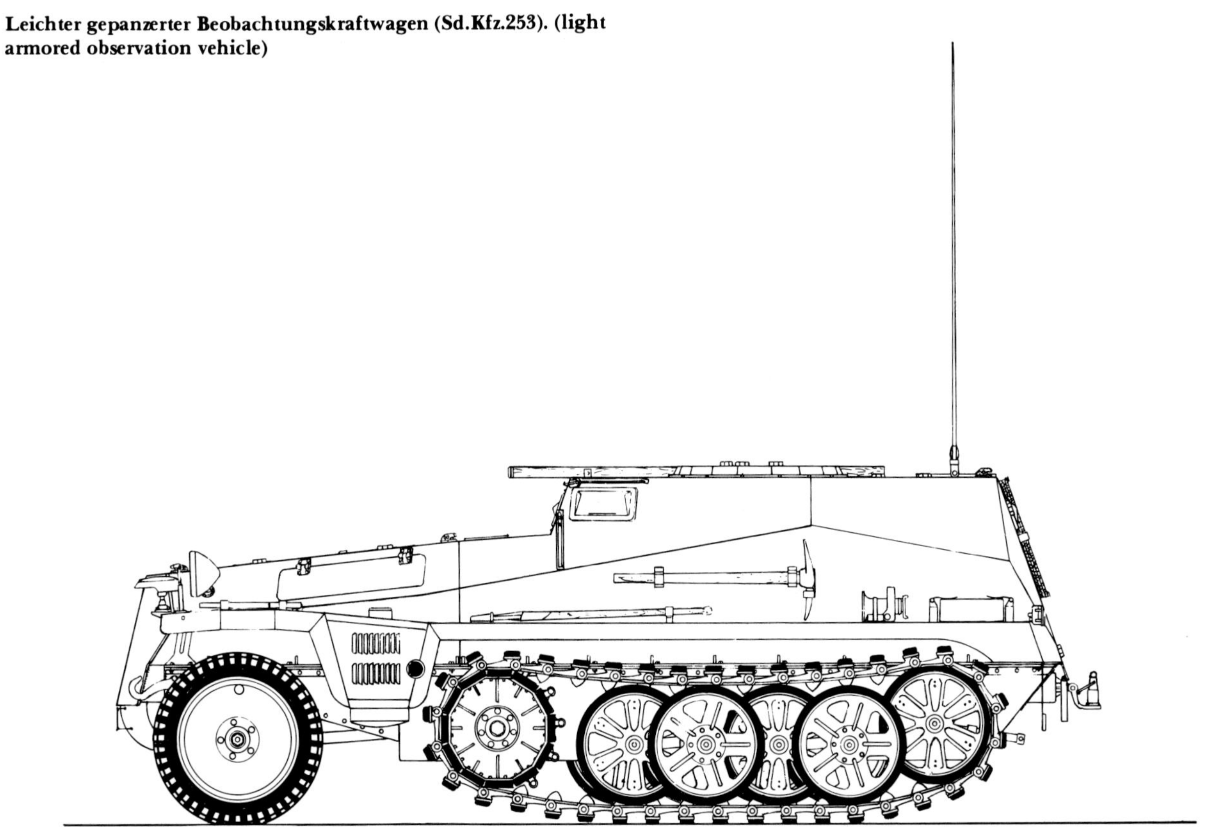

Leichter gepanzerter Beobachtungskraftwagen (Sd.Kfz.253). (light armored observation vehicle)

This overhead view of the Sd.Kfz.253 shows the enclosed superstructure.

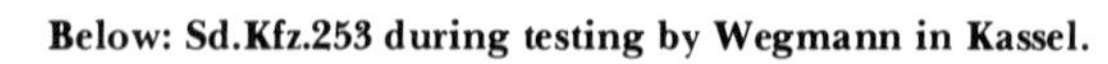

Below: Sd.Kfz.253 during testing by Wegmann in Kassel.

Sd.Kfz.253 — a platoon leader's vehicle in StuG-Abt.184.

Driver at the wheel. The vision block has been lowered.

Sd.Kfz.253 — Radio racks with installed radio equipment. (BA)

Replacement of the Assault Artillery Special Vehicles by the Leichten Schützenpanzerwagen (Sd.Kfz.250)

The first 39 leichten Schützenpanzerwagen (Sd.Kfz.250) were produced in June of 1941. It differed from its predecessors, the Sd.Kfz.252/253 in that it had a modified, open superstructure. The Sd.Kfz.250/6, Ausf.A served Sturmgeschütz units as a carrier for 7.5cm Kanone L/24 ammunition. Seventy rounds of this 75mm ammunition were carried in 35 ammunition cases.

The leichte Munitions-Panzerwagen (Sd.Kfz.250/6), Ausf.B served in units which possessed Sturmgeschütz with 7.5cm StuK 40 L/43 or L/48. 60 rounds of ammunition for the long 75mm StuK 40 were carried in canisters.

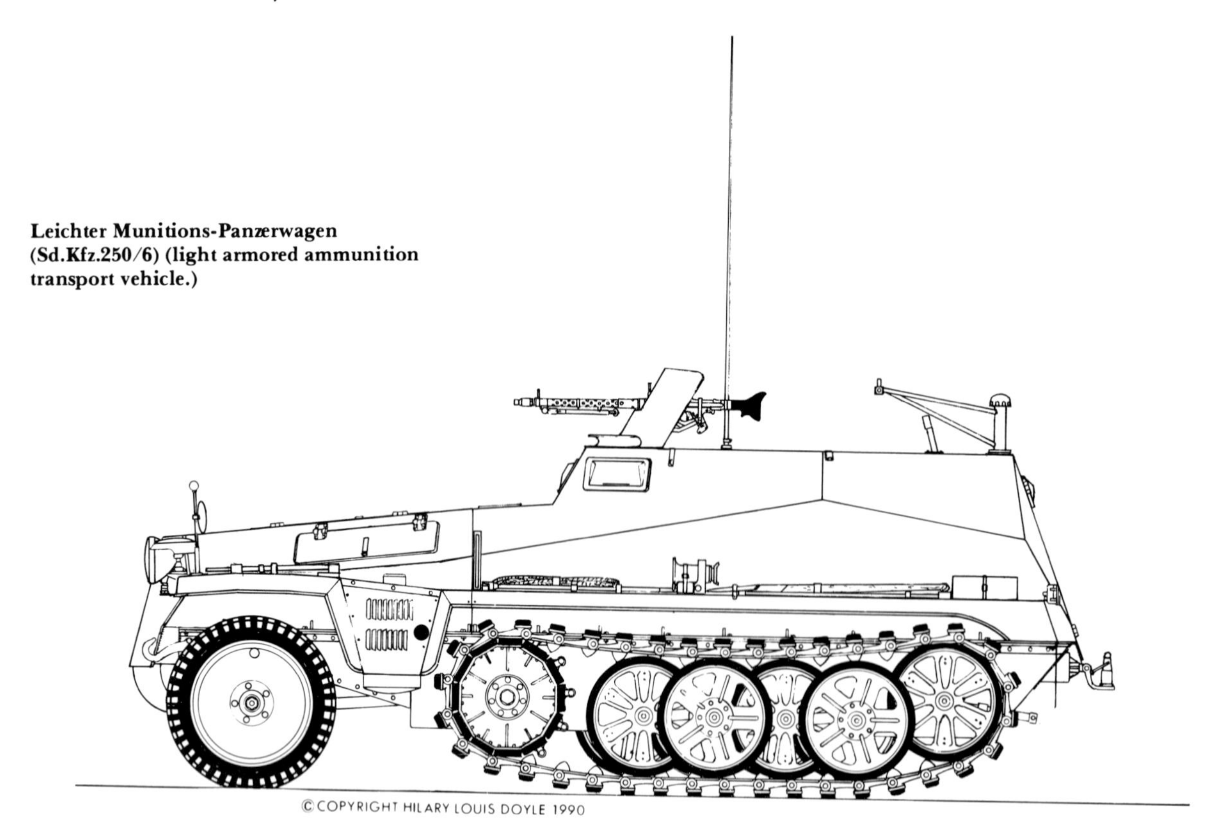

Leichter Munitions-Panzerwagen (Sd.Kfz.250/6) (light armored ammunition transport vehicle.)

Leichter Munitions-Panzerwagen (Sd.Kfz.250/6).

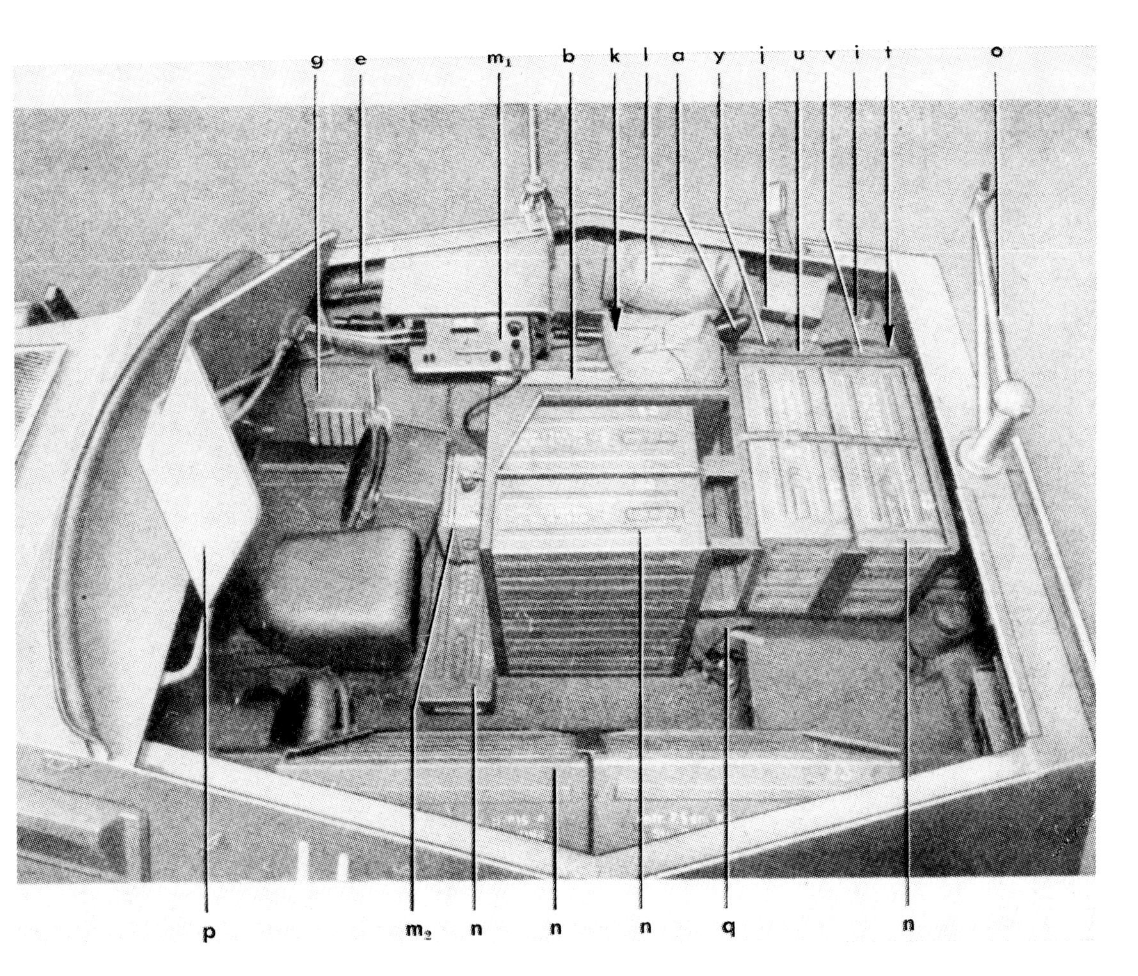

Sd.Kfz.250/6, Ausf.A — right side of the interior with 70 rounds of 7.5cm Kanone L/24 ammunition.

Left interior.

Sd.Kfz.250/6, Ausf.B — right side of the interior with 60 rounds of 7.5cm Sturmkanone 40 ammunition.

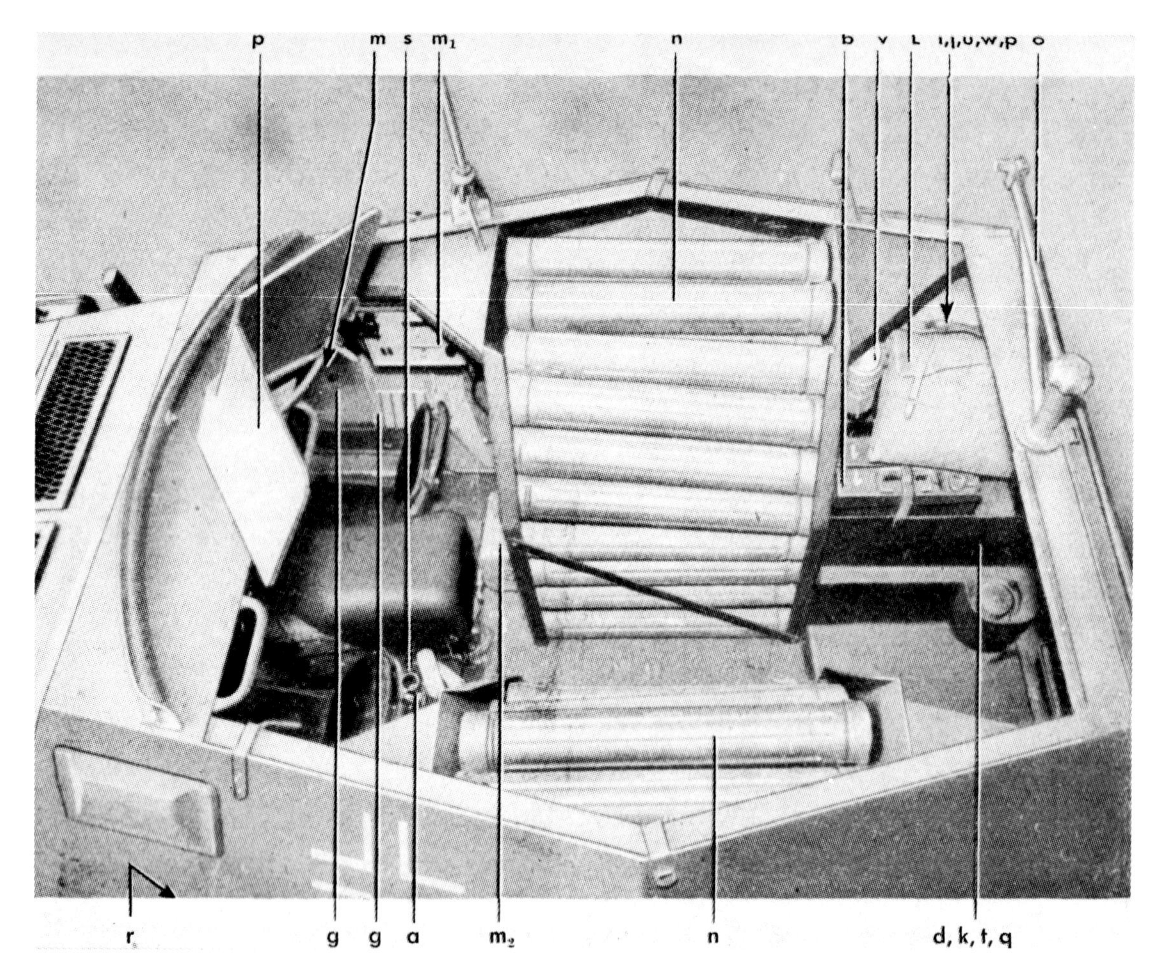

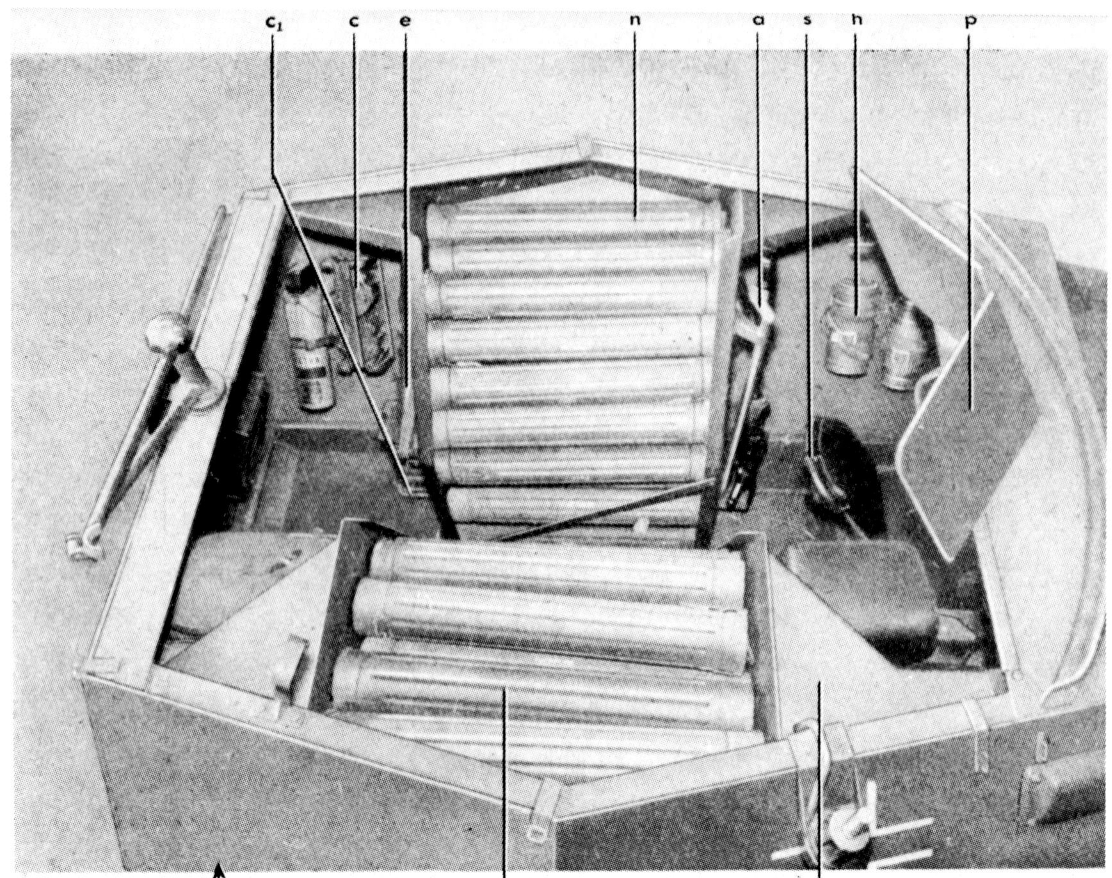

Left interior.

Munitionsanhänger (Sd.Anh.32) (ammunition trailer) —closed.

Munitionsanhänger (Sd.Anh.32) (ammunition trailer) — opened. The inserts were changed based on the type ammunition to be carried.

Wegmann developed a single-axle unarmored trailer (Sd.Anh.32/A) for the Sd.Kfz.252, which also was used by other units. The trailer had a maximum weight of 780kg, a tare weight of 330kg. There were four compartments, with measurements of: 415mm in length, 400mm in width, and a height of 410mm. The price per trailer was 668 RM. The trailers were often used for other than their intended purposes, and were towed behind a wide variety of vehicles.

Schwere Zugkraftwagen 18 T (Sd.Kfz.9)*

At the beginning of the war, the Sturmgeschütz units were reliant on outside support for major vehicle maintenance and overhaul. As the new branch grew, a call went out for its own maintenance organization, which led to the Sturmgeschütz being repaired by the maintenance services organic to the Sturmgeschütz Abteilung.

In cases where damage was heavy, a divisional Werkstatt-Kompanie or a Panzer-Werkstatt-Kompanie could be called upon.

When the damage could not be repaired at the unit level, the Sturmgeschütz could be sent to Panzer-Instandsetzungs-Dienst units attached to their Heeres-Gruppe or were turned over to the Panzer-Instandsetzungs-Kompanie-Werke. Those Sturmgeschütz vehicles which could not be repaired at these facilities were returned for major overhaul and rebuild to the Heimatsinstandsetzungsdienste (repair facilities in Germany).

— Sturmgeschütz from the Eastern Front and from the Southeast (Balkan) region were sent to the Heeres-Panzer-Nebenzeugamt in Oppeln/Oberschlesien

— all remaining regions were to send their vehicles to the Heeres-Panzer-Zeugamt in Königsborn/Magdeburg.

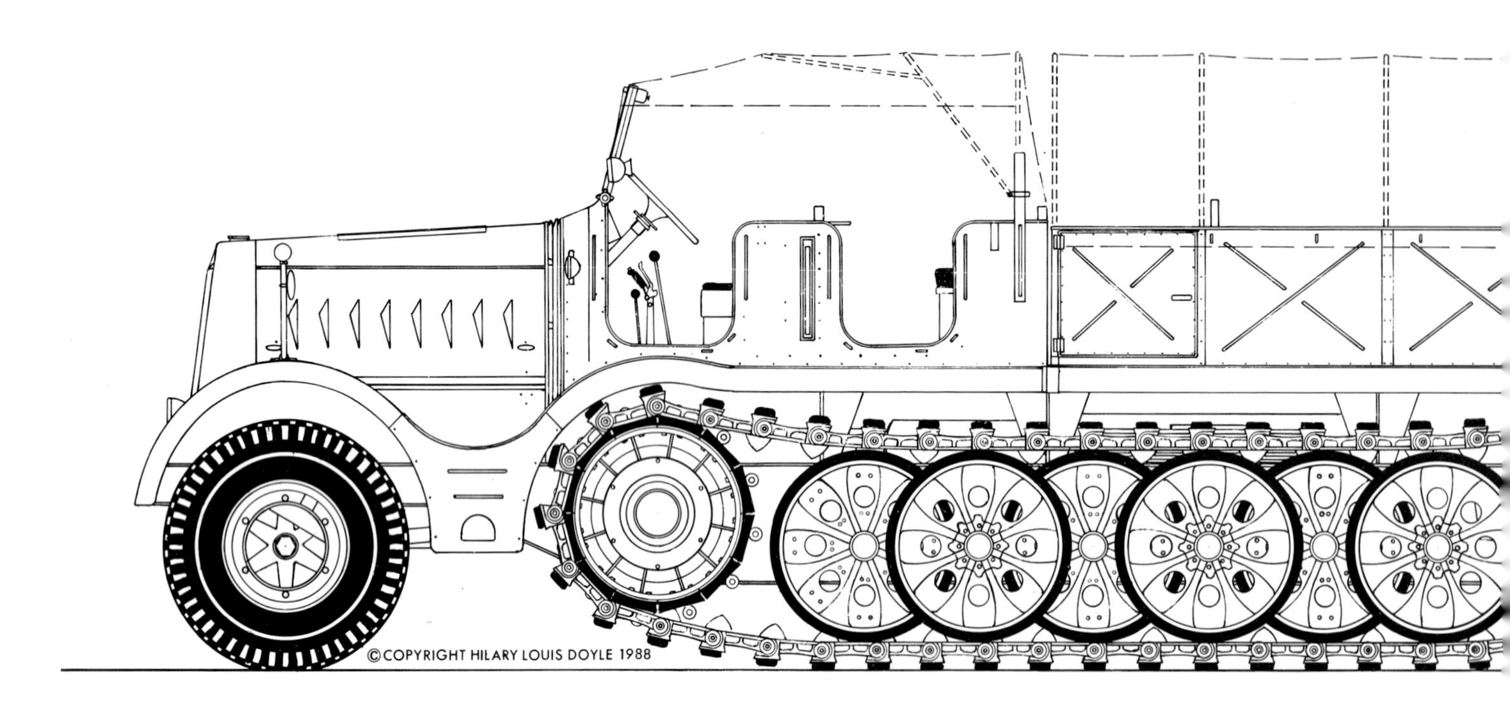

Important equipment the recovery of a broken down Sturmgeschütz was the schwere Zugkraftwagen 18 (heavy 18 ton half-track) working with the Tiefladeanhänger für Panzerkampfwagen (23t) (23 ton capacity, low-boy trailer for tanks).

Facts surrounding the creation of the 18-ton Zugkraftwagen are presented in a letter written to the author by Prof. Dr.-Ing. Rudolf Franke on 19 February 1980. This letter contained the following (excerpts):

"When I arrived in June 1933 at the WaPrüf 6 department of the Heereswaffenamt as first assistant to Herr Kniepkamp, my first assignment was to help in test the first KMZ 85 prototype at Krauss-Maffei, initially in the closed-down facilities at Hirschau. During the transfer of the vehicle to Kummersdorf, one track tore off and wrapped itself around behind the vehicle like a snail. The hardened pins of the Ritscher lubricated, needle-bearing track had developed hardening cracks which had not previously been recognized.

At the same time, at Daimler-Benz in Marienfelde, the prototype of a heavy half-tracked towing vehicle, predecessor to the Zgkw 12 t, was being driven with conventional running gear for the period, small road wheels and return rollers. The running gear was essentially inferior to the interleaved running gear of the KM towing vehicle at high road speeds, so that Kniepkamp was able to overcome major opposition to his "schachtellaufwerk" (interleaved running gear) design.

He found the name "ZKW", meaning Zugkraftwagen, to be analogous to Pkw and Lkw. It was then officially made the name of the vehicle, but unfortunately with the hard to pronounce abbreviation of "Zgkw." If I recall correctly, Kniepkamp received permission to issue a contract for developing a prototype of a 50 km/h tractor to KM. This was only approved as a "driveable testbed" because the project had previously been dropped as being unrealistic.

Then an entire family of Zkw's had to be developed, among which was a Zkw 18t for a very heavy long-barrelled gun from WaPrüf 4. This gun was to be hauled as three separate loads, each weighing 18 tons, behind three Zkw's.

I and my co-worker Spiess had to develop the design specification data for this "family" of vehicles. We often did not know, as in the case of the Zugkraftwagen 18t, the loads and equipment. We simply planned sensible and gradual increases for the future. Only when the first prototypes were presented to the "customers" did they become aware of their existence and want to have these vehicles. I still remember that we and FAMO were counting on a contract for about 10 vehicles. I was ecstatic when 60 vehicles were ordered. Interest in this vehicle for the purpose of recovery of immobile tanks was suddenly created. I turned the Zugkraftwagen development over to Herr Student in roughly 1937 and received more than enough work as the group director for tracked vehicle engines and transmissions."

As further confirmation of Herr Dr. Franke's statements, the following text of a document is printed, which, unfortunately has no date. It can, however, be safely

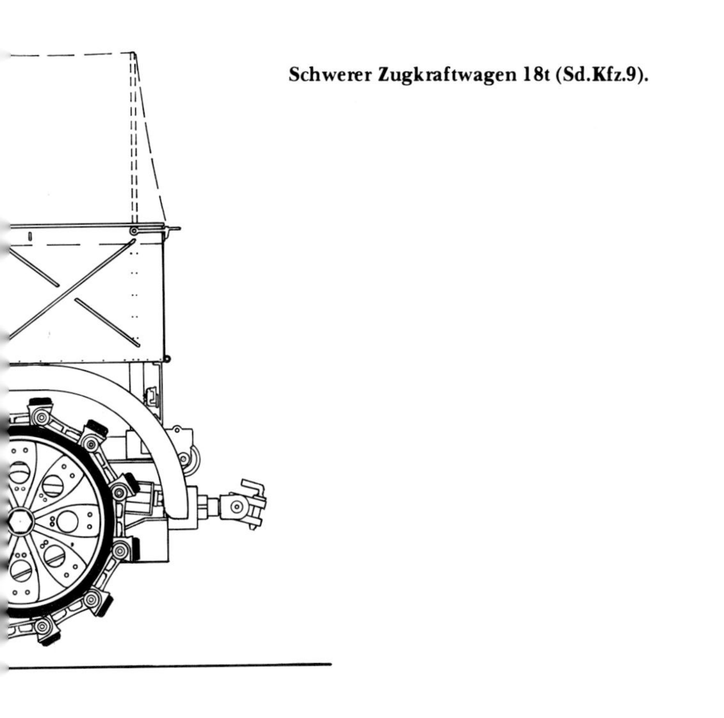

Schwerer Zugkraftwagen 18t (Sd.Kfz.9).

assumed that it is from the years 1934/1935:

Special case of "schweren-geländigängigen Kettenschlepper" (heavy cross-country tracked towing vehicles):

For especially heavy loads, a very heavy cross-country capable tracked towing vehicle is now being designed. No details of the vehicle will be given here before definitive test results are available.

The schwere, geländigängigen Kettenschlepper (Sd.Kfz.8) is sufficient for 16t loads, even in difficult terrain. A test has shown that this Kettenschlepper was able to carry a load of 50t up a 6 degrees incline on a plank trail (this consisted of loose boards laid next to one another in a potato field.) The rolling resistance (in poor terrain, loose sand) was about 300 kg per ton of trailer weight. In the previous case, a draw capacity of 16(t) x 4.6t is necessary.

This Kettenschlepper, however, has a towing capacity of 8.5 tons, and a large reserve of power, which can be used to climb slopes. Inclines of 10 degrees require about 175kg of pulling power per ton of trailer weight. Taking these values as a basis, this Kettenschlepper could climb a slope of about 14 degrees with a trailer load of about 16 tons.

In this specific case, though, the data for climbing ability data are conservatively listed.

These super-heavy half-tracked vehicles were developed and produced by FAMO, Fahrzeug- und Motorenwerke GmbH, formerly Machinenbau Linke-Hofmann, Breslau. FAMO produced, among other products, commercial wheeled and fully tracked tractors. In the final months of the war, the FAMO plant was relocated to Sachsen. FAMO affiliated facilities in Warsaw, Poland had also been involved in production since 1941.

The production of the Zugkraftwagen 18t (Sd.Kfz.9) began in 1935 with the "Typ FMgr 1", an 18-ton half-tracked towing vehicle with a total useful weight of 35.5 tons. The vehicle was originally intended for towing the 24cm-Kanone 3, which was developed by Krupp and was to be introduced in 1937. The price of the towing vehicle was 75,000 RM. In 1938, the "Typ F2" followed which only cost 60,000 RM, due to the increased number ordered. The Panzertruppe had intended this to be their standard armor recovery vehicle and at the same time the towing vehicle for the Tiefladeanhänger (23t). For this mission, the vehicles did not receive the usual Artillerie superstructure with benches for the gun crews, but had an open superstructure with rows of seats for the crew and an equipment platform in the rear.

The "Typ F3" appeared in 1939 and was produced, essentially unaltered, until the end of the war.

Vehicle Description

During development of half-tracked vehicles for the German Reichswehr (and later the Wehrmacht), Maybach had achieved an absolute monopoly on the development and production of the engines for these vehicles. A short look into the history of the power plants for the Zugkraftwagen 18t confirms this fact:

A 12-cylinder V carbureted engine with over 200 hp was planned for the experimental vehicles. In 1935 this led to the beginning of the design of "Typ HL 95" prototype engines.

A brief comparison makes the switch-over to the mass-produced HL 108 clear:

— HL 95 with d/s 94/114 P/n 230/2600 12 V (two test engines)

Schwerer Zugkraftwagen 18t (Sd.Kfz.9). (BA)

Drehkran (Hebekraft 6t) (rotating crane with 6-ton lifting capacity) from Bilstein on the s.Zgkw 18t (Sd.Kfz.9/1). (BA)

Drehkran (Hebekraft 10t) (rotating crane with 10-ton lifting capacity) from Demag on the s.Zgkw 18t (Sd.Kfz.9/2). (BA)

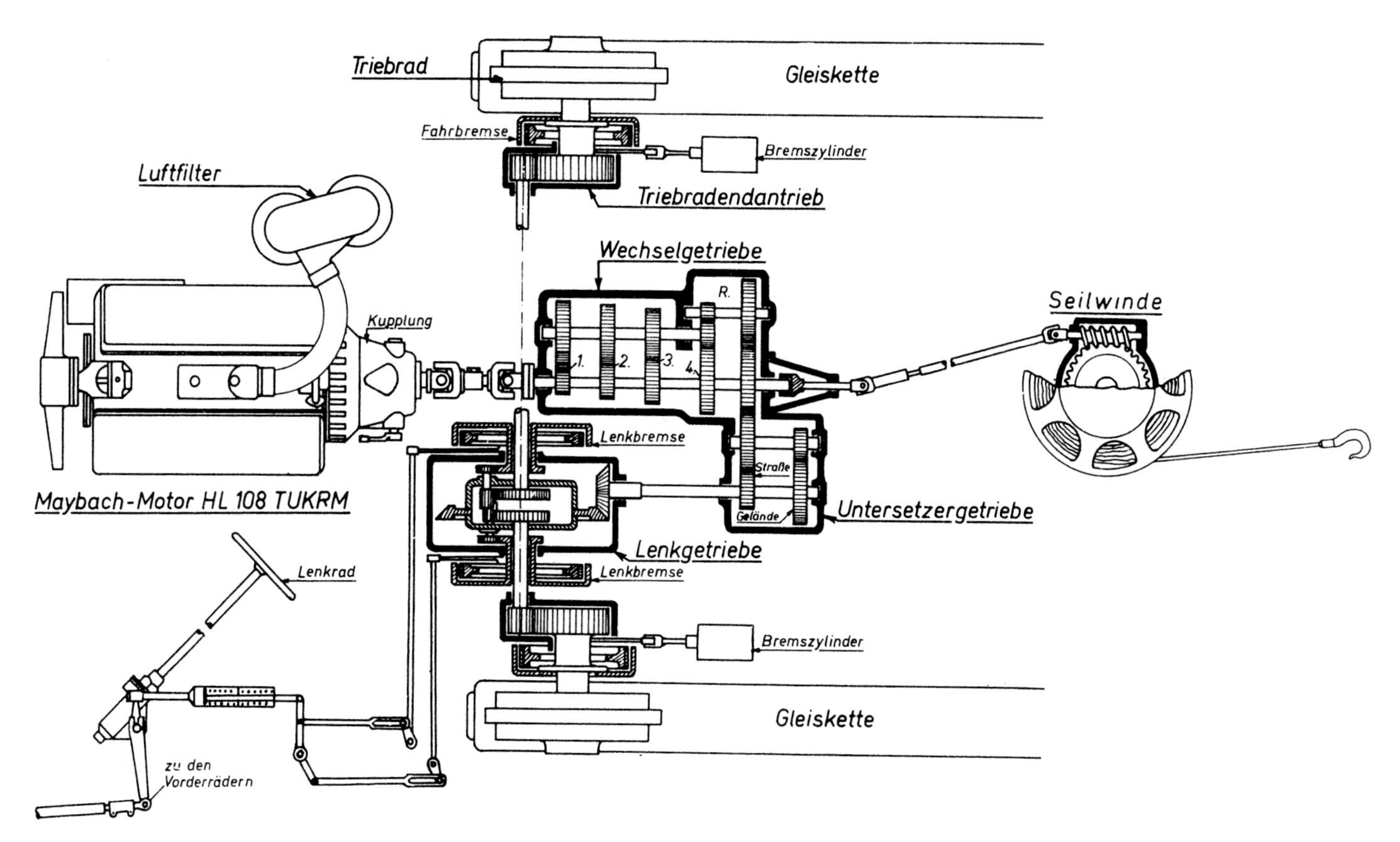

Zugkraftwagen 18t — transmission and drive train (including the drive for the winch).

— HL 98 with d/s 95/115 P/n 230/2600 12V (for military use)
— HL 108 with d/s 100/115 P/n 230/2600 12V (mass produced)

Further development in 1943 as model HL 109 with the same data. In 1940 the diesel test engine HL 174 with d/s 125/130, P/n 400/2600 12 V.

The Maybach HL 108 engine, which preceded the mass-produced HL 120, was also an important power plant system for the 0-Serie for the Sturmgeschütz.

The engine was situated forward in the frame, mounted on three rubber dampers. In view of the widely varied attitudes which the vehicle could attain, an oil pump was installed at the lowest point in the oil pan. Coolant circulation was achieved by a centrifugal pump. The air-fuel mixing was accomplished in two Solex two-stage downdraft cross-country carburetors.

The fuel (230 liters) was located in a fuel tank under the driver's seat and in a reserve tank (60 liters). For starting the engine, electrical and centrifugal starters were installed.

The main clutch was built into the flywheel and served to interrupt the power flow to the transmission. The transmission (Typ ZF G VL 65 VL 230) was a four-speed Aphon gearbox with one reverse speed. A common partitioned housing contained the variable-speed gears, reduction and steering gears. A additional transfer case with clutch for providing power for a winch was flangged to the transmission housing. The steering gears had a two-fold mission. They were effective both as a differential gear and also aided steering. When the steering wheel was turned 15 degrees, a connecting rod actuated the steering brakes. This caused the revolutions of the encumbered driveshaft to decrease and that of the opposite driveshaft to increase correspondingly. The drive wheels were propelled by the final drives. The running gear consisted of two drive wheels, road wheels, two idler wheels with track tensioners, and the track. Located within the drive wheels were the brake drums for the drive brakes. The running gear was set up as a schachtellaufwerk (interleaved running gear). The suspension for the roadwheels consisted of torsion bars.

The tracks (Zgw 50/440/260) consisted of 47 links, each with a rubber pad (Typ W 601), which were connected to one another by pins. Each track link contained a lubrication chambers, which had to be regularly replenished with grease. The forward axle was a single-piece solid axle. The front wheels were Trilexeinheitsräder (Trilex universal wheels) with tire sizes 12.75 — 20 extra.

The permissible total weight for the Zugkraftwagen was 18000 kilograms, the useful load (road and cross-country) was 2200 kilograms. Maximum speed was 50 km/h.

Production of the Zugkraftwagen 18t

Production commenced at FAMO, Breslau, beginning in 1938 and continued to February 1945. The chassis numbers series ran from 45001 through approximately 47500. Files have not been found to prove whether Ursus, Warsaw under the supervision of FAMO, produced completed vehicles or simply manufactured components for assembly at the FAMO facilities in Breslau.

An increase in production of these important vehicles was possible due to Vomag in Plauen beginning assembly in 1941 and continuing to early 1944. Vomag assembled 300 of the s.Zgkw 18t vehicles with the chassis number series 40001-40300.

On January 1, 1938, none of these vehicles were reported as available in the inventories of either the Army or the Luftwaffe. On April 1, 1938, four vehicles were reported to be in the Army inventory. In the time period from April to the end of December 1938, 34 Zugkraftwagen 18t vehicles were accepted by the Army. The remaining production consisted of:

1939	Accepted by Waffenamt
January	8
February	8
March	8
April	13
May	1
June	0
July	8
August	14
September	13
October	9
November	12
December	2

1940	Accepted by Waffenamt
January	12
February	14
March	15
April	20
May	19
June	21
July	26
August	27
September	27
October	25
November	25
December	9

1941	Accepted by Waffenamt
January	26 - 16 complete vehicles 10 chassis with Bilstein 6-ton crane
February	29 - 20 complete vehicles 7 chassis with Bilstein 6-ton crane
March	28 - 15 complete vehicles 13 chassis with Bilstein 6-ton crane
April	28 - 16 complete vehicles 12 chassis with Bilstein 6-ton crane
May	10
June	14
July	19
August	30
September	22
October	28
November	31
December	27 - 22 from FAMO, 5 from Vomag

1942	Accepted by Waffenamt	
	FAMO	VOMAG
January	19	5
February	19	0
March	22	7
April	14	10
May	25	10
June	25	9
July	23	8
August	25	11
September	22 in addition 3 chassis for cranes	10
October	24 in addition 3 chassis for cranes	10
November	28	11
December	26	11

1943	Accepted by Waffenamt	
	FAMO	VOMAG
January	30 in addition 5 complete vehicles for Japan	11
February	30	13
March	32	13
April	33 in addition 8 chassis for cranes	16
May	34	15
June	35	16
July	37	16
August	40 in addition 10 chassis	15
September	45 in addition 12 chassis	15
October	50 in addition 20 chassis	15
November	60 in addition 12 chassis	10
December	55 in addition 5 chassis one additional vehicle – without engine – to Tatra, Nesseldorf, for fitting with the air-cooled 12-cylinder diesel engine (Typ 103) The vehicle was transferred to Kummersdorf for testing (Order nr.65/44 of 3 Nov 1944 – Tatra Diesel 12 cylinder for Zgkw 12 and 18t)	20

In December 1943, FAMO delivered an Zugkraftwagen 18 t —without an engine — to the Tatra firm for the installation of an air-cooled 12-cylinder "Typ 103" diesel engine. These photos show the front and rear views of the converted vehicle.

1944	Accepted by Waffenamt
	FAMO **VOMAG**
January	80 15 the 15 vehicles reported by VOMAG were completed by FAMO in Breslau
February	65 - 55 complete vehicles 10 chassis with Bilstein 6-ton crane
March	70 - 51 complete vehicles 11 chassis with Bilstein 6-ton crane 8 chassis with Demag 12-ton crane
April	80 - 56 complete vehicle 24 chassis with Bilstein 6-ton crane
May	80 - 65 complete vehicles 15 chassis with Bilstein 6-ton crane
June	85 - 81 complete vehicles 4 chassis with Bilstein 6-ton crane
July	85 - 77 complete vehicles 8 chassis with Bilstein 6-ton crane
August	50 - 45 complete vehicles 6 chassis with Bilstein 6-ton crane Although 120 units were targeted for August, there was a failure in the delivery of trans missions caused by a moving the production shops back from Warsaw. The deliveries from France had stopped due to obvious reasons. 32 Zgkw 18 t were overhauled within industry circles.
September	52
October	63 - due to the bottleneck in transmission production, FAMO was only able to produce 25 complete transmissions itself
November	83 - 63 complete vehicles 20 chassis with Bilstein 6-ton crane
December	101 - plus 1 from August reported late from the production in France

1945	Accepted by Waffenamt
January	85 - deliveries from sub-contractors delayed as a result of the conditions near Breslau
February	24 - Based on war events, only 24 Zgkw 18 t vehicles were produced from a planned 110 units. Production ceased after the relocation of facilities from Breslau to Schönebeck.

The official designations:
- schwerer Zugkraftwagen 18 t (Sd.Kfz.9)
- Drehkraftwagen 6 t (Sd.Kfz.9/1) (Bilstein)
- Drehkraftwagen 10 t (Sd.Kfz.9/2) (Bilstein)
- schwere Zugkraftwagen 18 t mit 40-t Seilwinde (Sd.Kfz.9/6)

Recovery Vehicles – Authorized Within Sturmgeschütz Units

Ssturmgeschützbatterie (Mot.) (ZU 7 Geschütze)

K.St.N.446 dated 1 Nov 1941

Instandsetzungsgruppe
1 schwere Zugkraftwagen 18 t (Sd.Kfz.9)*
1 Tiefladeanhänger f. Pz.Kpfw., 22 t (Sd.Ah.116)*

* only applicable to independent batteries

Zugkraftwagen 18 t towing a Sturmgeschütz. (BA)

Stabsbatterie (Mot.) einer Sturmgeschütz-Abteilung mit Batterien ZU 7 Geschützen

K.St.N.588 dated 1 Nov 1941

Abschleppstaffel
2 schwere Zugkraftwagen 18 t (Sd.Kfz.9)
1 Drehkrankraftwagen (Hebekraft 3 t) (Sd.Kfz.100)
1 Tiefladeanhänger f. Pz.Kpfw., 22 t (Sd.Ah.116)

Sturmgeschütz Batterie (Mot.) (ZU 10 Geschütze)

K.St.N.446a dated 1 Nov 1942

Instandsetzungsgruppe
- 1 schwere Zugkraftwagen 18 t (Sd.Kfz.9)*
- 1 Tiefladeanhänger f. Pz.Kpfw., 22 t (Sd.Ah.116)*

---

* only applicable to independent batteries

Stabsbatterie einer Sturmgeschütz Abteilung mit Batterien ZU 10 Geschütze

K.St.N.558a dated 1 Dec 1942

Abschleppstaffel
- 2 schwere Zugkraftwagen 18 t (Sd.Kfz.9)
- 1 Drehkrankraftwagen (Hebekraft 3 t) (Sd.Kfz.100)
- 2 Tiefladeanhänger f. Pz.Kpfw., 22 t (Sd.Ah.116)

Stabsbatterie (Mot.) einer Sturmgeschütz Abteilung mit Batterien ZU 14 Geschütze

K.St.Nr.416b dated 1 Feb 1944

Abschleppstaffel
- 3 schwere Zugkraftwagen 18 t (Sd.Kfz.9)
- 1 Drehkrankraftwagen (Hebekraft 3 t) (Sd.Kfz.100)
- 3 Tiefladeanhänger f. Pz.Kpfw., 22 t (Sd.Ah.116)

Sturmgeschütz Abteilung (Panzerjäger Abteilung) (10 Oder 14 Geschütze)

K.ST.Nr.1149 dated 1 Feb 1944

Bergetrupp
- 1 schwere Zugkraftwagen 18 t (Sd.Kfz.9) or
- 1 Fahrgestell Pz.Kpfw.III oder IV

Stab- und Stabsbatterie (Mot.) einer Sturmgeschütz Brigade mit Batterien ZU 10 Oder 14 Geschütze

K.St.Nr.416 dated 1 Jun 1944

Bergetrupp
- 3 schwere Zugkraftwagen 18 t (Sd.Kfz.9)
- 1 Drehkrankraftwagen (Hebekraft 3 t) (Sd.Kfz.100)

Tieflaneanhänger für Panzerkampfwagen (23 T)

The Tiefladeanhänger für Panzerkampfwagen (low-boy trailer for tanks), "Typ Ba 41" through "Typ Ba 44" (Sd.Ah.116) was designed for a useful load of 23 tons (older trailers with mechanical winches had only a 22-ton capacity); it served to carry tanks or Sturmgeschütz and other cargo within the limits of its useful load and its size. A loading ramp could be made out of the chassis (this feature was not available on the last trailers produced).
The trailers with a 22t and 23t capacity had the following serial numbers:
Busch, beginning at 1130
Schenk, from 5992 to 6061, 5132 to 6171 and 6385 to 6424 and continuing again from 6456
Ackermann, from 2445 to 2544, and continuing on from 2760
Soga, from 2501 to 2700

This Tiefladeanhänger consisted of a forward chassis, the loading bed, and rear chassis. The chassis frame consisted of beams welded together both lengthwise and across. In the center of the frame were two mounts for the loading ramp. The load bed was secured by two removable pins. On the frame there were eight spring trestles as mountings for the load-bearing springs.

The forward axle was straight, the rear axle was offset. On each axle were two pivoting axle arms and the wheels. The wheels were interchangeable, tire size 13.50-20.

Steering was accomplished through pivoting axle linkage. The placement of the wheels permitted a steering deflection up to an angle of 35 degrees. The steering mechanism could be secured by means of a special device. When secured the trailer fork was prevented from turning to the side. This prevented the forward chassis from turning while backing the trailer. The towing fork could be removed and mounted on a crosspiece at the rear of the chassis.

The load bed was hung between the forward and rear chassis. On the older models, the forward portions of the

The Tiefladeanhänger für Panzerkampfwagen (23 t) (low-boy trailer for tanks, 23 ton capacity) also used with Sturmgeschütz units. The tractor was the Zugkraftwagen 18t.

A defective Sturmgeschütz loaded onto the Sd.Anh.116 low-boy trailer.

The rear chassis was removed to facilitate loading of damaged vehicles. The load bed also served as a loading ramp.

bed's sides could be folded up, thereby creating two loading ramps.

A loaded tank could be prevented from moving by securing the vehicle to the fore and aft brackets.

The rear chassis was similar to the forward in construction. Independent from the forward chassis, the rear chassis could be steered by means of the pivoting axle linkage. While the forward chassis was guided by the towing fork, the rear chassis was guided using the steering wheel provided. A power steering system using compressed air was installed to minimize the strength needed to steer the rear chassis. A meter was installed on the steering column to indicate the deflection of the wheels. Also on the rear chassis, a canopy with a removable foldable

A complete recovery system with tractor and low-boy trailer. (BA)

covering could be erected to protect the driver from the effects of weather. Two lifting winches were installed on earlier models for raising and lowering the load bed.

On later models, two hydraulic lifters were installed in their place.

The trailer had an airbrake system which worked on all eight wheels.

Electric power for electrical lights and signal system was provided from the towing vehicle. When on the move, coordination between the driver of the towing vehicle and the driver of the trailer was accomplished using the following signals:

1 long = attention
1 short, 1 long= ready to move (given by trailer driver)
2 short = stop
3 short = quick stop
1 long, 1 short= over to the right
(can be given by either driver)
2 long = steering blocked
(from the trailer driver)
unblock steering (from the driver of the towing vehicle

Driving the trailer cross-country was to be avoided as much as possible. An attempt was to be made to load tanks on a road.

Trailer Technical Data

Overall length with fork	14,100 mm
Overall length without fork	12,200 mm
Overall width	2,900 mm
Overall height	2,650 mm
Gauge, front and rear	2,480 mm
Axle base	1350+7400+1350 mm
Ground clearance, no load	530 mm
Ground clearance, loaded	480 mm
Turning radius, with towing vehicle and trailer steering	19 m
Length of load bed	5,700 mm
Width of load bed	2,990 mm
Height of load bed above ground, no load	770 mm
Height of load bed above ground, loaded	720 mm
Tires, front and rear	13.50-20
Tire pressure, front and rear,	7.8 atu
Wheel rims	11"-20
Compression depth	105 mm

	w/mech. winch	w/hydr. winch	w/hydr. lifter w/o load ramp
tare weight	14,800kg	13,800kg	13,140kg
useful load	22,000kg	23,000kg	23,000kg
Max loaded weight	36,800kg	36,800kg	36,140kg
Axle weight, no load	3,700kg	3,450kg	3,285kg
Axle weight, loaded	9,200kg	9,200kg	9,035kg

Max speed	40 km/h
Trailer fork, eye diameter	40 mm
Type suspension, front and rear	leaf springs
Shock absorbers, double action	"Stabilus Modell 70"
Brakes, driving and automatic	airbrakes, type BBR 100/140 for all eight wheels
Brake pressure	4 atu minimum
Hand brake	spindle brake, effective on four rear chassis wheels
Toe-in of all axles (no load)	8-10mm
Wheel arc	2 1/2 degrees
Expansion	2 degrees
Steering, forward chassis	axle pivot by towing fork
Steering, rear chassis	horse linkage, connected with air pressure steering, effecting axle linkage steering
Hauling device for load	cable guide for the cable cable from the Zugkraftwagen
Lifting device for load bed	mechanical lifting winch on rear chassis
Lubrication	central pressure and point lubrication

Bergepanzer III, Recovery Vehicle

The Bergepanzer III was a logical supplement to the motorpools of Sturmartillerie units. Chassis from Panzerkampfwagen III repaired in overhaul facilities were converted for use as Bergepanzer. In a speech given by Hitler on April 7th 1944 it was determined that the Bergepanzer III would not be permitted to be stricken from the production program in favor of the Sturmgeschütz. The prototype of the Bergepanzer III entered a thorough testing program at Kummersdorf on 1 March 1944.

Those 240 Panzerkampfwagen III which were currently in-country for major repairs were to converted into Bergepanzer.

Beginning April 15, it was determined that conversion of those Panzerkampfwagen III under repair into Bergepanzer vehicles would be accomplished in order to:
a) fill a gap in the production of the Zugkraftwagen 18t
b) help the Sturmgeschütz-Abteilungen in Infantry Divisions which had no towing vehicles.
Hitler put great emphasis on an increased output of Sturmgeschütz and Panzerjäger and ordered that the last 15 Panzerkampfwagen III coming out of overhaul be utilized as Sturmgeschütz chassis.

The following equipment was planned for the Bergepanzer III:
— rear coupling for towing damaged vehicles — anchor on a single-axle trailer — recovery equipment — to include a jib crane in the future.

On March 1st 1944 the vehicle testing branch at Kummersdorf began work under contract Nr. Pz.IIIf 1/44 by conducting recovery experiments with the Bergepanzer III. The tests were completed in December 1944. The testing extended to:
a) Driving the complete Bergepanzer in difficult terrain
b) Assessing the usefulness of the intended equipment for recovery purposes including the hatch fasteners and the box superstructure
c) Recovery attempts in various types of terrain
The Bergepanzer III delivered by the firm Daimler-Benz AG, Werk 40 (chassis number 74104) together with the recovery anchor had met expectations in various recovery and towing tests in Berka and Kummersdorf. Usable pulling power in first gear on firm ground was 10 to 15 tons. This was sufficient for recovery and towing purposes. Rigged with pulleys providing an increase by a factor of 6, in normal cases, 40 tons of force could be exerted when only 7 tons of force were provided by the towing vehicle. During tests on wet clay, however, a pulling force of only 3 to 4 tons was noted. The pulling power was also inadequate in snow. Tests with grippers on the tracks failed because of their inadequate design.

The centered, tow coupling was mounted too low, and too far forward of the rear corner of the Bergepanzer. This design was unsuitable and a new design was to be initiated.

Separation of the box superstructure into compartments was bothersome, as, among other things, the roller guides for the anchor could not be stored. The superstructure was too high, a headroom of 500mm seemed sufficient and would assist in loading. The hatch for the crew was to be omitted. The superstructure and a hatch were to be covered by a tarpaulin. The positions of the "Pilze" mounts for the 1-ton jib crane were relocated so that in addition to loading equipment into the superstructure, the crane could be used to remove and install the vehicle's own engine components.

During testing, the left final drive shaft broke twice, the right shaft once. This occurred within a short time when reversing under load. Because this type of damage had appeared only in later models, a material or production process defect was suspected. An investigation of the materials was to be conducted.

The Bergeanker (recovery anchor) in use.

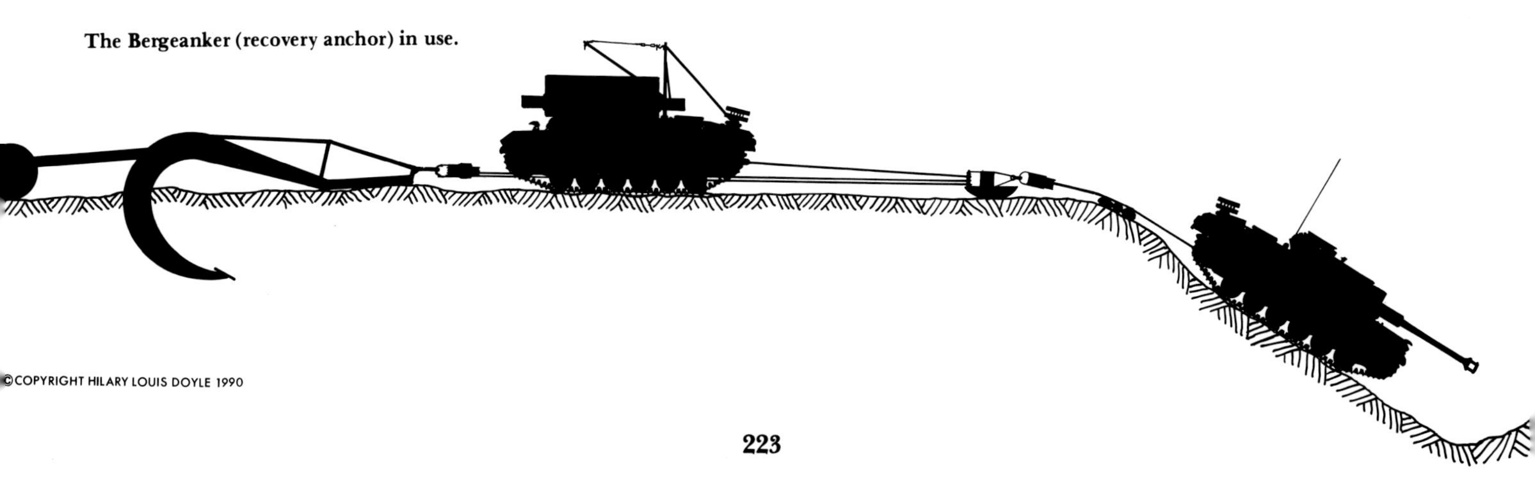

Bergepanzer III towing a Bergeanker (recovery anchor).

The Bergeanker (recovery anchor) in use.

The Bergepanzer III as seen from the right rear sporting Ostketten tracks.

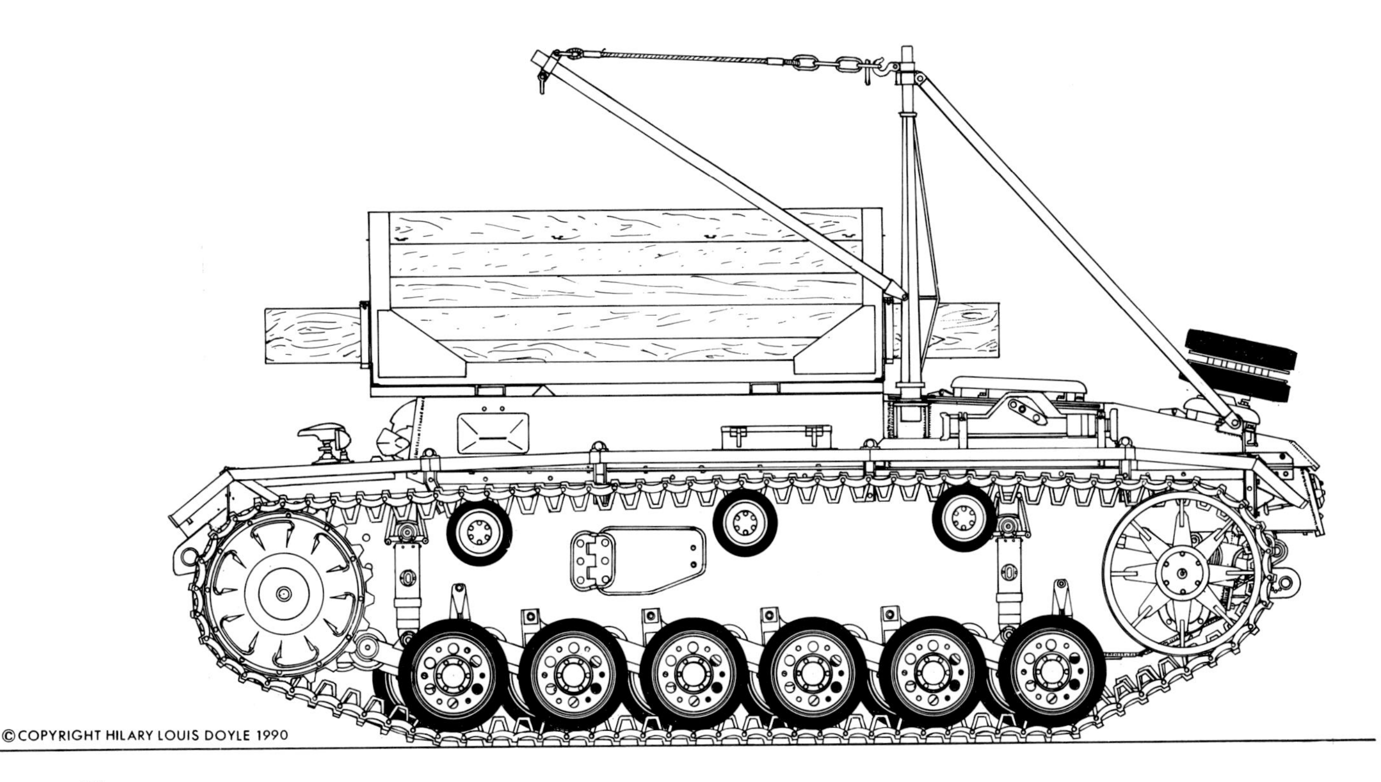

Bergepanzer III.

Summary

The Bergepanzer III in conjunction with the recovery anchor proved to be a good means of tank recovery after several modifications had been made. These included the a redesigned centered tow coupling on the hull rear and track grippers which actually held. The recovery of heavy tanks was limited to easier cases.

The conversions to the Bergepanzer was time and again prevented by the intervention of higher authorities. An order was issued to provide Panzerkampfwagen III Flammenwerfer to support an urgent need for a planned combat action. To accomplish this, finished Bergepanzer III were to be re-equipped with their old turrets mounting flame-throwers. A teletype message from Wegmann in Kassel to the Speer ministry dated 30 November 1944 stated among other things: Six 8./Z.W. chassis are available, which have been converted to Bergepanzer. Based on orders from OKH, these completed Bergepanzer are to be re-converted and used for this action.

Bergepanzer III (auf Panzerkampfwagen III Fahrgestell)*
Fabricated using rebuilds

January 1944	0
February 1944	0
March 1944	15
April 1944	11
May 1944	43
June 1944	20
July 1944	20
August 1944	18
September 1944	11
October 1944	17
November 1944	12
December 1944	0
January 1945	5
February 1945	1
March 1945	3
April 1945	0
Total	176

On January 15 1945 there were a total of 114 Bergepanzer III reported as on-hand with various combat units.

*Only from the conversion of overhauled Panzerkampfwagen III's

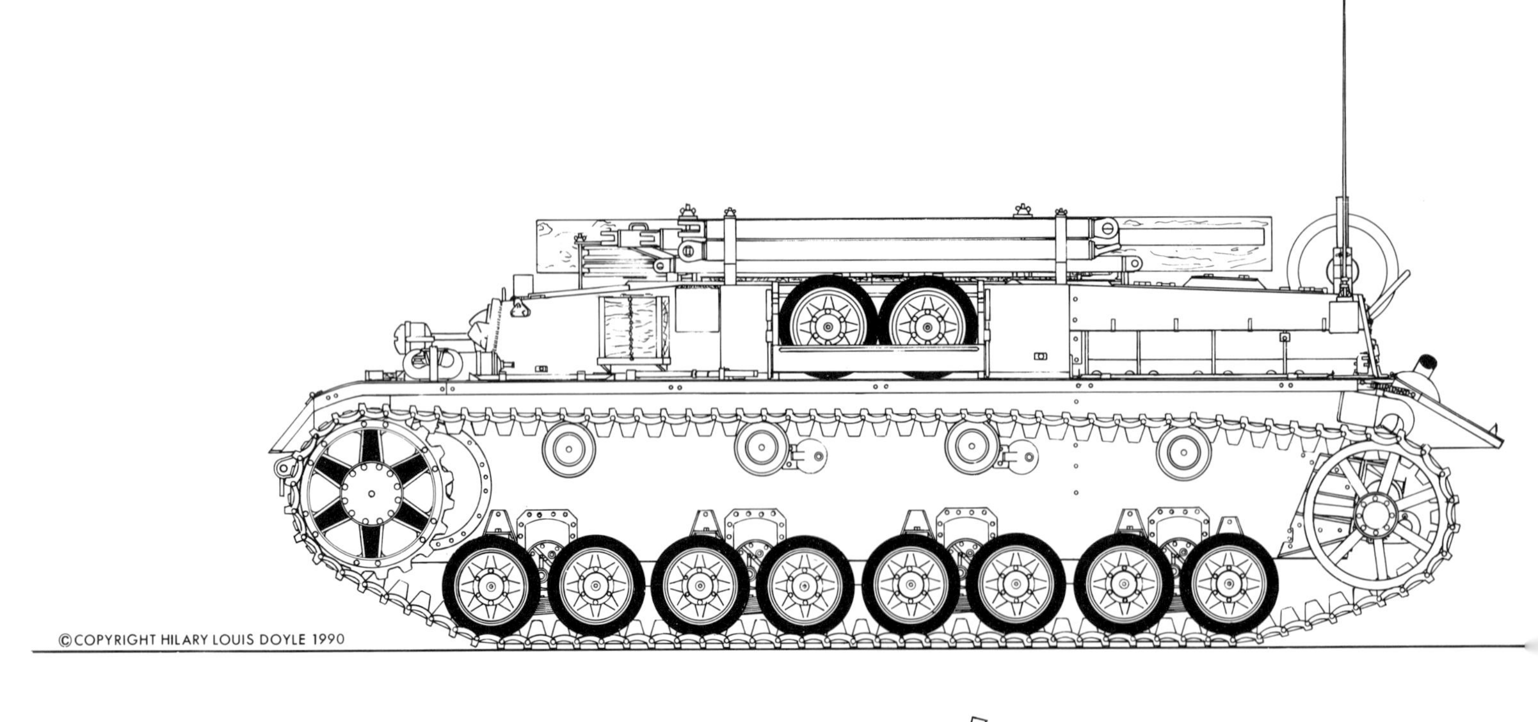

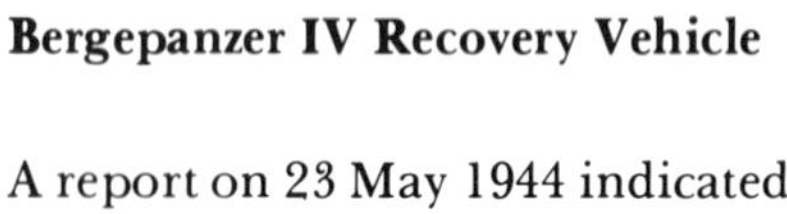

Bergepanzer IV Recovery Vehicle

A report on 23 May 1944 indicated that a Bergepanzer IV was to be designed following the example of the Bergepanzer III. Once again, chassis were to come from overhaul facilities, new chassis were not be used for this purpose. On 24 May 1944 it was declared that the final layout of the Bergepanzer IV had not been clarified. In the meantime, Daimler-Benz had sent a test vehicle to Kummersdorf.

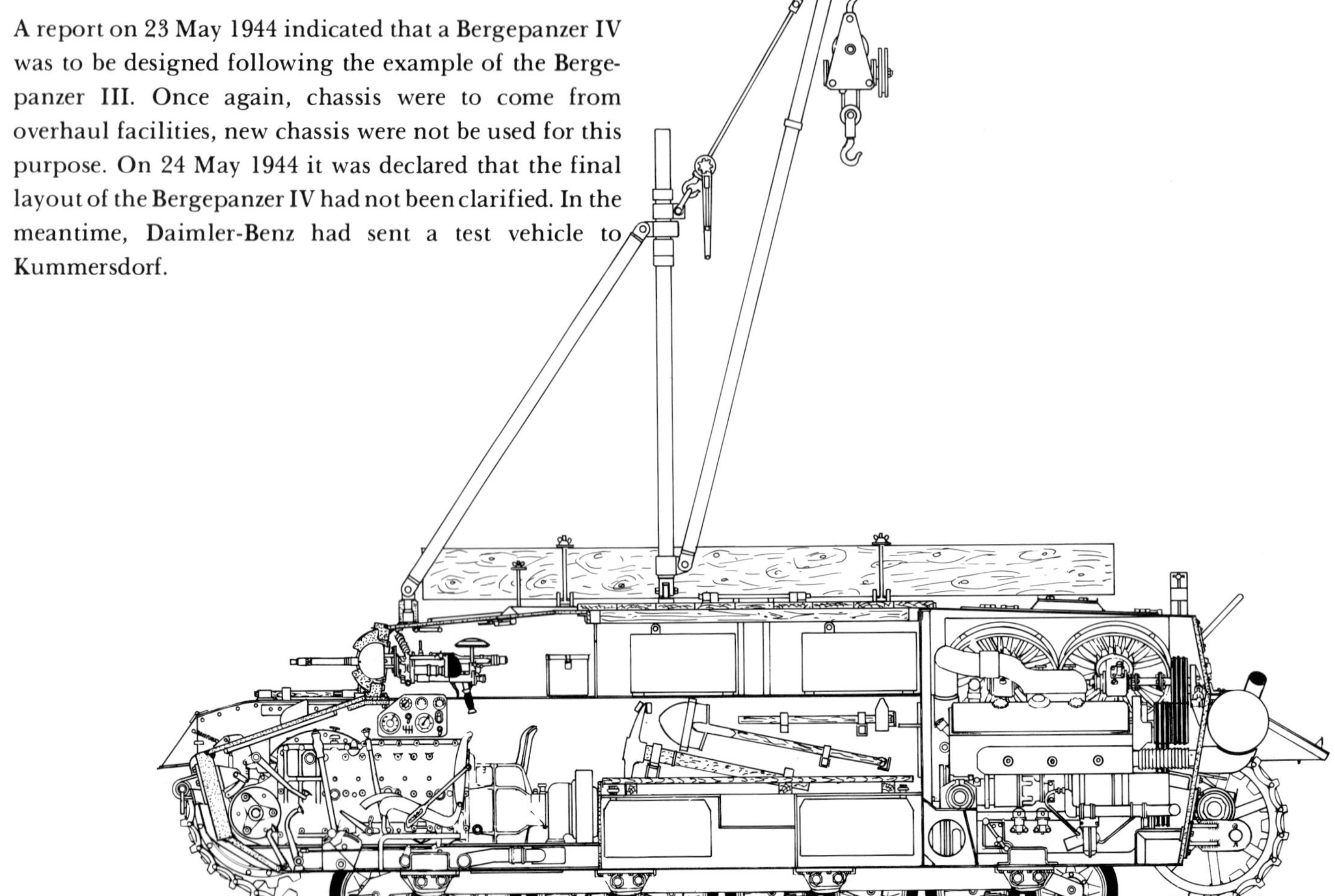

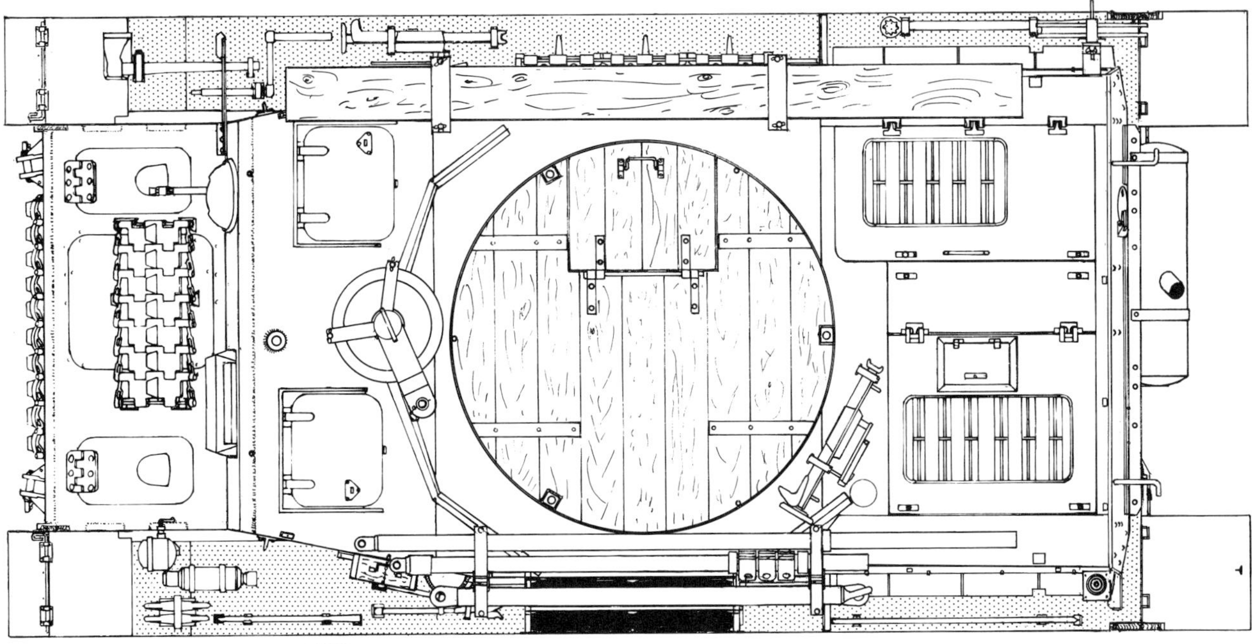

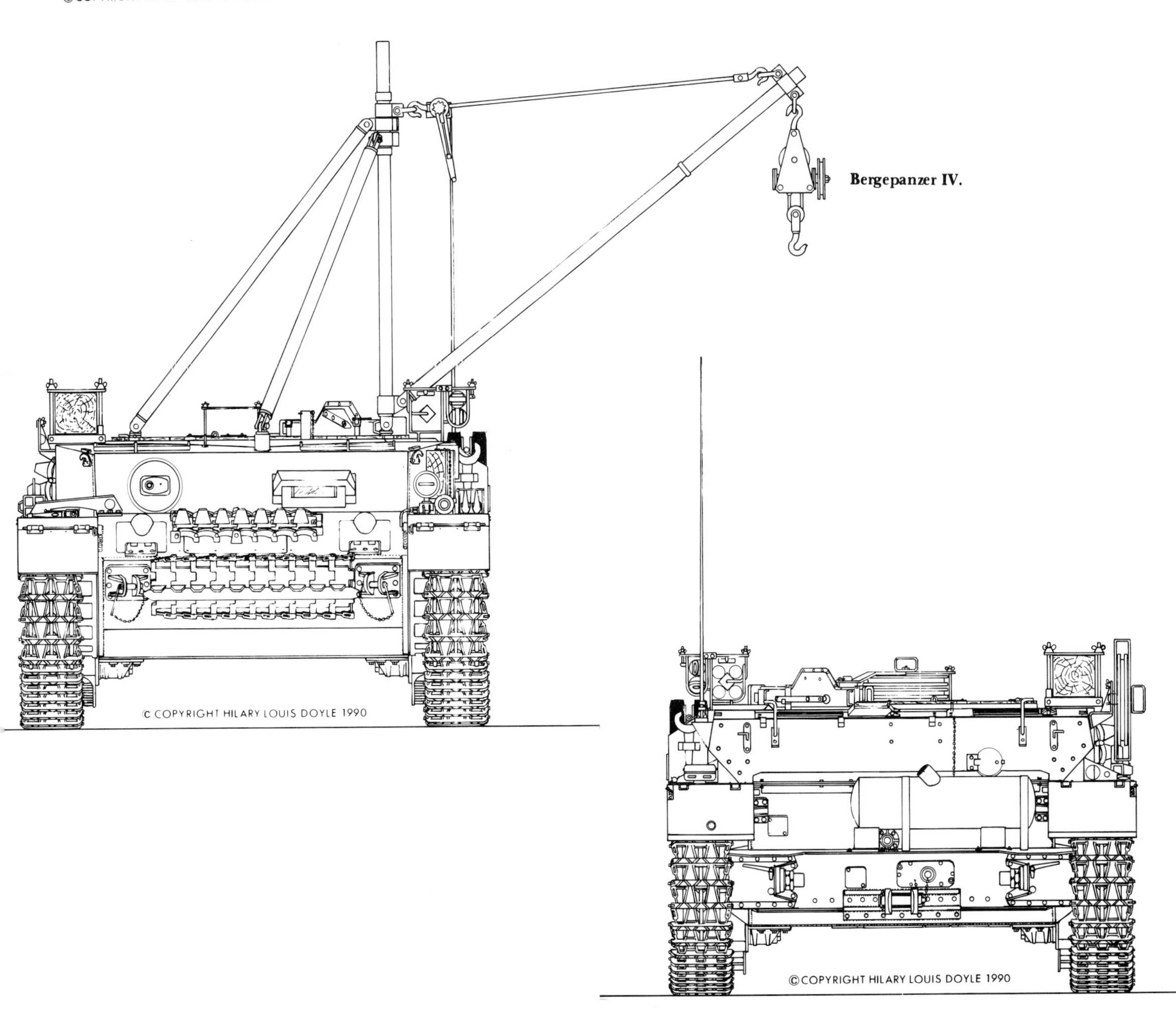

Bergepanzer IV.

Fortunately, original drawings from Fried.Krupp AG still exist which reveal details about the design of this vehicle. On the superstructure, the large opening inside the turret race was covered with wooden planks. A wooden hatch cover allowed access inside the hull. Krupp drawing SKA 6351 dated 20 August 1944 further indicates that wooden boards covered the floor of the crew compartment over the fuel tanks. A modification introduced on 8 October 1944 reduced from 1540mm to 1510mm, the distance between the fore and aft mounts for the jib crane.

A new Krupp blueprint SKB 6351 dated 14 December 1944 shows additional brackets on the right side of the vehicle for block and tackle pulleys and a double-pulley mounted on the superstructure roof.

The minutes of a meeting held in Kummersdorf on 13 and 14 January 1945 show that Krupp-Grusonwerk could not produce the winch for the Bergepanzer IV in Magdeburg. However, the test vehicle could be assembled and when available, the spade and winch installed. Initially, these components were only to be installed on the test vehicle. WaPrüf 6 would attempt to have the winch produced (under the urgency code "Zl") by the firm of Baumgarten in Neuruppin. The winch from Demag was altered in such a way that the cable unwound to the rear. The proposal for installation of the winch (SKB 6675) was delivered to WaPrüf 6 by Krupp. Demag was to plan to deliver one more winch.

Wa Prüf 6 handed over to Krupp data on:
— a device which shorted out the engine after sensing that the maximum allowable tow line had been fed out,
— a device for cleaning the cable as it rewound on the winch, and
— the design of a cable end connector.

It could be gathered from the files still available from Zahnradfabrik Friedrichshafen (ZF) concerning the gearbox AK 5-80 with capstan drive, that the gearbox could not be installed in the Panzerkampfwagen IV in this form. Krupp suggested, instead of feeding the capstan drive to the side, to lead it out over the countershaft of the gearbox. In this way the gearbox could be installed and connection to the winch would be more favorable. A test with the Bergepanzer 38 showed that the winch attached in this fashion was indeed more favorable. ZF was called upon to expedite completion of a drawing for the altered version of the gearbox.

The design work to be done by the Krupp firm was to be credited to the contract for the Bergepanzer IV.

A discussion on 6 April 1945 at Grusonwerk in Magdeburg resulted in the following memorandum about the test vehicle:Bergepanzer IV, chassis number 83945
The work could not be carried out at the planned pace due to personnel leaving for work in the trenches. All parts are available to meet scheduled completion date in mid-April. Krupp is to decide where the vehicle is to be sent for completion. It is suggested that the assembly work be completed at Grusonwerk. Those parts still needed must be delivered as quickly as possible.

Bergepanzer 38, with winch and spade, served as an example for the Bergepanzer IV.

Bergepanzer IV, chassis number 84557
Difficulties still exist in obtaining supplies of oxygen and acetylene so that the welding work has not been finished. Section 6 will attempt to complete the work at the boiler makers' plant. The vehicle must be outfitted with new mounts. The tracks are still missing, as is a set of drive wheels for the final drives.
The last sentence of this memorandum seems quite ironic when reading it today:
The planned production deadline for this vehicle is approximately mid-May 1945, unforeseen events notwithstanding.

On 7 May 1945, the German Reich ceased to exist.

The reason so few Panzer IV's were converted to Bergepanzer vehicles (compared to the Panzer III) was that most of the Panzer IV which were overhauled in the major repair facilities were returned as battle tanks. Authorized conversions were limited in number.

Bergepanzer IV (auf Panzerkampfwagen IV Fahrgestell)*
Fabricated using rebuilds

October 1944	5
November 1944	11
December 1944	1
January 1945	1
February 1945	2
March 1945	1
April 1945	0
Total	21

***only from the conversion of overhauled Panzerkampfwagen IV's**

On 15 January 1945 there were a total of 5 Bergepanzer IV's present with various combat units.

Sturmgeschütz Export Model

Despite the rather strained situation concerning raw materials and the limited production capacity, the German Reich had entered into agreements to deliver war material to allied and friendly nations.

The office responsible for the conclusion of such agreements was the:
Ausfuhrgemeinshaft für Kriegsgerät
bei der Reichsgruppe Industrie
in Berlin-Wannsee, Dreilindenstrasse 47/49.

On July 4th 1944 the A.G.K. wrote to its member firms:
All claims against foreign States, which have come to exist by means of the direct or indirect export of war material, must be reported to the Reichsbank by means of an individual monthly EVE-collection, independent of the provisional export currency declaration made when crossing the border, in the month the bill was charged.

For filling out the EVE-forms, we are formulating an overall view of the war material which belong to those in our area of concern, coordinated between examination stations and data from the test station numbers.

Equipment which was added to the list of war material permitted for import and export by decree of the Reichskommissar beginning 2 September 1940 are:

Nr. of the Examination station
2: Examination Station for iron industry Armor Plating
Equipment, which was declared in accordance with the law of 6 November 1935 governing import and export:
Nr. of the Examining station
6: Examination for machine construction complete guns over 20mm for Army, Navy, Marines and Air Force, incl. complete ammunition
7: Examination for vehicle industry: Panzerkampfwagen, Panzerspähwagen, Sturmgeschütze, tractors, towing vehicles, fully and half-tracked vehicles over 50 hp or 12 km/h, special military vehicles, two and multi-axle drive vehicles.

(The list was shortened by the author to those fields relevant to this book)
The export trade of these arms is guided exclusively by the A.G.K. via the corresponding place of registration. Contracts permitting export are to be submitted to the A.G.K.

On 14 July 1944 the A.G.K. reported that the A.G.K. price examination station connected up until now with the place of registration, would, for organizational reasons, now be conducted as an independent department with headquarters in Berlin.
The following enclosure
to the circular Igb. Nr. 120006/G/44-I-602-Allg
Subj: Summary of surcharges for war material producers to countries named — situation as of 1 July 1944
Rumania: surcharge of 27% on export prices in effect beginning 1 January 1944
Bulgaria: surcharge of up to 50% on export prices in effect beginning 15 February 1941. Special rules apply to program tax.
Hungary: surcharge of 20% on export prices in effect beginning 1 July 1943
Croatia: surcharge of 900% on pre-war export prices to the former Yugoslavia. If no comparative price exists, figure the Wehrmacht price plus 10 to 60%.
Finland: surcharge of 50% on the Wehrmacht price
Slovakia: surcharge of 20 to 30% on pre-war export prices
Sweden: surcharge of 20 to 30% on pre-war export prices
Italy: surcharge of 30 to 50% on pre-war export prices
Portugal: commensurate with every obtainable price; lowest price level is that in 1942
Spain: inasmuch as not comparable equipment exists 100% surcharge on 1939 export prices, otherwise commensurate export prices, at least 60 to 90% over Wehrmachts price.
Turkey: surcharge of 250% on 1940 export prices
Japan: normal export prices. For models or special special equipment affix special pricing.

The export prices are to be established according to the stipulation of the firms own costs; only when these figures are not available are the Wehrmachts prices to be used.

Delivery of the Sturmgeschütz to Foreign Lands

On 17 August 1943, on Hitler's orders, 10 Sturmgeschütz per month in June, July and August were to be allocated to the Finnish Army together with four ammunition rations.

A total of 59 Sturmgeschütz vehicles were received by Finland, The last of these were retained in service until the end of 1966.

Based on a meeting with the Head of the Bulgarian General Staff on 1 February 1943 at Supreme Headquarters, Generalstab des Heeres was to incorporate into their plans the following assignments of war materials to Bulgaria: Allocation was to occur through the "Chef H. Rüt u. BdE" in direct agreement with the Wirtschaftsamt and military attache in Sofia within the framework of this shipment schedule: A total of 20 7.5cm Sturmgeschütz, at a rate of five per month, beginning in March 1943.

According to a report dated 25 April 1943, 25 Sturmgeschütz were to be delivered (instead of 20) by the General Quartiermeister to Bulgaria.

Situation report for Bulgaria effective 12 June 1943:
Sturmgeschütz 40 L/48
On-hand with army: 0
On-hand, Reserves: 0
In delivery from Germany: 25

(It is evident from these files that after the Heereszeugamt made the shipments by rail, there were delays in delivering the Sturmgeschütz to the Bulgarian Army by the Military Attache in Sofia.)

In August 1943 it was determined that an additional 10 Sturmgeschütz per month in October, September and October 1943 were to be delivered to Bulgaria.

Situation report for Bulgaria effective 31 December 1943:
Sturmgeschütz 40
Ordered: 55
Delivered: 25
Situation report for Bulgaria effective 22 February 1944:
Sturmgeschütz 40
Total to be delivered: 55
Delivered: 45
Still to be delivered: 10
Situation in August 1944:

25 Sturmgeschütz in the Bulgarian 1st Sturmgeschütz Abteilung in Sofia, and 25 Sturmgeschütz vehicles in the

Sturmgeschütz in use by foreign armies. Finland received 59 Sturmgeschütz, Ausf.G.

Above and below: Sturmgeschütz were given to the Italian Army in May 1943 (Ausf.G, MIAG chassis number 95102 completed in April 1943).

A Rumanian crew familiarize themselves with the Sturmgeschütz.

Bulgarian 2nd Sturmgeschütz Abteilung in Plovdiv.

Sturmgeschütz delivered to Bulgaria did not see action against the Red Army, but rumored to have were used by the Bulgarian Army against the German Wehrmacht.

Month	Country	Sturm-geschütz	Shipped by rail from HZA
1943			
February	Bulgaria	5	
March	Bulgaria	5	
April	Bulgaria	5	
May	Bulgaria	10	
August	Bulgaria	10	
September	Bulgaria	10	
December	Bulgaria	10	16 Dec 1943
Total:		55	
1943			
June	Finland	10	
July	Finland	10	
August	Finland	10	
1944			
March	Finland	7	5 Apr 1944 (not delivered)
June	Finland	15	22 Jun 1944
July	Finland	15	9 Jul 1944
August	Finland	15	Not delivered*
Total:		59	

* Cease-fire between Russia and Finland on 9 September 1944

Month	Country	Sturm-geschütz	shipped by rail from HZA
1943			
May	Italy	5	
1943			
October	Spain	10	
1943			
November	Rumania	4	20 Nov 1943
December	Rumania	2	28 Jan 1944
1944			
January	Rumania	2	6 Feb 1944
February	Rumania	12	26 Feb 1944
March	Rumania	10	6 Apr 1944
April	Rumania	10	1 May 1944
May	Rumania	20	28 May 1944
June	Rumania	20	14 Jul 1944
July	Rumania	20	24 Jul 1944
August	Rumania	20	12 Aug 1944*
Total:		120	

*Coup d'Etat on 23 August 1944

Month	Country	Sturm-geschütz	shipped by rail from HZA
1944			
July	Hungary	10	1 Aug 1944
July	Hungary	10	26 Aug 1944
August	Hungary	10	28 Aug 1944
September	Hungary	10	7 Sep 1944
Total:		40	
1942			
November	Country unk.	2	(Ausf.F/8)

The Sturmartillerie in Action

Western Campaign beginning 10 May 1940

At the start of the campaign in the West only two Sturmgeschütz batteries were available, 16.Sturmbatterie of Infanterie-Regiment "Grossdeutschland" (formerly Sturmbatterie 640) and Sturmbatterie 659. These were followed by Sturmbatterie 660 (on 13 May 1940) and Sturmbatterie 665 (on 10 June 1940). Each battery was authorized to possess 6 Sturmgeschütz, 5 leichten, gepanzerten Beobachtungswagen (Sd.Kfz.253) and 6 leichten, gepanzerten Munitions-Transportkraftwagen (Sd.Kfz.252) in accordance with K.St.N.445 dated 1 November 1939.

Batterie 7.5cm Sturmgeschütz (6 Geschütze) (mot.S) K.ST.N 445 dated 1 November 1939

Due to production delays, Sturmbatterie 640 received 4 MTW (Sd.Kfz.251) (medium armored personnel transport vehicles) instead of the Sd.Kfz.253. Instead of the Sd.Kfz.252, a platoon from Munitions-Transport-Kolonne 601, outfitted with Panzerkampfwagen I, Ausf.A ohne Aufbau als Munitionspanzer (Panzer I chassis without turret, converted to ammo carrier), was attached to Sturmbatterie 660.

Batterie 7,5cm Sturmgeschütz (6 Geschütze) (mot.S), K.St.N.445 dated 1.November 1939

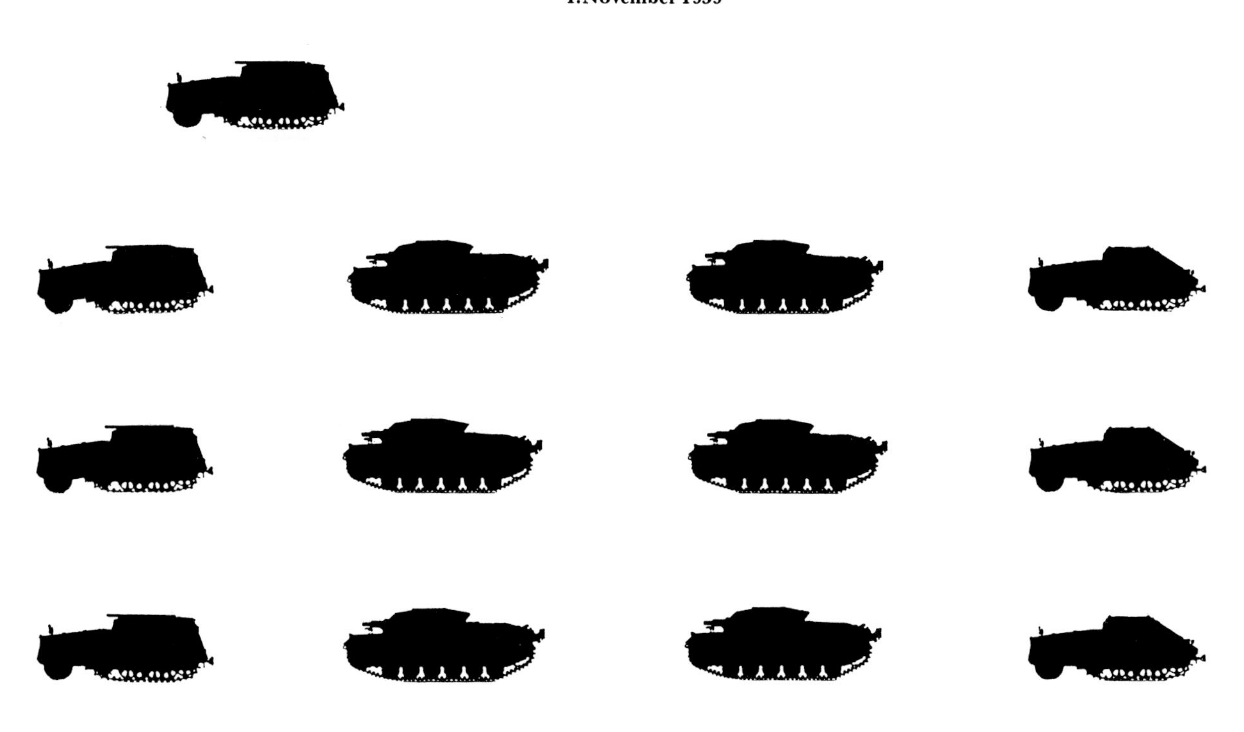

Balkan Campaign beginning 6 April 1941

Only three Sturmgeschütz Abteilungen (184, 190 and 191) and the 16.Sturmgeschütz-Kompanie/Infanterie Regiment "Grossdeutschland" were employed in the campaign in Greece and Yugoslavia. There were three Sturmbatterien in each Abteilung which possessed a total of 18 Sturmgeschütz.

A new version of K.St.N.445 had been published on 1 February 1941, but the Batterie still only had 6 Sturmgeschütz. The title of the unit was officially changed to "Sturmgeschütz Batterie" when K.St.N.445 was replaced by K.St.N.446 dated 18 April 1941.

26 March 1941 — Photo of the entire 3.Batterie/Sturmgeschütz-Abteilung 197.

The Sturmgeschütz as originally intended — in the role of an infantry support vehicle.

Operation "Barbarossa" beginning 22 June 1941

At the start of the campaign in the East, the Sturmartillerie fielded 7 Sturmgeschütz-Batterien and 11 Sturmgeschütz-Abteilungen. Each Sturmgeschütz-Abteilung consisted of three Sturmgeschütz-Batterien. Although several of the Batterien still possessed only 6 Sturmgeschütz, the majority had been converted to the supplemental K.St.N.446 dated 18 April 1941.

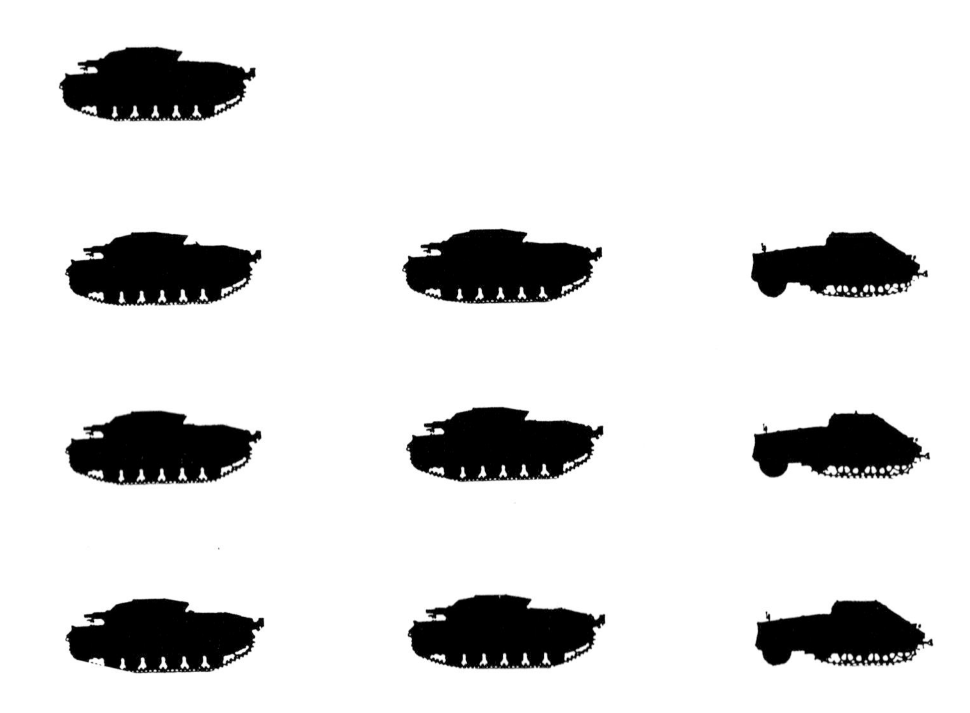

Sturmgeschütz Batterie (7 Gesch.) (mot.) K.St.N.446 dated 18 April 1941 (supplemental) and 1 November 1941

Sturmgeschütz Batterie (7 Geschütze) (mot.) K.St.N.446 dated 18 April 1941 (supplemental) and dated 1 November 1941

By this time, a sufficient number of Sturmgeschütz had been produced to issue one to the Batterie-Chef in most of the Sturmgeschütz-Batterien. In addition, the Zug-Führer (platoon leader) were mounted in a Sturmgeschütz instead of the thinly armored, leichten, gepanzerten Beobachtungswagen (Sd.Kfz.253). The Oberkommando des Heeres, Organization Abteilung issued an order dated 9 August 1941 that all Sturmgeschütz-Batterien were to be outfitted with 7 Sturmgeschütz.

Many of the Sturmgeschütz-Batterien had been initially created under K.St.N.445 or 446 and still retained their leichten, gepanzerten Beobachtungswagen (Sd.Kfz.253) even though they had subsequently been issued a seventh Sturmgeschütz.

There was a total of 272 Sturmgeschütz with Sturmartillerie units (including StuG Lehr and Grossdeutschland) at the front on 22 Juni 1941 at the start of the campaign.

Summer Offensive in the East beginning June 1942

At the start of the major offensives in southern Russia that culminated with a drive toward Stalingrad, there were 19 Sturmgeschütz-Abteilungen (including Grossdeutschland) and one Batterie at the front, of which 13 were under Heeres Gruppe Süd (Army Group South). Having entered production in March 1942, Sturmgeschütz with the long 7.5cm Sturmkanone 40 L/43 (followed in June 1942 by the Sturmkanone 40 L/48) were issued to the Sturmgeschütz-Abteilungen starting in April 1942. By December of 1942, due both to increased production and attrition, most of the Sturmgeschütz-Abteilungen at the front possessed Sturmgeschütz with the long Sturmkanone 40 instead of the short 7.5cm Kanone L/24.

Sturmartillerie Status on the Eastern Front

Date	Nr. of Abt.	Combat Ready	Under Repair	Total
18 Jun	18	166	44	210
25 Jul	19	258	42	300
23 Aug	19	277	75	352
25 Sep	20	294	115	409
16 Oct	19	295	120	415
26 Nov	20	347	101	448
28 Dec	27	315	127	442

Throughout the war, most of the Sturmartillerie units were deployed on the Eastern Front. Only two units were sent to North Africa in 1942. Three Sturmgeschütz, Ausf.D with the 5.Panzerjägerkompanie/Sonderverband 288 in April 1942 and four Sturmgeschütz, Ausf.F/8 with Sturmgeschütz-Batterie 90 of the 10.Panzer Division in November 1942.

In late 1942, revised K.St.N. were again published reflecting the increased production and availability of Sturmgeschütz for issue.

Sturmgeschütz Batterie (10 Geschütz) (mot.) K.St.N.446a dated 1 November 1942

The leichten, gepanzerten Munitionskraftwagen (Sd.Kfz.252 and 250/6) were no longer authorized by this organization chart, having been replaced in each Batterie by two munitions trucks. Although, units previously issued these Sd.Kfz.252 or 250/6 retained them in service through the end of the war, their numbers gradually decreased through attrition.

For the first time the Abteilung Kommandeur was officially authorized to possess a Sturmgeschütz under K.St.N.416 dated 1 December 1942 for the organization of the Stab Sturmgeschütz Abteilung (mot.). Even though the K.St.N. officially authorizing the Abteilung Kommandeur a Sturmgeschütz was not published until December 1942, as early as April 1942, many units were authorized to possess 22 Sturmgeschütz (7 for each Batterie plus 1 for the Kommandeur).

When a new K.St.N. was published, it did not automatically become effective for all of the units. Most of the

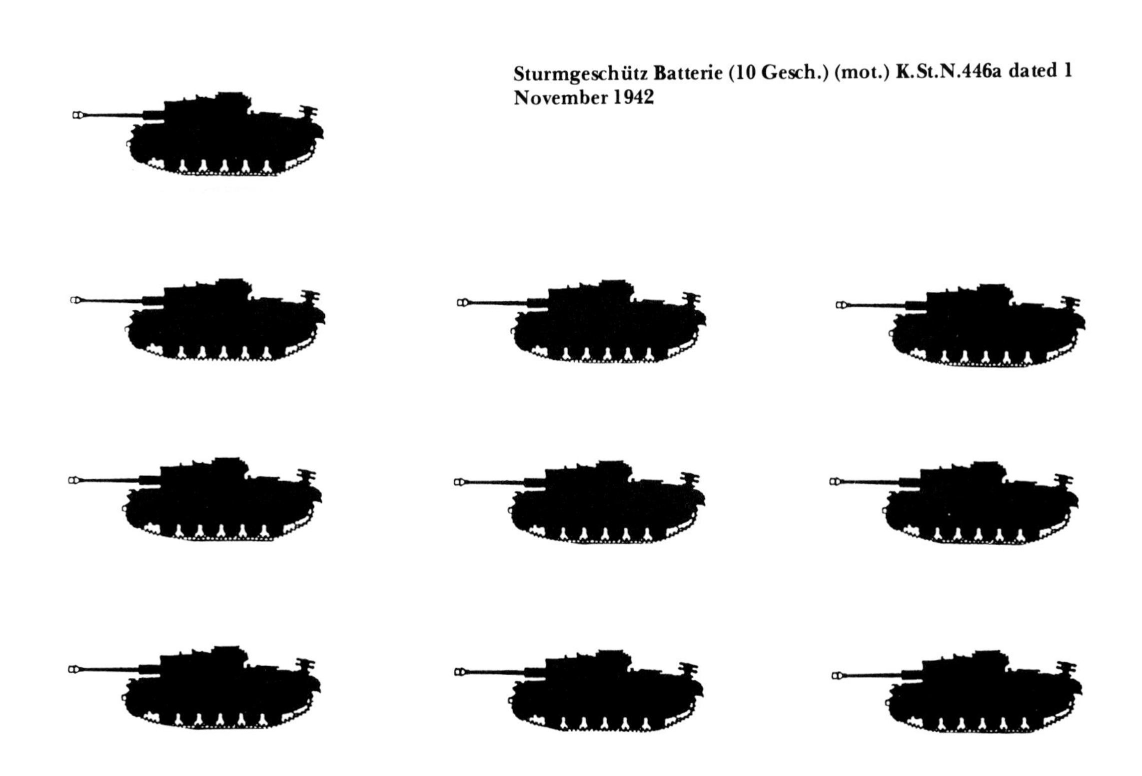
Sturmgeschütz Batterie (10 Gesch.) (mot.) K.St.N.446a dated 1 November 1942

units outfitted with Sturmgeschütz in late 1942 and early 1943 were still under the older K.St.N.446, which only authorized 7 Sturmgeschütz per Batterie.

On March 2nd 1943, Oberkommando des Heeres Organization Abteilung issued order number 2156 g., which reflected the production of the Sturmhaubitze by authorizing each Batterie of the Sturmgeschütz-Abteilung to possess either seven 7.5cm Sturmgeschütz (Sd.Kfz.142/1) and three 10.5cm Sturmhaubitze (Sd.Kfz.142/2) or 10 7.5cm Sturmgeschütz (Sd.Kfz.142/1) per Batterie.

In the East beginning July 1943, during the time of Operation "Zitadelle"

At the start of the major summer offensive near Kursk, three Sturmgeschütz-Abteilungen and two Batterien were assigned to Heeres Gruppe Nord (Army Group North). Twelve Sturmgeschütz-Abteilungen were assigned to Heeres Gruppe Mitte. Nine Sturmgeschütz-Abteilung and one Batterie were assigned to Heeres Gruppe Süd, and two additional Sturmgeschütz-Abteilung to Heeres Gruppe A. Only nine Sturmgeschütz-Abteilung units had been issued Sturmhaubitze (one in Heeres Gruppe Nord, six in Heeres Gruppe Mitte and two in Heeres Gruppe Süd).

Sturmgeschütz Batterie (14 Gesch.) (mot.) K.St.N.446b dated 1 February 1944

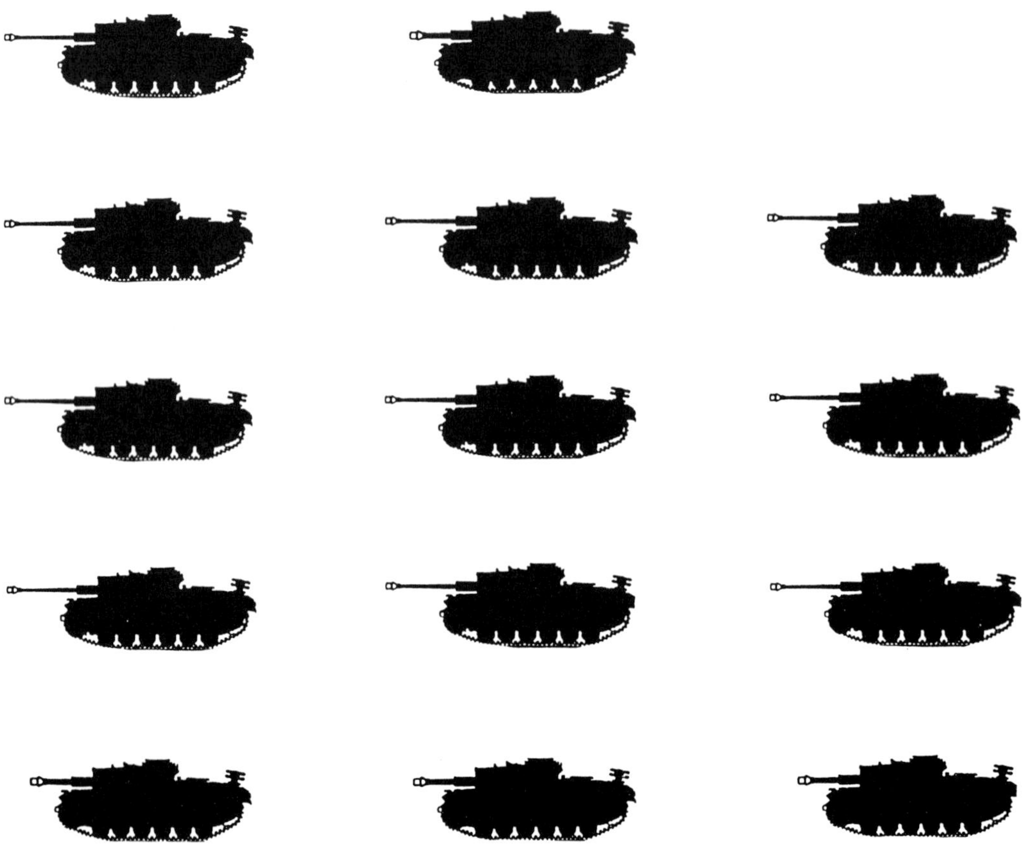

Sturmartillerie Status on the Eastern Front in 1943

Date	# Abt.	StuG (short) *Rep.*	Oper.	StuG (long) Rep.	Oper.	StuH Rep.	Total	
31Mar43	20	39	16	297	86	5	0	443
30Apr43	21	34	16	426	91	13	1	581
31May43	22	33	15	548	84	13	1	694
30Jun43	26	29	6	657	70	57	2	821
31Jul43	29	23	5	543	202	46	21	840
31Aug43	30	16	12	487	235	49	14	813
30Sep43	31	14	11	354	375	27	37	818
31Oct43	34	8	12	423	331	37	29	840
30Nov43	37	5	18	388	452	33	27	923
31Dec43	39	10	6	539	393	37	21	1006

*Oper. = OperationalRep. = Under repair

In February 1944, a new organization under K.St.N.446b was issued for the Sturmgeschütz-Batterie authorizing 14 Sturmgeschütz per Batterie.

Sturmgeschütz Batterie (14 Gesch.) (mot.) K.St.N.446b dated 1 February 1944

The new organization K.St.N.416b dated 1 February 1944 for the Stab und Stabsbatterie (mot.) einer Sturmgeschütz-Abteilung mit Batterien zu 14 Geschützen authorized the Stab to possess 3 Sturmgeschütz III für 7.5cm Sturmkanone 40 (L/48) (Sd.Kfz.141/1).

The older K.St.N.446 and 446a remained in effect, so that at this point in time it was possible to have a Sturmgeschütz-Abteilung authorized to possess 22, 31 or 45 Sturmgeschütz, depending on which K.St.N. it had been ordered to follow. In actual fact, starting in early 1943 all Sturmgeschütz-Abteilungen were authorized to have 31 Sturmgeschütz, and every newly created Abteilung was in possession of their full complement before being sent into action.

The OKH planned to eventually outfit all Sturmgeschütz-Abteilungen with 45 Sturmgeschütz. But, not until specific orders were issued was an Abteilung authorized to change to a new K.St.N. By the end of the war, only four units had actually been authorized by orders to posses and had been issued 45 Sturmgeschütz:

Sturmgeschütz-Brigade 259 in May 1944 (Eastern Front)
Sturmgeschütz-Brigade 341 in May 1944 (West)
Sturmgeschütz-Brigade 278 in Nov.1944 (Eastern Front)
Sturm-Artillerie-Brigade 303 in Dec.1944 (Eastern Front)

An OKH Befehl von 25.Feb.44 entitled, Renaming Sturmgeschütz Units, converted the names of Heeres Sturmgeschütz-Abteilungen into Heeres Sturmgeschütz-Brigaden. Each Brigade was to retain the same number as the Abteilung. The organization of each Brigade remained exactly as it had been when it was known as an Abteilung.

The reason for renaming the Sturmartillerie organizations was to distinguish between these Heeres-Sturmgeschütz units and those units with Sturmgeschütz that had been assigned to the Infanterie-, Jäger- and Gebirgs-Divisionen. Their designations were also changed with the same order (dated 25.Feb.44) from "Pz.Jg.Kp. (Sturmgeschütz)" within the Pz.Jg.Abt. of the Infanterie-, Jäger- and Gebirgs-Divisions to "Sturmgeschütz-Abteilungen." In February 1944, two new organizations were developed to provide additional support to the Sturmgeschütz Abteilungen. "Sturmgeschütz Begleit Batterie für Sturmgeschütz Abteilung" K.St.N.448 (supplemental) dated 8 February 1944 was established to provide an infantry unit directly under the control of the Abteilung Kommandeur. These infantry were to accompany the Sturmgeschütz during attacks and provide guards and outposts in the defense. These units effectively relieved the Sturmgeschütz crews of these additional duties, enabling them to carry out the necessary repairs and maintenance of the vehicles that was so often neglected when the crews were required to be on guard or provide their own security at night. The Inspektor der Artillerie had requested that the Infanterie in this Begleit Batterie be mounted in m.SPW (Sd.Kfz.251) but these mSPW were never issued to the Begleit Batterie or authorized by the K.St.N. reissued as "Begleit Grenadier Batterie zur Sturm Artillerie Brigade" K.St.N.448 dated 1 December 1944.

The second new unit, "Begleit Panzer Batterie (Panzer II) für Sturmgeschütz Abteilung" K.St.N.447 (supplemental) dated 8 February 1944, was established to provide reconnaissance, scouting and anti-aircraft duties. Each Begleit Panzer Batterie was to consist of 14 Pz.Kpfw.II (2 in the Batterietrupp and 4 in each of the 3 Zügen. Only four of these Begleit Panzer Batterien were formed in April and June 1944. All four, 5.Pz. Begl.Battr./Sturm Art.Brig.236, 5.Pz.Begl.Battr./Sturm Art.Brig.239, and 5. u. 6.Pz.Begl.Battr./Sturm Art.Brig.667 were cancelled by an order in November 1944 although

Sturmgeschütz Batterie (10 Gesch.) (mot.) K.St.N.446 Ausf.A dated 1 June 1944

Sturmgeschütz Batterie (14 Gesch.) (mot.) K.St.N.446 Ausf.B dated 1 June 1944

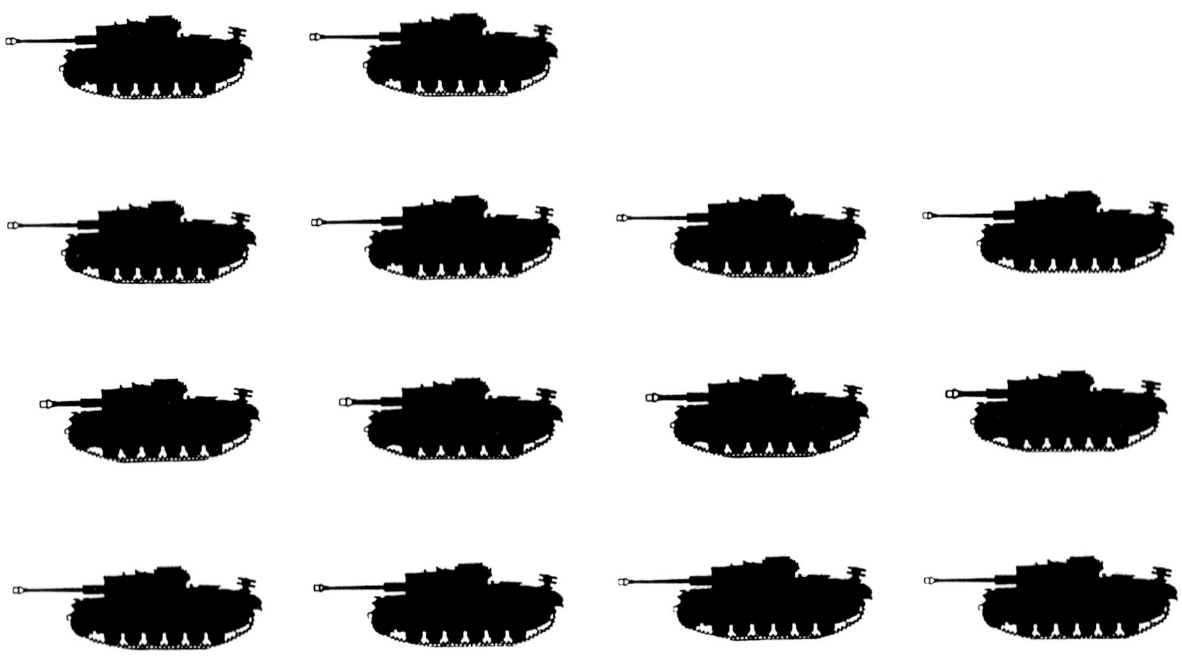

Sturm Art.Brig.239 still possessed 4 Pz.Kpfw.II on 15.Mar.45.

The Heeres Sturmgeschütz units without Begleit Batterie were officially named Sturmgeschütz-Brigade, and the units with Begleit Batterie were named Sturmartillerie-Brigade. With the exception of the Begleit Batterie, the organization of the Stab, Stabsbatterie and Sturmgeschütz Batterien remained the same for both the Sturmgeschütz-Brigade and the Sturmartillerie-Brigade.

Beginning on 1 June 1944, the previously existing Batterie organizations were consolidated into K.St.N.446 with Ausf.A for Batterien with 10 Sturmgeschütze and Ausf.B for Batterien with 14 Sturmgeschütz.

Sturmgeschütz Batterie (10 Gesch.) (mot.) K.St.N.446 Ausf.A dated 1 June 1944

Sturmgeschütz Batterie (14 Gesch.) (mot.) K.St.N.446 Ausf.B dated 1 June 1944

In accordance with K.St.N.416 dated 1 June 1944, Stab und Stabsbatterie (mot.) einer Sturmgeschütz-Brigade were authorized to possess one Sturmgeschütz III if it had three 10-gun batteries and three Sturmgeschütz III's if it had three 14-gun batteries. These organizations for Batterie and Brigade Stab remained in effect until the end of the war.

The Great Russian Offensive starting 22 June 1944 and the Allied Landing in the West

In June 1944, prior to the Great Offensive on the Eastern Front, there were:

- 5 Sturmgeschütz Brigade (184, 226, 303, 909 and 912) under Heeres Gruppe Nord;
- 9 Sturmgeschütz Brigade (177, 185, 189, 190, 244, 245, 281 and 901) and one Sturmartillerie Brigade (667) under Heeres Gruppe Mitte;
- 9 Sturmgeschütz Brigade (210, 237, 249, 270, 300, 301, 311, 322 and 500) under Heeres Gruppe Nordukraine; and
- 9 Sturmgeschütz Brigade (228, 243, 259, 278, 286, 325, 905, 911 and Grossdeutschland) under Heeres Gruppe Südukraine.

At the same time in June 1944 the remaining Heeres-Sturmartillerie units were assigned as follows:

- In the west: Sturmgeschütz Brigade 341 and Sturmgeschütz-Lehr-Brigade 902
- In the southeast: Sturmgeschütz Brigade 201
- In the southwest: Sturmgeschütz Brigade 242, 907 and 904
- In the Germany for replenishment:

Sturmgeschütz Brigade 191, 202, 203, 209, 232, 261, 276, 277, 279 and 280) and Sturmartillerie Brigades 236 and 239.

Status of the Sturmartillerie in 1944 and 1945

Date	StuG Nr.of. Brig.	Oper.	StuH Rep.	Oper.	Rep.	Total
In the East:						
1Jun44	32	615	158	95	25	893
1Jul44	32	718	108	123	16	965
1Aug44	31	463	175	116	32	786
1Sep44	28	383	173	135	57	748
1Oct4427	427	103	184	50	764	
1Nov44	29					826
1Dec44	30				859	
15Jan45	34	558	100	272	66	996
15Mar45	30	364	154	141	49	708
In the West:						
1Jun44	2	44	6	9	0	59
1Jul44	3	94	1	12	0	107
1Aug44	3	65	7	15	0	87
1Sep44	3	18	0	0	0	0
1Oct44	4	43	25	19	18	105
1Nov44	6					136
1Dec44	6					126
15Jan45	9	94	62	36	25	217
15Mar45	7	83	43	18	12	156

The following reports attest to the Sturmgeschütz's effectiveness in combat:

Report covering tank combat in the battles from 12 January through 31 March 1943 in Heeres Gruppe Nord by Lodoga Sea.

Deployment of Sturmgeschütz-Abteilung 226 during the entire period:

a) Number of deployed Sturmgeschütz: 41
(including new arrivals)
b) Number of enemy tanks destroyed: 210c) Losses
Personnel (killed, wounded, missing): 87
Material (not repairable): 13 StuG
Deployment of Sturmgeschütz Batteries of the 1st, 10th, 12th and 13th Luftwaffe-Feld-Divisionen
a) Number of deployed Sturmgeschütz: 20

(including new arrivals)
b) Number of enemy tanks destroyed: 17
c) Losses
Personnel (killed, wounded, missing): 32
Material (not repairable): 5 StuG

During this entire period, deployment of the Sturmgeschütz units of the Luftwaffe was done in support of and cooperation with the units of the Sturmgeschütz-Abteilungen in order to give them the combat experience they lacked, and because the shortage of technical expertise and equipment necessitated this.

Experiences of the Infanterie-Division "Grossdeutschland" in armor combat during the winter battle around Charkov

In the period from 7 to 20 March 1943, the following were knocked out:
230 T-34's
16 T60's or T70's
3 KWI's.

The kills were accomplished with the following weapons:
188 by 71 Pz.Kpfw.IV, long
41 by 35 Sturmgeschütz 75mm, long
30 by 9 Pz.Kpfw.VI (Tiger)

Trip report of the Commander of the Sturmgeschütz-Ersatz und Ausbildungs-Abteilung to the Eastern Front from 30 August to 22 September 1943

While Heeres Gruppe Nord remained in reinforced positions without significant engagements with the enemy, Heeres Gruppen Mitte and Süd were involved in defensive and retrograde actions, and retired from the Kuban bridgehead with well-planned and unopposed disengagements. The retrograde and break-off actions succeeded according to plan, and were accepted calmly and good naturedly by the troops. Panic retreat or abandoned equipment was never observed. The troops regret having to give up what was in part, at least, good agricultural land, but were convinced that the Russians would one day be decisively defeated. Not very much is thought of Russian combat effectiveness.

The entire Russian infantry is quite poor, and generally only goes into combat under armor escort.

The Russian tanks have become poorer. The crews are well-chosen and cared for, but poorly trained. When faced with anti-tank weapons, they very often panic. According to captured orders, the tank crews have been forbidden to combat German Sturmgeschütz.

The Russian anti-tank gun has been improved with larger caliber guns, that are skillfully used. They are feared by the Sturmgeschütz crews. Aside from these, mines and anti-tank rifles are the most unpleasant means of defense used against the Sturmgeschütz. The anti-tank rifles can even penetrate the commander's cupola.

In the current combat situation, German tanks are worthless in comparison to the Sturmgeschütz.
The following facts support this:
1. The tank operates in large areas, its strengths are mobility, evasion and flanking actions. It does not seek to engage russian tanks. It doesn't suit it to fight in co-operation with the German Infantry.
2. Its armor and optics are inferior to that of the Sturmgeschütz.
3. The tank is substantially higher than the Sturmgeschütz, therefore easier to knock out.

A high-placed commander was heard to say: I would rather have one Sturmgeschütz-Abteilung than an entire Panzer Division. A Regimental commander stated: I prefer two Sturmgeschütz to ten Panzerkampfwagen.

The Sturmgeschütz has very likely become the most valued weapon this summer. Every infantry commander speaks with enthusiasm about and recognizes the capabilities of the Sturmartillerie. Aside from defending against enemy infantry assaults, the Sturmartillerie can claim a large portion of the high number of enemy tanks destroyed. The number of enemy tanks claimed in August 1943 by 11 Sturmgeschütz-Abteilungen was 423, opposed to their own losses of only 18 Sturmgeschütz that are beyond repair.

Material Observations

The Sturmkanone as well as the Sturmhaubitze have completely proven themselves. The latter is indispensable against infantry targets. A ratio of 7 to 3 in each Batterie is viewed as correct.

The caliber of 75mm is sufficient for all armor encountered up to this point. If a russian tank appears which is the equal to the German Tiger tank, a caliber of 88mm is recommended.

Schürzen side skirts have proven valuable, are effective

against anti-tank rifles and light anti-tank guns, but their mountings are bad.

The commander's cupola is too lightly armored.

The loader's machine gun has not proven effective. It is suggested that it be internally mounted for the driver or the gunner.

The 30-watt radio is good. Ranges of up to 200 kilometers have been achieved.

Shaped charge ammunition (HEAT rounds) have also proven to be effective. However, Panzergranaten (capped, armor piercing, with explosive filler, shells) are preferred. The high-explosive shell is indispensable against infantry.

The on-board intercom has proven worthwhile.

The engine and chassis are weak. Automotive and cross-country capability are still inferior to the T-34.

An overview of claimed tank kills, ranked in accordance with the various weapons (Eastern Front) by the Chef of the Generalstab des Heeres dated 29 May 1944:

	Jan	Feb	Mar
Reported number of tanks knocked out:	4679	2189	2563
Cause of destruction known:	2652	1219	893

In Detail:

Close-combat weapons	185: 8%	62: 5.1%	105: 11.8%
Anti-Tank Guns	769: 29%	291: 23.8%	305: 34.2%
StuG and Pz.Jäger	634: 23%	355: 29.2%	256: 28.6%
Panzerkampfwagen	822: 31%	397: 32.5%	122: 13.6%
Other weapons (artillery, mines, etc)	242: 9%	114: 9.4%	105: 11.8%

Claims for Knocked-Out Tanks (missing reports from the Western Front) from the Inspektor der Artillerie in September 1944

Results achieved by the Sturmartillerie in 1944:

	Enemy tanks destroyed	Operational Sturmgeschütze	Our own non-repairable losses
January	860	671	61
February	429	718	71
March	578	511	177
April	542	533	121
May	147	732	15
June	245	757	34
July	1019	909	138
August	847	654	96

Total number of tanks destroyed since 22 June 1941 = 18261

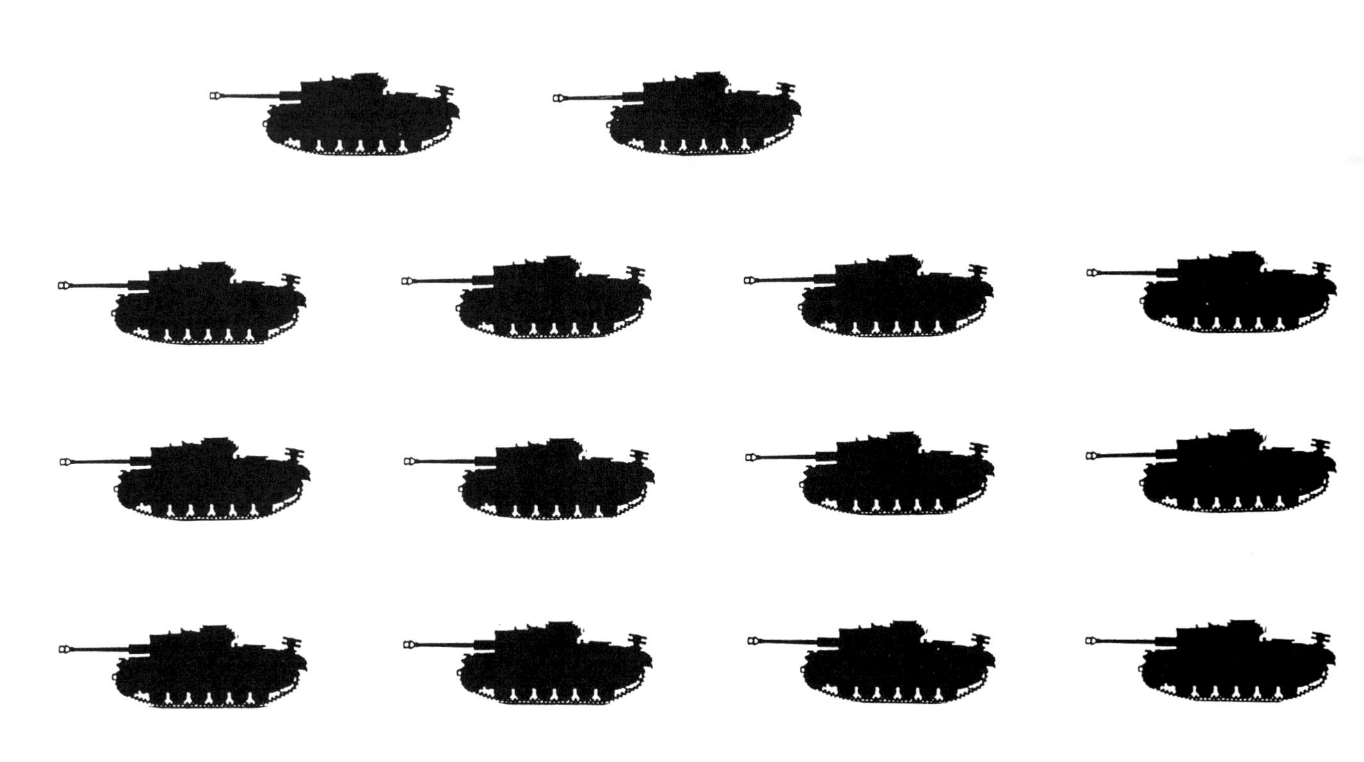

Panzer Sturmgeschütz Kompanie (14 Gesch.) (f.Pz.StuG-Abt.) K.St.N.1159 dated 20 June 1943 and 1 November 1943

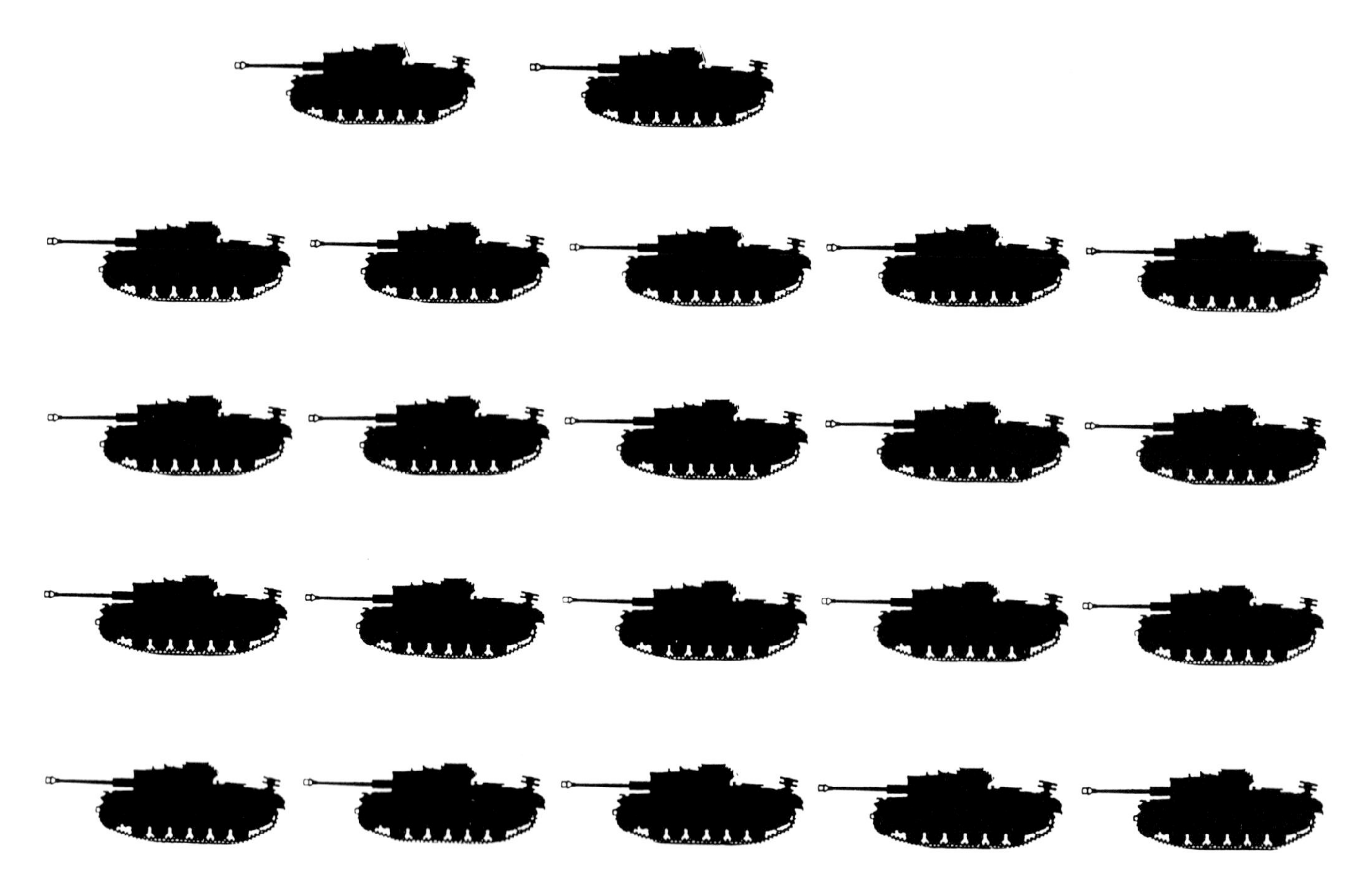

Panzer Sturmgeschütz Kompanie (22 Gesch.) (f.Pz.Abt.) K.St.N.1158 dated 20 April 1943 and 1 November 1943

Sturmgeschütze in the Panzertruppe

A shortage of tanks occurred in early 1943, when the Panzerkampfwagen III ceased production, the Panzerkampfwagen IV production wasn't sufficient to meet the needs and the Panther production was experiencing continuous problems. As a temporary measure, it was decided to issue Sturmgeschütz to the Panzer Truppen. Starting in May 1943, the General Inspekteur der Panzer Truppen obtained control of the issue of 100 Sturmgeschütz per month. These were initially used to outfit the newly reformed Panzer-Abteilungen for the Panzer- and Panzer-Grenadier Divisions that had been lost at Stalingrad. Following this, it became the general practice to outfit the Panzer-Abteilung of Panzer-Grenadier Divisions with Sturmgeschütz instead of the Panzerkampfwagen IV.

The units authorized 14 Sturmgeschütz per company were:

Unit	Belonging to	Shipped from HZA
1.-3.Kp./Pz.Abt.129	29.Pz.Gr.Div.	42 StuG III May43
1.-3.Kp./Pz.Abt.103	3.Pz.Gr.Div.	42 StuG III May43
9.-11.Kp./Pz.Rgt.2	16.Pz.Div.	42 StuG III Jun43
1.-3.Kp./Pz.Abt.FHH	Pz.Gr.Div.FHH	42 StuG III Jul/Aug43
1.Kp./Pz.Jg.Abt.228	16.Pz.Gr.Div.	14 StuG III Sep43
1.-3.Kp./Pz.Abt.5	25.Pz.Gr.Div.	42 StuG III Sep43
1.-3.Kp./Pz.Abt.7	10.Pz.Gr.Div.	42 StuG III Sep/Oct43
1.-3.Kp./Pz.Abt.8	20.Pz.Gr.Div.	42 StuG III Sep/Oct43
1.-3.Kp./Pz.Abt.118	18.Pz.Gr.Div.	42 StuG III Nov/Dec43
1.-3.Kp./Pz.Abt.190	90.Pz.Gr.Div.	42 StuG IVFeb44

Those units authorized 22 Sturmgeschütz per Kompanie formed mixed Abteilungen with two companies of Panzerkampfwagen IV and two companies of Sturmgeschütz.

Unit	Belonging to	Shipped from HZA
9.u.11.Kp./Pz.Rgt.24	24.Pz.Div.	44 StuG III May/Jun43
10.u.12.Kp./Pz.Rgt.36	14.Pz.Div.	44 StuG III Jun/Jul43

Later in 1944, production shortages again occurred and Sturmgeschütz III were issued to the Panzer Truppen in place of the Panzerkampfwagen IV in the Panzer-Regiments and in place of the Jagdpanther and Panzer IV/70(V) in the Panzer-Jäger-Abteilung as follows:

Unit	Belonging to	Shipped from HZA
2 Kpn./Pz.Jg.Abt.559	Heeres Truppen	28 StuG III Aug44
2 Kpn./Pz.Jg.Abt.519	Heeres Truppen	28 StuG III Sep44
4.Pz.Jg.Kp./Pz.Abt.2111	Pz.Brig.111	10 StuG III Sep44
4.Pz.Jg.Kp./Pz.Abt.2112	Pz.Brig.112	10 StuG III Sep44
4.Pz.Jg.Kp./Pz.Abt.2113	Pz.Brig.113	10 StuG III Sep44
2 Kpn./Pz.Jg.Abt.38	2.Pz.Div.	21 StuG III Oct-Nov44
5.u.6.Kp./Pz.Rgt.3	2.Pz.Div.	28 III Nov44
1 Kp./I./Pz.Rgt.33	9.Pz.Div.	14 StuG III Dec44
5.Kp./Pz.Rgt.33	116.Pz.Div.	14 StuG III Dec44
3 Kpn./Pz.Jg.Abt.510	Heeres Truppen	30 StuG III Feb45
II.Abt./Pz.Rgt.2	Heeres Truppen	31 StuG III Mar45

This list was limited to newly formed or reformed units and does not include the numerous cases in which Panzer Truppen units were issued Sturmgeschütz as replacements or "inherited" Sturmgeschütz from other units. From mid-1944 until the end of the war, all but four Panzer- and Panzer-Grenadier Divisions had acquired from one through 21 Sturmgeschütz, either in their Panzer- or Panzer-Jäger units.

After-Action-Account on the Deployment of Sturmgeschütz within the framework of a Panzer-Abteilung in Panzer-Regiment 36 dated 7 December 1943

For its deployment to the Eastern Front, the III./36 was equipped in the following manner:
2 companies and reconnaissance platoon with Pz.Kpfw.IV = 49 Pz.Kpfw.IV
2 companies with Sturmgeschütz = 44 StuG

The Abteilung first entered combat on 28 October 1943 and by 1 December 1943, had engaged in 16 days of combat.
The employment of the Sturmgeschütz within the Panzer-Abteilung and in cooperation with the Panzer-Grenadiers was accomplished in four different combat situations:
1. The Sturmgeschütz in the first wave of the attack.
2. The Sturmgeschütz in the second wave for flank protection.
3. The Sturmgeschütz in cooperation with the Panzer-Grenadiers.
4. The Sturmgeschütz in the defense.

All four possibilities were tested in the six-week deployment and the following experience was obtained:
Point 1: The deployment of the Sturmgeschütz in the attack on the first wave had one advantage — that it offers a somewhat smaller target than the Pz.Kpfw.IV. The disadvantages are: the tank can maintain its direction of thrust of the battle by use of its revolving turret, even when it must engage targets appearing to the left or right side. The Sturmgeschütz must always turn its front toward the enemy, so that for example to engage an enemy target to its left front, it must first turn half-left. These turns delay the engagement of enemy targets and lengthen the attack of the Panzer-Gruppe. Particularly difficult and inhibitive is the engagement of targets on the flanks in difficult terrain during the muddy season.

When breaking through enemy infantry positions, which are mostly very strongly outfitted with anti-tank rifles, the lack of a machine gun under armor protection on the Sturmgeschütz is a noteworthy disadvantage. The protective machine gun shield on the Sturmgeschütz is not effective against anti-tank rifles or infantry fire received from the sides.
Point 2: The Sturmgeschütz is somewhat better utilized in the second wave and in protecting the flanks, because the Sturmgeschütz in protected positions can allow an enemy armor attack to advance.
During the attack to eliminate a flanking threat, however, the same weaknesses are seen as identified in Point 1 above. The Sturmgeschütz only needs to turn to one side or the other to engage targets, but then has more problems when faced with enemy tanks approaching from its flanks, than a Pz.Kpfw.IV would.
Point 3: The most valued deployment of the Sturmgeschütz was in cooperation with the Panzer-Grenadier. The Sturmgeschütz give the Panzer-Grenadier a strong sense of morale, especially against enemy armor attacks. With the Panzer-Grenadier attacking, the Sturmgeschütz can engage the enemy's heavy weapons, such as anti-tank guns, tanks, artillery, infantry guns etc., while the enemy infantry and anti-tank rifles are kept away from the Sturmgeschütz by our own Panzer-Grenadiers.
Point 4: The Sturmgeschütz was a valuable weapon in the defense. It can, as mobile tank destroyer, from previously reconnoitered firing positions behind the forward lines, successfully engage the enemy. It has been shown to be very worthwhile to scout out firing positions in advance on foot.

Sturmgeschütz of the Waffen-SS.

Sturmgeschütz from SS-Das Reich, loaded on railcars. (BA)

Successes and Losses:

Knocked-Out

	Tanks	Anti-tank	Guns
Pz.Kpfw.IV:	136	117	20
Sturmgeschütz	75	59	36

Non-Repairable Losses:
20 Pz.Kpfw.IV by enemy fire and 13 due to mechanical failure.
19 StuG by enemy fire and six by mechanical failure.

A maintenance platoon was able to overhaul the following in 35 work-days:
52 Pz.Kpfw.IV and 74 Sturmgeschütz
In summary it can be said that in a combined Panzer-Abteilung, the Pz.Kpfw.IV has proven itself superior to the Sturmgeschütz, particularly in the attack.

Infanterie Sturmgeschütz Kompanie (Sfl.) K.St.N.190 dated 1 November 1941

Sturmgeschütz in the Waffen SS

The fifth Batterie that was formed with the initial 30 Sturmgeschütz, Ausf.A, belonged to the LSSAH instead of the Heeres Sturmartillerie. Formed too late to see action in the West, this Batterie and the Batterie for Das Reich fought in the Balkans in April 1941. Two additional SS Sturmgeschütz Batterien were formed in 1941 and all four were engaged on the Eastern Front in 1941/42. The Sturmgeschütz-Batterien in the SS used the same K.St.N. as the Heeres, but the orders for their creation normally authorized them to automatically upgrade to the latest K.St.N. as soon as new K.St.N were published. In late 1941, a specific K.St.N. was published for Sturmgeschütz-Batterien that were an integral part of Infanterie units. This was K.St.N.190 dated 1 November 1941 for the Infanterie-Sturmgeschütz-Kompanie (Sfl.).

When the Sturmgeschütz-Abteilung for the LSSAH was formed in February 1942, the orders specified that the unit be organized in accordance with K.St.N.190. Later in 1942, when Sturmgeschütz-Abteilungen were being formed for Das Reich and Totenkopf the orders specified the normal K.St.N.446. By the spring of 1943, all three Sturmgeschütz-Abteilungen had reorganized under K.St.N.446a. During upgrade and expansion of the SS, additional Sturmgeschütz-Batterien and Abteilungen were created for the SS in 1942 and 1943. Starting in 1943 with SS-Pz.Abt.11 and continuing through 1944 and 1945, just as with the Heeres Panzer Truppen, the SS-Panzer-Abteilungen and Panzer-Jäger-Abteilung were issued Sturmgeschuetz III und IV due to the shortage of Pz.Kpfw.IV, Jagdpanzer IV and Panzer IV/70(V).

SS-Units equipped with the Sturmgeschütz

Formed in	Unit	Belonged to	Organization
May 1940	StuG Battr.	LSSAH	6 StuG
Feb 1941	StuG Battr.	Das Reich	6 StuG
Jun 1941	StuG Battr.	Totenkopf	6 StuG
Sep 1941	StuG Battr.	Wiking	7 StuG
Feb 1942	1.-3 Battr./StuG Abt.1	1.SS Div.	7 StuG/Battr.
Mar 1942	StuG Battr.	7.SS Div.	7 StuG/Battr.
Sep 1942	StuG Battr.	5.SS Div.	7 StuG/Battr.
by Oct 1942	StuG Battr.	8.SS Div.	7 StuG
Oct 1942	1.-3.Battr./StuG Abt.2	2.SS Div.	7 StuG/Battr.
Oct 1942	1.-3.Battr./StuG Abt.3	3.SS Div.	7 StuG/Battr.
by Apr 1943	StuG Battr.	16.SS Brig	10 StuG
by Jul 1943	1.-3.Battr./StuG Abt.16	16.SS Brig	10 StuG/Battr.
Jul 1943	StuG Battr.	Prinz Eugen	10 StuG
by Aug 1943	StuG Battr.	1.SS Brig	10 StuG
by Aug 1943	1.-3.Kp./Pz.Abt.11	11.SS Div.	14 StuG/Battr.
Oct 1943	StuGBattr./Pz.Jg.Abt.54	4.SS Brig	10 StuG
by Nov 1943	1./StuG Abt.4	4.SS Brig	10 StuG
by Dec 1943	StuG Battr.6	6.SS Brig	10 StuG
by Jan 1944	1.-3 Kp./Pz.Abt.17	17.SS Div	14 StuG/Kp.
Jan 1944	2 Kpn./I./Pz.Rgt.5	5.SS Div	22 StuG/Kp.
Jan 1944	2 Kpn./II/Pz.Rgt.9	9.SS Div	22 StuG/Kp.
Jan 1944	2 Kpn./II/Pz.Rgt.10	10.SS Div	22 StuG/Kp.
Feb 1944	1.-3.Kp./Pz.Abt.4	4.SS Div	14 StuG/Kp.
Feb 1944	1.-3.Kp./Pz.Abt.18	18.SS Div	14 StuG/Kp.
by Mar 1944	2 Kpn./Pz.Jg.Abt.16	16.SS Div	10 StuG/Kp.
by Apr 1944	1.-3.Kp./Pz.Abt.16	16.SS Div	14 StuG/Kp.
Nov 1944	2 Kpn./II/Pz.Rgt.2	2.SS Div	14 StuG/Kp.
by Dec 1944	1.-3.Kp./Pz.Jg.Abt.17	17.SS Div	10 StuG/Kp.
by Feb 1945	1.-3.Kp./Pz.Jg.Abt.18	18.SS Div	10 StuG/Kp.
by Feb 1945	1.-3.Kp./Pz.Jg.Abt.54	4.SS Brig	10 StuG/Kp.

Sturmgeschütz in Luftwaffe Field Units

On 1. Juni 1942, the Luftwaffe ordered the formation of a Sturmgeschütz platoon with 4 StuG L/24. This unit was to become the basis as a training unit for all of the Luftwaffe Sturmgeschütz units that were to follow.

In October 1942, the III./Sturmgeschütz/Artillerie Regiment Hermann Göring was formed and received 21 Sturmgeschütz. Ausf.F/8 from the Heeres Zeugamt at Magdeburg in early November 1942. Renamed the III.(Sturmgeschütz) Abteilung of Fallschirm Panzer Regiment Hermann Göring, the unit fought in Sicily in July 1943 and then in Italy until it was reorganized as a Panzer-Jäger-Abteilung in April 1944 and issued Jagdpanzer IV.

Starting in late 1942 and early 1943, a Sturmgeschütz Batterie with 4 Sturmgeschütze within the Panzer-Jäger-Abteilung (Numbered 1101 through 1121) were formed for each of the Luftwaffe Feld Divisionen. In 1944, the Sturmgeschütz Batterie for Pz.Jg.Abt.1104, 1106, 1112, 1116, 1117, 1118, 1119 and 1121 were authorized to expand to 10 Sturmgeschütz per Batterie.

The Luftwaffe also formed two Sturmgeschütz Brigaden, Fallschirm Sturmgeschütz Brigade XI and XII with a normal complement of 31 Sturmgeschütz und Sturmhaubitze. Both units fought on the Western Front in 1944 and were replenished for the major offensives in December and January "Wacht am Rhein" und "Nordwind."

With the expansion of the Luftwaffe Feld Truppen to form a Panzer Korps in 1944, a Panzer-Abteilung was formed for the Fallschirm Panzer Grenadier Division 2 Hermann Göring and issued 45 Sturmgeschuetz III in October 1944. This unit fought on the Eastern Front along with the replenished Fallschirm Panzer Division Hermann Göring.

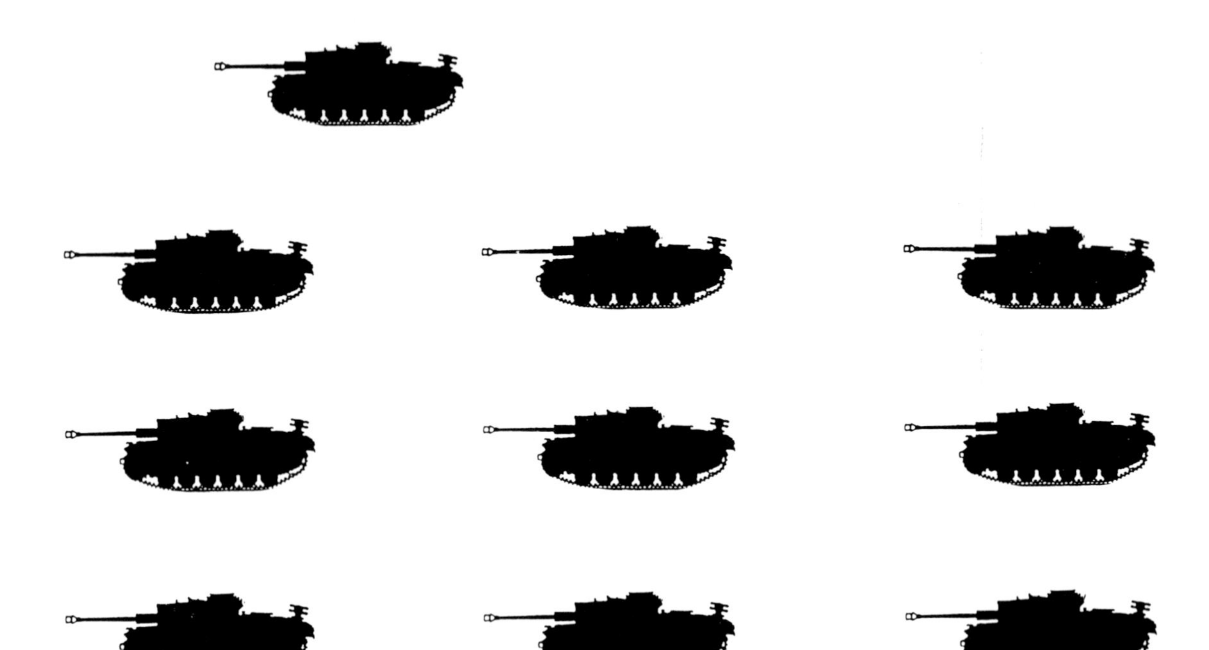

Sturmgeschütz Abteilung (10 Gesch.) (in Pz.Jg.Abt.) K.St.N.1149 Ausf.A dated 1 February 1944

Sturmgeschütz in Infantry Units

The Heeres Sturmgeschütz-Abteilungen and Batterien were normally assigned to support Infanterie Regiments and Battalions at the Front. As the reputation of the Sturmgeschütz as an effective tank destroyer increased, and the Russian tank threat became greater, the infantry wanted there own mobile anti-tank protection without having to depend on an outside unit coming to their aid. Starting in 1943, it was decided to outfit one Kompanie of the Panzer-Jäger-Abteilung of the Infanterie-, Gebirgs-, Jäger-, and Volks-Grenadier Divisions with Sturmgeschütz. In 1943, each Sturmgeschütz Panzer-Jäger-Kompanie was authorized to possess 14 Sturmgeschütz. This was decreased to 10 Sturmgeschütz in 1944. The applicable K.St.N. for these units were: Sturmgeschütz Abteilung (10 Gesch.) (in Pz.Jg.Abt.) K.St.N.1149 Ausf.A dated 1 February 1944
Sturmgeschütz Abteilung (14 Gesch.) (in Pz.Jg.Abt.) K.St.N.1149 Ausf.B dated 1 February 1944

Sturmgeschütz Abteilung (14 Gesch.) (in Pz.Jg.Abt.) K.St.N.1149 Ausf.B dated 1 February 1944

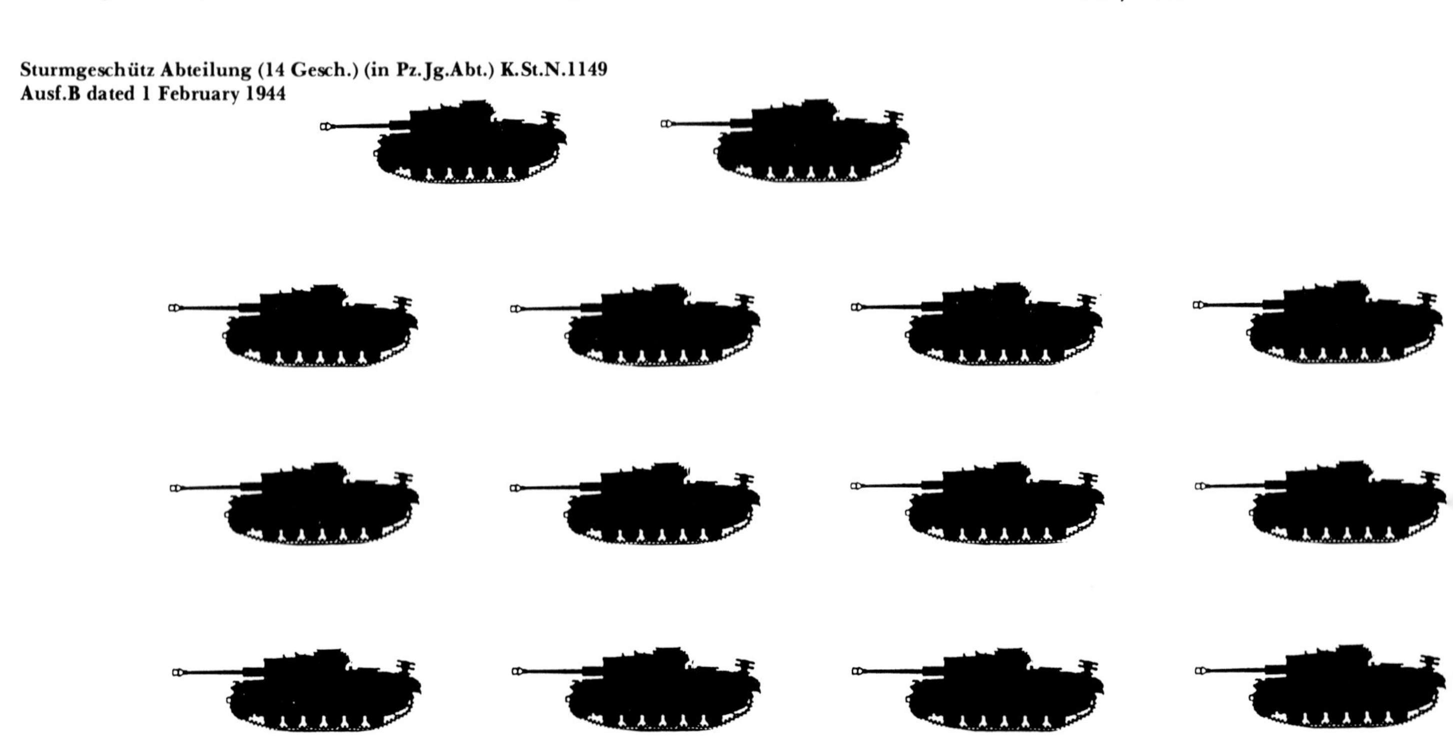

Not including Sturmgeschütz shipped as replacements to units at the front, the following Sturmgeschütz III and IV were issued to Panzer-Jäger-Kompanien and Sturmgeschütz-Abteilungen with Infanterie-, Gebirgs-, Jäger-, and Volks-Grenadier Divisionen:

Month	Issued	Units
Jul.43	42 StuG III	3 Pz.Jg.Kp.
Sep.43	28 StuG III	2 Pz.Jg.Kp.
Oct.43	28 StuG III	2 Pz.Jg.Kp.
Jan.44	40 StuG III	4 Pz.Jg.Kp.
Feb.44	80 StuG III and 50 StuG IV	13 StuG Abt.
Mar.44	60 StuG IV	6 StuG Abt.
Apr.44	30 StuG III and 80 StuG IV	14 StuG Abt.
May.44	90 StuG IV	9 StuG Abt.
Jun.44	10 StuG III and 70 StuG IV	8 StuG Abt.
Jul.44	114 StuG III and 94 StuG IV	20 StuG Abt. (2 have 14)
Aug.44	110 StuG III and 50 StuG IV	16 StuG Abt.
Sep.44	110 StuG III and 70 StuG IV	18 StuG Abt.
Oct.44	80 StuG IV	8 StuG Abt.
Nov.44	50 StuG III and 30 StuG IV	8 StuG Abt.
Dec.44	40 StuG III and 40 StuG IV	8 StuG Abt.
Jan.45	30 StuG III and 30 StuG IV	6 StuG Abt.

The following observations of the effectiveness of Sturmgeschütz units of the Infantry Divisions were published in the Nachrichtenblatt der Panzertruppen, an armor branch pamphlet, in December 1944:

After-Action-Report from Panzerjäger-Kompanie 1045 with the Sturmgeschütz III

The company was prepared as divisional reserves. The enemy attacked one morning after a half-hour pummeling of artillery preparatory fire with heavy air support and about 30 new T-34 tanks and mechanized infantry deployed on a wide front. The enemy tried to force a breakthrough with portions of 5 or 6 divisions. The terrain was unusually favorable for the enemy. Above all, the forested areas provided him with suitable firing positions and assembly areas.

The company went into action with 9 Sturmgeschütz, and on the first day was able to knock-out or destroy the following within three hours:
16 T-34 (new): 2 mortars
1 KW I: 2 observation points w/ radios
2 T-34 (made immobile): 1 anti-tank gun
17 machine guns: 1 infantry gun

On the second day:
2 T-34: 3 anti-tank weapons
1 self-propelled gun: 2 grenade launchers
21 machine guns: 2 anti-tank guns

The tanks were knocked-out at ranges of 600 to 800 meters. In a period of 15 minutes, one Sturmgeschütz was able to shoot five tanks out of a column.

The enemy tanks did not fire a single aimed round. The remaining T-34 tanks were individually hunted down. One T-34 was knocked out at a range of 1000 meters with 3 rounds.

Equipping the Sturmgeschütz with machine guns and sub-machine guns has proven to be very beneficial, because these weapons were able to keep the enemy tank-hunter-teams as well as the infantry at bay. In sum it should be said that after the enemy's high losses of tanks, it no longer utilized them during infantry attacks but rather dug them into defensive positions.

After-Action-Report from Panzer-Jäger-Kompanie 1253 with the Sturmgeschütz IV

Due to the situation, the company, almost without exception, was deployed in platoon strength for the duration of the operation. This will continuously be necessary for Panzer-Jäger-Kompanien of Infanterie Divisions. Therefore, the main emphasis, more than before, must be placed on small-team operations and platoon training.

The company had no time to practice together with the Infantry due to its overly hasty deployment and transport to the front. The individual situations did not permit a thorough briefing by the commander, above all the necessary contact with the platoon leaders.

Our own attacks were, from the start, greatly affected by the decreased combat effectiveness of the Infantry and the lack of training in cooperation during combat.

The enemy, tanks and anti-tank guns, evaded sweeping counterattacks and then only showed determined resistance when the infantry did not follow.

The company was equipped with Sturmgeschütz that were missing the old machine gun shield and the new all-round firing mount had not been installed. Because the enemy can be suppressed in many cases with machine gun fire, the lack of this weapon is often not replaceable by firing an increased number of high-explosive shells.

Units of the Heeres Sturmartillerie

Abteilung	Established	Deployed on Front	Remarks
Lehr 111	January 1945	East from January 1945	Sturmartillerie Brigade from January 1945
177	June 1941	East from Sept. 1941	redesignated Pz.Jg.Abt. 69 (August1944)
184	August 1940	East from June 1941	Sturmartillerie-Brigade from April 1945
185	August 1940	East from June 1941	
189	July 1941	East from August 1941	redesignated Pz.Jg.Abt. 70 (August 1944)
190	October 1940	East from June 1941	
191	October 1940	East from June 1941, Southeastern from July 1944	
192	November 1940	East from June 1941	redesignated StuG Abt. G.D.
197	November 1940	East from June 1941	redesignated s.Pz.Jg.Abt. 653
200	July 1943	West from Dec. 1944, East from Feb. 1945	initially with self-propelled guns on captured chassis
201	March 1941	East from June 1941	
202	September 1941	East from September 1941	
203	February 1941	East from June 1941	
209	December 1941	East from January 1942	
210	March 1941	East from June 1941	Sturmartillerie-Brigade from August 1945
226	April 1941	East from June 1941	
228	November 1942	East from December 1942	
232	November 1942	East from December 1942	
236	June 1943	East from July 1943	Sturmartillerie-Brigade from April 1944
237	July 1943	East from Sept. 1943	Sturmartillerie Brigade from April 1944
239	July 1943	East from August 1943	redesignated Sturm-Pz.Abt. 219
242	November 1942	East from January to February 1943, Southwest from June 1943	
243	May 1943	East from June 1941, West from Dec. 1944	Sturmartillerie Brigade from February 1945
244	June 1941	East from July 1941, West from December 1944	
245	June 1941	East from July 1941	
249	January 1942	East from Feb. 1942	Sturmartillerie Brigade from February 1945
259	July 1943	East from September 1943	
261	August 1943	East from August 1943	Sturmartillerie-Brigade from November 1944
270	December 1942	East from Dec. 1942	redesignated Pz.Jg.Abt. 152 (August 1944)
276	August 1943	East from October 1943	
277	August 1943	East from September 1943	
278	September 1943	East from October 1943	
279	September 1943	East from October 1943	
280	September 1943	East from November 1943, West from September 1944	
281	October 1943	East from January 1944	redesignated H.Art.Pak Abt 1052 (August 1944)
286	October 1943	East from Dec. 1943	Sturmartillerie-Brigade from April 1945
300	October 1943	East from January 1944	Strumartillerie-Brigade from March 1945
301	November 1943	East from February 1944	Sturmartillerie-Brigade from April 1945
303	November 1943	East from January 1944	Sturmartillerie-Brigade from November 1944
311	November 1943	East from March 1944	Sturmartillerie-Brigade from April 1945
322	December 1943	East from March 1944	
325	December 1943	East from April 1944	Sturmartillerie-Brigade from April 1945
341	December 1943	West from June 1944	
393	March 1944	East from July 1944	
394	March 1944	West from July 1944	
600	March 1942	East from Batterien 660, 665, 666	
667	June 1942	East from July 1942, West from Dec. 1944	Sturmartillerie-Brigade from June 1944
902	December 1943	Western from June 1944	
904	December 1942	East from March 1943	
905	December 1942	East from April 1943	Sturmartillerie-Brigade from November 1944
907	December 1943	Southwest from January 1944	
909	January 1943	East from March 1943	
911	January 1943	East from March 1943, West from Dec. 1944	Sturmartillerie-Brigade from November 1944
912	March 1943	East from April 1943	
914	January 1944	Southwest from Sept. 1944	at first with Italian StuG
920	July 1944	East from July 1944	
1170	April 1945	Germany from April 1945	Sturmartillerie-Brigade from April 1945
G.D.	April 1942	East from June 1943	created from StuG Abt. 192

Abteilung	Established	Deployed on Front	Remarks
90	November 1942	Tunisia from Nov. 1942 to May 1943	redesignated from 1./242
247	March 1943	Southwest from July 1943	disbanded (October 1943)
287	August 1942	East from October 1942 to April 1943	redesignated 1.Kp./Pz.Abt. Rhodos
288		North Africa March to October 1942	
393	November 1942	East from January to September 1943	disbanded (October 1943)
395	November 1942	East from January to March 1943	disbanded (March 1943)
640/G.D.	November 1939	West 1940, Balkans East 1941	redesignated to 16.Sturmbattr. in & "G.D." (April 1940)
659	March 1940	West 1940, East from June 1941	redesignated StuG Battr.287 (August 1942)
660	March 1940	West 1940, East from June 1941	redesignated 1.Battr./StuG Abt.600 (March 1942)
665	May 1940	West 1940, East from June 1941	redesignated 2.Battr./StuG Abt.600 (March 1942)
666	May 1940	East from June 1941	redesignated 1.Battr./StuG Abt.184 (March 1942)
667	July 1940	East from June 1941	redesignated 1.Battr./StuG Abt.667 (June 1942)
741	February 1943	East from July 1943 early 1944	dissolved into StuG Brig.394 (May to 1944)
742	February 1943	East from July to October 1943	dissolved into 2.Battr./StuG Abt.201 (October 1943)
Lehr 900	May 1941	East from June 1941 to April 1942	dissolved (May 1942)
Lehr 901	December 1942	East from Dec. 1942 to April 1943	used for the new 8.ALR (May 1943)

Sturmgeschütz Technical Data

1 Manufacturer	Daimler Benz	Daimler Benz	Alkett	Alkett	Alkett
2 **Ausführung**	**0-series**	**model A**	**model B**	**model C**	**model D**
3 Type	2/ZW	5/ZW	/ZW	/ZW	/ZW
4 Number	5	30	120	100	150
5 Engine: manufacturer, type	Maybach HL 108 TR	Maybach HL 120 TR	Maybach HL 120 TRM	Maybach HL 120 TRM	Maybach HL 120 TRM
6 Nr of cylinders, layout	12, 60° V	12, 60° V	12, 60° V	12, 60°	12, 60°
7 Bore/stroke (mm)	100 x 115	105 x 115	105 x 115	105 x 115	105 x 115
8 Capacity	10838	11867	11867	11867	11867
9 Compression ratio 1:	6.5	6.5	6.5	6.5	6.5
10 Power (metric HP)	250	300	300	300	300
11 RPM	3000	3000	3000	3000	3000
12 Power-to-weight ratio (HP/t)	15.3	13.6	13.6	13.6	13.6
13 Valve location	hanging	hanging	hanging	hanging	hanging
14 Crankshaft bearing	7-roller	7-roller	7-roller	7-roller	7-roller
15 Carburetor type, no.	2 Solex 40 JFF II	2 Solex 40 JFF II	2 Solex 40 JFF II	2 Solex JFF 40 II	2 Solex 40 JFF II
16 ignition sequence	1-8-5-10-3-7-6-11-2-9-4-12	1-12-5-8-3-10-6-7-2-11-4-9	1-12-5-8-3-10-6-7-2-11-4-9	1-12-5-8-3-10-6-7-2-11-4-9	1-12-5-8-3-10-6-7-2-11-4-9
17 Starter	Bosch (BNG) 4/24 CRS 178	Bosch BNG 4/24 ARS 129	Bosch BNG 4/24 ARS 129	Bosch BNG 4/24 ARS 129	Bosch BNG 4/24 ARS 129
18 Generator	Bosch GQL 600/12-900	Bosch GTLN 700/12-1500	Bosch GTLN 700/12-1500	Bosch GTLN 700/12-1500	Bosch GTLN 700/12-1500
19 Batteries, nr.	2, 12V, 105 Ah	2, 12V, 105 Ah	2, 12V, 105 Ah	2, 12V, 105 Ah	2, 12V, 105 Ah
20 Fuel delivery	2 Solex EP 100 pumps	2 Solex and 1 elec. pump	2 Solex and 1 elec. pump	2 Solex and 1 elec. pump	2 Solex and 1 elec. pump
21 Cooling	liquid	liquid	liquid	liquid	liquid
22 Clutch	triple plate	hydraulic with accelerator	triple plate	triple plate	triple plate
23 Transmission	ZF SFG 75 Schub-	Maybach SRG 328145-Variorex	ZF SSG 77 Aphon	ZF SSG 77 Aphon	ZF SSG 77 Aphon
24 Number of gears (F/R)	5/1	10/1	6/1	6/1	6/1
25 Drive	front	front	front	front	front
26 Drive reduction	4.4	4	4	4	4
27 Max speed (km/h)	35	30	40	40	40
28 Range: road/terrain (km)	165/95	160/100	165/95	165/95	165/94
29 Steering type	DB/Wilson clutch	DB/Wilson clutch	DB/Wilson clutch	DB/Wilson clutch	DB/Wilson clutch
30 Turning radius	5.85m	5.85m	5.85m	5.85m	5.85m
31 Suspension	leaf spring	cross torsion bar	cross torsion bar	cross torsion bar	cross torsion bar
32 Brake system	mechanical	hydraulic with servo	hydraulic with servo	hydraulic with servo	hydraulic with servo
33 Running gear type	roadwheels and return rollers	roadwheels and return rollers	roadwheels and return rollers	roadwheels and return rollers	roadwheels and return rollers
34 Track gauge (mm)	2376	2490	2510	2510	2510
35 Track length (mm)	3224	2860	2860	2860	2860
36 Track type, sections per track	TKgs 6100/380/120	Kgs 61/400/120-93	Kgs 61/400/120-93	Kgs 61/400/120-93	Kgs 61/400/120-93
37 Track width (mm)	360	360	400	400	400
38 Ground clearance (mm)	375	385	375	375	375
39 Length x width x height (mm)	5665 x 2810 x	5380 x 2920 x 1950	5400 x 2950 x 1960	5400 x 2950 x1960	5400 x 2950 x 1960
40 Firing height (mm)		1500	1500	1500	1500
41 Ground pressure (kg/cm²)		0.93	0.93	0.93	0.93
42 Combat weight (kg)	16000	19600	22000	22000	22000
43 Crew	4	4	4	4	4
44 Fuel supply (liters)	300	310	310	310	310
45 Fuel consumption (l/100km)	280	187	195	195	195
46 Climb/Step/Ford	30°/600/800	30°/600/800	30°/600/800	30°/600/800	30°/600/800
47 Armor: Hull (mm)	14.5	50	50	50	50
48 Superstructure (mm)	14.5 (soft steel)	50	50	50	50
49 Armament: Gun	7.5cm StuK L/24	7.5cm StuK L/24	7.5cm StuK L/24	7.5cm StuK L/24	7.5cm StuK L/24
50 Machine Gun	1 (loose)	-	1 (loose)	1 (loose)	1 (loose)
51 Submachine gun	1	2 (loose)	1	1	1
52 Gun elevation (degrees)		-10 to+20	-10 to+20	-10 to+20	-10 to+20
53 Gun Traverse (degrees)		+12 to -12	24 total	24 total	24 total
54 Gun sight	Rundblickfernrohr	Rundblickfernrohr 32 4 x 10°	Rundblickfernrohr	Sfl-Zielfernrohr	Sfl-Zielfernrohr
55 Ammunition supply		44	44	44	44
56 Radio equipment		1 VHF receiver	1 HF receiver	1 VHF receiver	1 VHF receiver
57 Remarks	superstructure of soft steel				on-board intercom

1 Manufacturer	Alkett	Alkett	Alkett	Alkett/MIAG	Alkett	Krupp-Gruson
2 **Ausführung**	**model E**	**model F**	**model F/8**	**model G**	**10.5cm StuH 42**	**Sturmgeschütz IV**
3 Type	/ZW	/ZW	/ZW	/ZW	/ZW	7/BW
4 Number	284	364	250			
5 Engine: manufacturer, type	Maybach HL 120 TRM	Maybach HL 120 TRM	Maybach HL 120 TRM	Maybach HL 120 TRM	Maybach HL 120 TRM	Maybach HL120TRM
6 Nr of cylinders, layout	12, 60° V	12, 60° V	12, 60° V	12, 60°	12, 60°	12, V
7 Bore/stroke (mm)	105 x 115	105 x 115	105 x 115	105 x 115	105 x 115	105 x 115
8 Capacity	11867	11867	11867	11867	11867	11867
9 Compression ratio 1:	6.5	6.5	6.5	6.5	6.5	6.5
10 Power (metric HP)	300	300	300	300	300	300
11 RPM	3000	3000	3000	3000	3000	3000
12 Power-to-weight ratio (HP/t)	13.6	12.3	12.3	11.1	11.1	10.6
13 Valve location	hanging	hanging	hanging	hanging	hanging	hanging
14 Crankshaft bearing	7-roller	7-roller	7-roller	7-roller	7-roller	7-roller
15 Carburetor type, no.	2 Solex 40 JFF II	2 Solex 40 JFF II	2 Solex 40 JFF II	2 Solex JFF 40 II	2 Solex 40 JFF II	2 Solex 40 JFF II
16 ignition sequence	1-12-5-8-3-10-6-7-2-11-4-9	1-12-5-8-3-10-6-7-2-11-4-9	1-12-5-8-3-10-6-7-2-11-4-9	1-12-5-8-3-10-6-7-2-11-4-9	1-12-5-8-3-10-6-7-2-11-4-9	1-12-5-8-3-10-6-7-2-11-4-9
17 Starter	Bosch (BNG) 4/24 ARS 129	Bosch BNG 4/24 ARS129	Bosch BNG 4/24 ARS 129	Bosch BNG 4/24 ARS129	Bosch BNG 4/24 ARS 129	Bosch BNG 4/24 ARS129
18 Generator	Bosch GTLN 700/12-1500	Bosch GTLN700/12-1500	Bosch GTLN 700/12-1500	Bosch GTLN700/12-1500	Bosch GTLN 700/12-1500	Bosch GTLN 700/12-1500
19 Batteries, nr.	2, 12V, 105 Ah	2, 12V, 105 Ah	2, 12V, 105 Ah	2, 12V, 105 Ah	2, 12V, 105 Ah	2, 12V, 105 Ah
20 Fuel delivery	2 Solex and 1 elec. pump	2 Solex and 1 elec. pump	2 Solex and 1 elec. pump	2 Solex and 1 elec. pump	2 Solex and 1 elec. pump	2 Solex and 1 elec. pump
21 Cooling	liquid	liquid	liquid	liquid	liquid	liquid
22 Clutch	triple plate	hydraulic with accelerator	triple plate	triple plate	triple plate	triple plate
23 Transmission	ZF SSG 77 Aphon	ZF SSG 77 Aphon	ZF SSG 77 Aphon	ZF SSG 77 Aphon	ZF SSG 77 Aphon	ZF SSG 77 Aphon
24 Number of gears (F/R)	6/1	6/1	6/1	6/1	6/1	6/1
25 Drive	front	front	front	front	front	
26 Drive reduction	4	4	4	3.88	4	3.23
27 Max speed (km/h)	40	40	40	40	40	38
28 Range: road/terrain (km)	165/95	165/95	165/95	155/95	155/95	210/130
29 Steering type	DB/Wilson clutch	DB/Wilson clutch	DB/Wilson clutch	DB/Wilson clutch	DB/Wilson clutch	Krupp/Wilson clutch
30 Turning radius	5.85m	5.85m	5.85m	5.85m	5.85m	5.92
31 Suspension	cross torsion bar	cross torsion bar	cross torsion bar	cross torsion bar	cross torsion bar	leaf spring
32 Brake system	hydraulic with servo	hydraulic with servo	hydraulic with servo	hydraulic with servo	hydraulic with servo	mechanical
33 Running gear type	roadwheels & return rollers	roadwheels and return rollers	roadwheels and return rollers	roadwheels and return rollers	roadwheels and return rollers	roadwheels and return rollers
34 Track gauge (mm)	2510	2510	2510	2510	2510	2510
35 Track length (mm)	2860	2860	2860	2860	2860	3520
36 Track type, sections per track	TKgs 61/400/120-93	Kgs 61/400/120-93	Kgs 61/400/120-93	Kgs 61/400/120-93	Kgs 61/400/120-93	Kgs 61/400/120-99
37 Track width (mm)	400	400	400	400	400	400
38 Ground clearance (mm)	375	390	390	390	390	400
39 Length x width x height (mm)	5400 x 2950 x 1960	6310 x 2920 x 2150	6770 x 2950 x 1850	6770 x 2950 x 2160	6000 x 2950 x 1850	6700 x 2950 x 2200
40 Firing height (mm)	1500	1550	1550	1570	1550	1550
41 Ground pressure (kg/cm²)	0.93	0.93	0.93	0.93	0.93	0.8
42 Combat weight (kg)	22000	23200	23400	23900	23900	23000
43 Crew	4	4	4	4	4	4
44 Fuel supply (liters)	300	310	310	310	310	310
45 Fuel consumption (l/100km)	280	187	195	195	195	430
46 Climb/Step/Ford (mm)	30°/600/800	30°/600/800	30°/600/800	30°/600/800	30°/600/800	30°/600/1200
47 Armor: Hull (mm)	50	50+30	50+30	50+30 (later 80)	50+30 (later 80)	80
48 Superstructure (mm)	50	50+30	50+38	50+30	50+30	80
49 Armament: Gun	7.5cm StuK L/24	7.5cm StuK L/24	7.5cm StuK L/24	7.5cm StuK L/24	10.5cm StuH 42 L/28	7.5cm StuK L/24
50 Machine Gun	1 (loose)	IMG 34		1 (2) MG	IMG 24	IMG 24
51 Submachine gun	1	IMP				2 MP
52 Gun elevation (degrees)	-10 to +20	-6 to +20	-6 to +20	-6 to +20	+20 to -6	+20 to -6
53 Gun traverse (degrees)	24 total	20 total	20 total	20 total	20 total	20 total
54 Gun sight	RbF 32/SFl Zf1/SF 14Z	SFl Zf1a/RbF 36	Sfl Zf1a/RbF 36	SFl ZF1a/RbF 36	SFl ZF1a/Rab/F 36	SFR SFl Zf 1a
55 Ammunition supply	50	44	54	54	36	63
56 Radio equipment	10WS h/VHF receiver	Fu 15 or Fu 16	Fu 15 or Fu 16	Fu 15 or Fu 16	Fu 15 or Fu 16	Fu 15 or Fu 16
57 Remarks	intercom	intercom	intercom	interco m	intercom	intercom

Bibliography

Adonyl-Naredy, F.v., *Ungarns Armee im Zweiten Weltkrieg*

Boelcke, W.A., *Deutschlands Rüstung im Zweiten Weltkrieg*

Doyle, H.L., Private archives

Doyle/Chamberlain/Jentz, *Encyclopedia of German Tanks*

Guderian, Heinz, *Erinnerungen eines Soldaten*

Haak, E., *Die Geschichte der deutschen Instandsetzungstruppe*

Jentz, T., Private archives

Kantakoski, P., *Suomalaiset panssarivaunujoukot*

Kurowski F./Tornau G., *Sturmartillerie*

Oswald, W., *Kraftfahrzeuge und Panzer der Reichswehr, Wehrmacht und Bundeswehr.*

Spielberger, W.J., *Der Panzerkampfwagen III und seine Abarten*

Spielberger, W.J., *Die Motorisierung der deutschen Reichswehr*

Thomas F./Wegmann G., *Die Ritterkreuzträger der Deutschen Wehrmacht, Teil I:Sturmartillerie*

Tornau G./Kurowski F., *Sturmartillerie - Fels in der Brandung*